W9-AAW-980

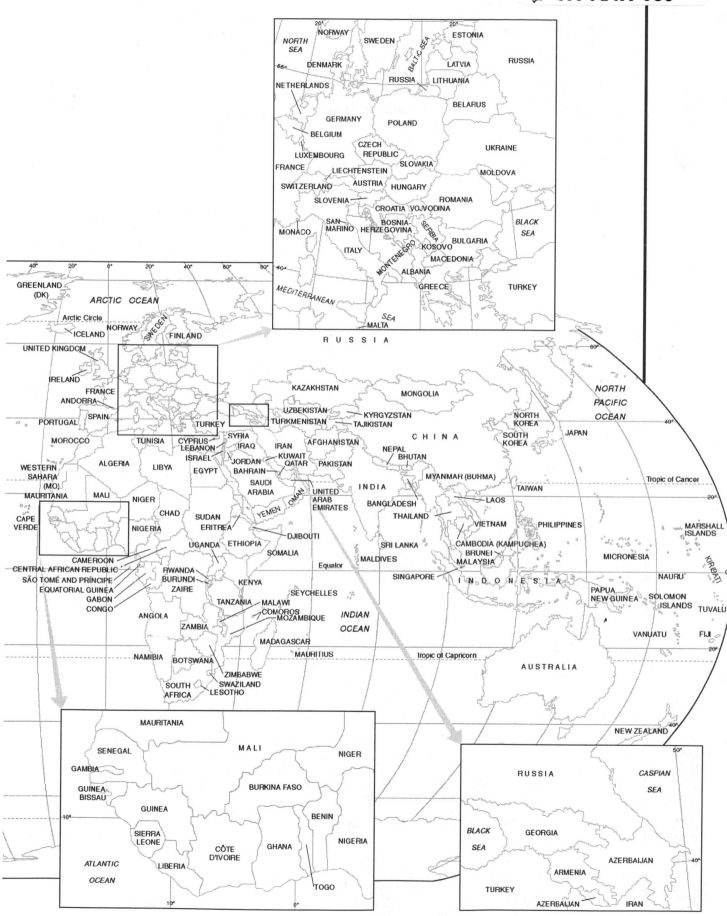

iii

Credits

1. Geography in a Changing World
Unit photo—Courtesy of New York Public Library.
2. Human-Environment Relationships
Unit photo—United Nations photo by Rick Grunbaum.
3. The Region
Unit photo—United Nations photo by Saw Lwin.
4. Spatial Interaction and Mapping
Unit photo—Courtesy of Digital Stock.
5. Population, Resources, and Socioeconomic Development
Unit photo—United Nations photo by J. K. Isaac.

Copyright

Cataloging in Publication Data
Main entry under title: Annual Editions: Geography. 2000/2001.
  1. Geography—Periodicals. 2. Anthropo-geography—Periodicals. 3. Natural resources—Periodicals. I. Pitzl, Gerald R., *comp.* II. Title: Geography.
ISBN 0–07–236551–X   87–641715   ISSN 1091–9937   910'.5

Fifteenth Edition

Cover image © 2000 PhotoDisc, Inc.

Printed in the United States of America   1234567890BAHBAH543210   Printed on Recycled Paper

06 18

GEOGRAPHY

00/01

Fifteenth Edition

EDITOR

Gerald R. Pitzl
Macalester College

Gerald R. Pitzl, professor of geography at Macalester College, received his bachelor's degree in secondary social science education from the University of Minnesota in 1964 and his M.A. (1971) and Ph.D. (1974) in geography from the same institution. He teaches a wide array of geography courses and is the author of a number of articles on geography, the developing world, and the use of the Harvard case method.

Dushkin/McGraw-Hill
Sluice Dock, Guilford, Connecticut 06437

Visit us on the Internet
http://www.dushkin.com/annualeditions/

This map has been developed to give you a graphic picture of where the countries of the world are located, the relationship they have with their region and neighbors, and their positions relative to the superpowers and power blocs. We have focused on certain areas to more clearly illustrate these crowded regions.

To the Reader

In publishing ANNUAL EDITIONS we recognize the enormous role played by the magazines, newspapers, and journals of the public press in providing current, first-rate educational information in a broad spectrum of interest areas. Many of these articles are appropriate for students, researchers, and professionals seeking accurate, current material to help bridge the gap between principles and theories and the real world. These articles, however, become more useful for study when those of lasting value are carefully collected, organized, indexed, and reproduced in a low-cost format, which provides easy and permanent access when the material is needed. That is the role played by ANNUAL EDITIONS.

New to ANNUAL EDITIONS is the inclusion of related World Wide Web sites. These sites have been selected by our editorial staff to represent some of the best resources found on the World Wide Web today. Through our carefully developed topic guide, we have linked these Web resources to the articles covered in this ANNUAL EDITIONS reader. We think that you will find this volume useful, and we hope that you will take a moment to visit us on the Web at *http://www.dushkin.com* to tell us what you think.

The articles in this fifteenth edition of *Annual Editions: Geography* represent the wide range of topics associated with the discipline of geography. The major themes of spatial relationships, regional development, the population explosion, and socioeconomic inequalities exemplify the diversity of research areas within geography.

The book is organized into five units, each containing articles relating to geographical themes. Selections address the conceptual nature of geography and the global and regional problems in the world today. This latter theme reflects the geographer's concern with finding solutions to these serious issues. Regional problems, such as food shortages in the Sahel and the greenhouse effect, concern not only geographers but also researchers from other disciplines.

The association of geography with other fields is important, because expertise from related research will be necessary in finding solutions to some difficult problems. Input from the focus of geography is vital in our common search for solutions. This discipline has always been integrative. That is, geography uses evidence from many sources to answer the basic questions, "Where is it?" "Why is it there?" and "What is its relevance?" The first group of articles emphasizes the interconnectedness not only of places and regions in the world but of efforts toward solutions to problems as well. No single discipline can have all of the answers to the problems facing us today; the complexity of the issues is simply too great.

The writings in unit 1 discuss particular aspects of geography as a discipline and provide examples of the topics presented in the remaining four sections. Units 2, 3, and 4 represent major themes in geography. Unit 5 addresses important problems faced by geographers and others.

Annual Editions: Geography 00/01 will be useful to both teachers and students in their study of geography. The anthology is designed to provide detail and case study material to supplement the standard textbook treatment of geography. The goals of this anthology are to introduce students to the richness and diversity of topics relating to places and regions on Earth's surface, to pay heed to the serious problems facing humankind, and to stimulate the search for more information on topics of interest.

I would like to express my gratitude to Barbara Wells-Howe for her continued help in preparing this material for publication. Her typing, organization of materials, and many helpful suggestions are greatly appreciated. Without her diligence and professional efforts, this undertaking could not have been completed. Special thanks are also extended to Ian Nielsen for his continued encouragement during the preparation of this new edition and to Addie Raucci for her enthusiasm and helpfulness. A word of thanks must go as well to all those who recommended articles for inclusion in this volume and who commented on its overall organization. Sarah Witham Bednarz, Miriam Helen Hill, Artimus Keiffer, Peter O. Muller, Ray Oldokowski, Tom Paradise, John A. Vargas, and Randy W. Widdis were especially helpful in that regard.

In order to improve the next edition of *Annual Editions: Geography,* we need your help. Please share your opinions by filling out and returning to us the postage-paid *article rating form* on the last page of this book. We will give serious consideration to all your comments.

Gerald R. Pitzl
Editor

Contents

UNIT 1

Geography in a Changing World

Eight articles discuss the discipline of geography and the extremely varied and wide-ranging themes that define geography today.

The concepts in bold italics are developed in the article. For further expansion please refer to the Topic Guide and the Index.

UNIT 2

Human-Environment Relationships

Ten articles examine the relationship between humans and the land on which we live. Topics include the destruction of the rain forests, desertification, pollution, and the effects of human society on the global environment.

The concepts in bold italics are developed in the article. For further expansion please refer to the Topic Guide and the Index.

UNIT 3

The Region

Nine selections review the
importance of the region as a
concept in geography and as an
organizing framework for research.
A number of world regional trends,
as well as the patterns of area
relationships, are examined.

The concepts in bold italics are developed in the article. For further expansion please refer to the Topic Guide and the Index.

ix

The concepts in bold italics are developed in the article. For further expansion please refer to the Topic Guide and the Index.

UNIT 4

Spatial Interaction and Mapping

Seven articles discuss the key
theme in geographical analysis:
place-to-place spatial interaction.
Human diffusion, transportation systems, urban growth,
and cartography are some
of the themes examined.

UNIT 5

Population, Resources, and Socioeconomic Development

Nine articles examine the effects of population growth on natural resources and the resulting socioeconomic level of development.

The concepts in bold italics are developed in the article. For further expansion please refer to the Topic Guide and the Index.

The concepts in bold italics are developed in the article. For further expansion please refer to the Topic Guide and the Index.

Topic Guide

This topic guide suggests how the selections and World Wide Web sites found in the next section of this book relate to topics of traditional concern to geography students and professionals. It is useful for locating interrelated articles and Web sites for reading and research. The guide is arranged alphabetically according to topic.

The relevant Web sites, which are numbered and annotated on pages 6 and 7, are easily identified by the Web icon (☉) under the topic articles. By linking the articles and the Web sites by topic, this ANNUAL EDITIONS reader becomes a powerful learning and research tool.

5

DUSHKIN ONLINE

● AE: Geography

The following World Wide Web sites have been carefully researched and selected to support the articles found in this reader. If you are interested in learning more about specific topics found in this book, these Web sites are a good place to start. The sites are cross-referenced by number and appear in the topic guide on the previous two pages. Also, you can link to these Web sites through our DUSHKIN ONLINE support site at *http://www.dushkin.com/online/*.

The following sites were available at the time of publication. Visit our Web site—we update DUSHKIN ONLINE regularly to reflect any changes.

General Sources

1. The Association of American Geographers
http://www.aag.org
Surf this site of the Association of American Geographers to learn about AAG projects and publications, careers in geography, and information about related organizations.

2. National Geographic Society
http://www.nationalgeographic.com
This site provides links to National Geographic's huge archive of maps, articles, and other documents. Search the site for information about worldwide expeditions of interest to geographers.

3. The New York Times
http://www.nytimes.com
Browsing through the archives of the *New York Times* will provide you with a wide array of articles and information related to the different subfields of geography.

4. Social Science Internet Resources
http://www.wcsu.ctstateu.edu/library/ss_geography.html
This site is a definitive source for geography-related links to universities, browsers, cartography, associations, and discussion groups.

5. U.S. Geological Survey
http://www.usgs.gov
This site and its many links are replete with information and resources for geographers, from explanations of El Niño, to mapping, to geography education, to water resources. No geographer's resource list would be complete without frequent mention of the USGS.

Geography in a Changing World

6. Alternative Energy Institute
http://www.altenergy.org
The AEI will continue to monitor the transition from today's energy forms to the future in a "surprising journey of twists and turns." This site is the beginning of an incredible journey.

7. Geological Survey of Sweden: Other Geological Surveys
http://www3.sgu.se/links/Surveys_e.shtml
This site provides links to the national geographical surveys of many countries in Europe and elsewhere, including Brazil, South Africa, and the United States. It makes for very interesting and informative browsing.

8. Mission to Planet Earth
http://www.earth.nasa.gov
This site will direct you to information about NASA's Mission to Planet Earth program and its Science of the Earth System. Surf here to learn about satellites, El Niño, and even "strategic visions" of interest to geographers.

9. Public Utilities Commission of Ohio
http://www.puc.ohio.gov/consumer/gcc/index.html

PUCO aims for this site to serve as a clearinghouse of information related to global climate change. Its extensive links provide for explanation of the science and chronology of global climate change, acronyms, definitions, and more.

10. Santa Fe Institute
http://acoma.santafe.edu/sfi/research/
This home page of the Santa Fe Institute—a nonprofit, multidisciplinary research and education center—will lead you to a plethora of valuable links related to its primary goal: to create a new kind of scientific research community, pursuing emerging science. Such links as Evolution of Language, Ecology, and Local Rules for Global Problems are offered.

Human-Environment Relationships

11. The North-South Institute
http://www.nsi-ins.ca/info.html
Searching this site of the North-South Institute—which works to strengthen international development cooperation and enhance gender and social equity—will help you find information on a variety of development issues.

12. United Nations Environment Programme
http://www.unep.ch
Consult this home page of UNEP for links to critical topics of concern to geographers, including desertification and the impact of trade on the environment. The site will direct you to useful databases and global resource information.

13. World Health Organization
http://www.who.int
This home page of the World Health Organization will provide you with links to a wealth of statistical and analytical information about health in the developing world.

The Region

14. AS at UVA Yellow Pages: Regional Studies
http://xroads.virginia.edu/~YP/regional/regional.html
Those interested in American regional studies will find this site a gold mine. Links to periodicals and other informational resources about the Midwest/Central, Northeast, South, and West regions are provided here.

15. Can Cities Save the Future?
http://pan.cedar.univie.ac.at/habitat/press/press7.html
This press release about the second session of the Preparatory Committee for Habitat II is an excellent discussion of the question of global urbanization.

16. IISDnet
http://iisd1.iisd.ca
The International Institute for Sustainable Development, a Canadian organization, presents information through gateways entitled Business and Sustainable Development, Developing Ideas, and Hot Topics. Linkages is its multimedia resource for environment and development policymakers.

17. NewsPage
http://pnp1.individual.com
Individual, Inc., maintains this business-oriented Web site. Geographers will find links to much valuable information

about such fields as energy, environmental services, media and communications, and health care.

18. Telecommuting as an Investment: The Big Picture—John Wolf
http://www.svi.org/telework/forums/messages5/48.html
This page deals with the many issues related to telecommuting, including its potential role in reducing environmental pollution. The site discusses such topics as employment law and the impact of telecommuting on businesses and employees.

19. The Urban Environment
http://www.geocities.com/RainForest/Vines/6723/urb/index.html
Global urbanization is discussed fully at this site, which also includes the original 1992 Treaty on Urbanization.

20. Virtual Seminar in Global Political Economy//Global Cities & Social Movements
http://csf.colorado.edu/gpe/gpe95b/resources.html
This Web site is rich in links to subjects of interest in regional studies, such as sustainable cities, megacities, and urban planning. Links to many international nongovernmental organizations are included.

21. WWW-LARCH-LK Archive: Sustainability
http://www.clr.toronto.edu/ARCHIVES/HMAIL/larchl/0737.html
This site gives you the opportunity to read and respond to a discourse on sustainability, with many different opinions and viewpoints represented.

Spatial Interaction and Mapping

22. Edinburgh Geographical Information Systems
http://www.geo.ed.ac.uk/home/gishome.html
This valuable site, hosted by the Department of Geography at the University of Edinburgh, provides information on all aspects of Geographic Information Systems and provides links to other servers worldwide. A GIS reference database as well as a major GIS bibliography is included.

23. GIS Frequently Asked Questions and General Information
http://www.census.gov/ftp/pub/geo/www/faq-index.html
Browse through this site to get answers to FAQs about Geographic Information Systems. It can direct you to general information about GIS as well as guidelines on such specific questions as how to order U.S. Geological Survey maps. Other sources of information are also noted.

24. International Map Trade Association
http://www.maptrade.org
The International Map Trade Association offers this site for those interested in information on maps, geography, and mapping technology. Lists of map retailers and publishers as well as upcoming IMTA conferences and trade shows are noted.

25. PSC Publications
http://www.psc.lsa.umich.edu/pubs/abs/abs94-319.html
Use this site and its links from the Population Studies Center of the University of Michigan for spatial patterns of immigration and discussion of white and black flight from high immigration metropolitan areas in the United States.

26. U.S. Geological Survey
http://www.usgs.gov/research/gis/title.html
This site discusses the uses for Geographic Information Systems and explains how GIS works, addressing such topics as data integration, data modeling, and relating information from different sources.

Population, Resources, and Socioeconomic Development

27. African Studies WWW (U.Penn)
http://www.sas.upenn.edu/African_Studies/AS.html
This site will give you access to rich and varied resources that cover such topics as demographics, migration, family planning, and health and nutrition.

28. Human Rights and Humanitarian Assistance
http://info.pitt.edu/~ian/resource/human.htm
Through this site, part of the World Wide Web Virtual Library, you can conduct research into a number of human-rights topics in order to gain a greater understanding of the issues affecting indigenous peoples in the modern era.

29. Hypertext and Ethnography
http://www.umanitoba.ca/faculties/arts/anthropology/tutor/ aaa_presentation.new.html
This site, presented by Brian Schwimmer of the University of Manitoba, will be of great value to people who are interested in culture and communication. He addresses such topics as multivocality and complex symbolization, among many others.

30. Research and Reference (Library of Congress)
http://lcweb.loc.gov/rr/
This research and reference site of the Library of Congress will lead you to invaluable information on different countries. It provides links to numerous publications, bibliographies, and guides in area studies that can be of great help to geographers.

31. Space Research Institute
http://arc.iki.rssi.ru/Welcome.html
Browse through this home page of Russia's Space Research Institute for information on its Environment Monitoring Information Systems, the IKI Satellite Situation Center, and its Data Archive.

We highly recommend that you review our Web site for expanded information and our other product lines. We are continually updating and adding links to our Web site in order to offer you the most usable and useful information that will support and expand the value of your Annual Editions. You can reach us at: *http://www.dushkin.com/annualeditions/.*

www.dushkin.com/online/

Unit Selections

1. **Rediscovering the Importance of Geography,** Alexander B. Murphy
2. **The Four Traditions of Geography,** William D. Pattison
3. **The American Geographies,** Barry Lopez
4. **World Prisms: The Future of Sovereign States and International Order,** Richard Falk
5. **Human Domination of Earth's Ecosystems,** Peter M. Vitousek, Harold A. Mooney, Jane Lubchenco, and Jerry M. Melillo
6. **The Role of Science in Policy: The Climate Change Debate in the United States,** Eugene B. Skolnikoff
7. **California Fumes Over Oil: Offshore-Drilling Decision Gives New Life to an Old Battle,** Betsy Streisand
8. **Lead in the Inner Cities,** Howard W. Mielke

Key Points to Consider

❖ Why is geography called an integrating discipline?

❖ How is geography related to earth science? Give some examples of these relationships.

❖ What are area studies? Why is the spatial concept so important in geography? What is your definition of geography?

❖ Why do history and geography rely on each other? How have humans affected the weather? The environment in general?

❖ How is individual behavior related to our understanding of the environment?

❖ Discuss whether or not change is a good thing. Why is it important to anticipate change?

❖ What does interconnectedness mean in terms of places? Give examples of how you as an individual interact with people in other places. How are you "connected" to the rest of the world?

❖ What will the world be like in the year 2010? Tell why you are pessimistic or optimistic about the future. What, if anything, can you do about the future?

 **Links**   **www.dushkin.com/online/**

6. **Alternative Energy Institute**
   http://www.altenergy.org
7. **Geological Survey of Sweden: Other Geological Surveys**
   http://www3.sgu.se/links/Surveys_e.shtml
8. **Mission to Planet Earth**
   http://www.earth.nasa.gov
9. **Public Utilities Commission of Ohio**
   http://www.puc.ohio.gov/consumer/gcc/index.html
10. **Santa Fe Institute**
    http://acoma.santafe.edu/sfi/research/

These sites are annotated on pages 6 and 7.

What is geography? This question has been asked innumerable times, but it has not elicited a universally accepted answer, even from those who are considered to be members of the geography profession. The reason lies in the very nature of geography as it has evolved through time. Geography is an extremely wide-ranging discipline, one that examines appropriate sets of events or circumstances occurring at specific places. Its goal is to answer certain basic questions.

The first question—Where is it?—establishes the location of the subject under investigation. The concept of location is very important in geography, and its meaning extends beyond the common notion of a specific address or the determination of the latitude and longitude of a place. Geographers are more concerned with the relative location of a place and how that place interacts with other places both far and near. Spatial interaction and the determination of the connections between places are important themes in geography.

Once a place is "located," in the geographer's sense of the word, the next question is, Why is it here? For example, why are people concentrated in high numbers on the North China plain, in the Ganges River Valley in India, and along the eastern seaboard in the United States? Conversely, why are there so few people in the Amazon basin and the Central Siberian lowlands? Generally, the geographer wants to find out why particular distribution patterns occur and why these patterns change over time.

The element of time is another extremely important ingredient in the geographical mix. Geography is most concerned with the activities of human beings, and human beings bring about change. As changes occur, new adjustments and modifications are made in the distribution patterns previously established. Patterns change, for instance, as new technology brings about new forms of communication and transportation and as once-desirable locations decline in favor of new ones. For example, people migrate from once-productive regions such as the Sahel when a

disaster such as drought visits the land. Geography, then, is greatly concerned with discovering the underlying processes that can explain the transformation of distribution patterns and interaction forms over time. Geography itself is dynamic, adjusting as a discipline to handle new situations in a changing world.

Geography is truly an integrating discipline. The geographer assembles evidence from many sources in order to explain a particular pattern or ongoing process of change. Some of this evidence may even be in the form of concepts or theories borrowed from other disciplines. The first article of this unit stresses the importance of geography as a discipline and proclaims its "rediscovery." The next two articles of this unit provide insight into both the conceptual nature of geography and the development of the discipline over time.

Throughout its history, four main themes have been the focus of research work in geography. These themes or traditions, according to William Pattison in "The Four Traditions of Geography," link geography with earth science, establish it as a field that studies land-human relationships, engage it in area studies, and give it a spatial focus. Although Pattison's article first appeared over 30 years ago, it is still referred to and cited frequently today. Much of the geographical research and analysis engaged in today would fall within one or more of Pattison's traditional areas, but new areas are also opening for geographers. In a particularly thought-provoking essay, Barry Lopez discusses local geographies and the importance of a sense of place. Richard Falk then reports on the geopolitical situation of sovereign states and the international order. In "Human Domination of the Earth's Ecosystems," we see how human activity is significantly altering the globe. This theme continues with Eugene Skolnikoff's analysis of the global climate change debate. "California Fumes Over Oil" raises an old issue involving economic development and the environment. Finally, Howard Mielke details the problem of lead content in inner-city soils.

POINT OF VIEW

Rediscovering the Importance of Geography

By Alexander B. Murphy

AS AMERICANS STRUGGLE to understand their place in a world characterized by instant global communications, shifting geopolitical relationships, and growing evidence of environmental change, it is not surprising that the venerable discipline of geography is experiencing a renaissance in the United States. More elementary and secondary schools now require courses in geography, and the College Board is adding the subject to its Advanced Placement program. In higher education, students are enrolling in geography courses in unprecedented numbers. Between 1985–86 and 1994–95, the number of bachelor's degrees awarded in geography increased from 3,056 to 4,295. Not coincidentally, more businesses are looking for employees with expertise in geographical analysis, to help them analyze possible new markets or environmental issues.

In light of these developments, institutions of higher education cannot afford simply to ignore geography, as some of them have, or to assume that existing programs are adequate. College administrators should recognize the academic and practical advantages of enhancing their offerings in geography, particularly if they are going to meet the demand for more and better geography instruction in primary and secondary schools. We cannot afford to know so little about the other countries and peoples with which we now interact with such frequency, or about

the dramatic environmental changes unfolding around us.

From the 1960s through the 1980s, most academics in the United States considered geography a marginal discipline, although it remained a core subject in most other countries. The familiar academic divide in the United States between the physical sciences, on one hand, and the social sciences and humanities, on the other, left little room for a discipline concerned with how things are organized and relate to one another on the surface of the earth—a concern that necessarily bridges the physical and cultural spheres. Moreover, beginning in the 1960s, the U.S. social-science agenda came to be dominated by pursuit of more-scientific explanations for human phenomena, based on assumptions about global similarities in human institutions, motivations, and actions. Accordingly, regional differences often were seen as idiosyncrasies of declining significance.

Although academic administrators and scholars in other disciplines might have marginalized geography, they could not kill it, for any attempt to make sense of the world must be based on some understanding of the changing human and physical patterns that shape its evolution—be they shifting vegetation zones or expanding economic contacts across international boundaries. Hence, some U.S. colleges and universities continued to teach geography, and the discipline was often in the background of many policy is-

sues—for example, the need to assess the risks associated with foreign investment in various parts of the world.

By the late 1980s, Americans' general ignorance of geography had become too widespread to ignore. Newspapers regularly published reports of surveys demonstrating that many Americans could not identify major countries or oceans on a map. The real problem, of course, was not the inability to answer simple questions that might be asked on *Jeopardy!*; instead, it was what that inability demonstrated about our collective understanding of the globe.

Geography's renaissance in the United States is due to the growing recognition that physical and human processes such as soil erosion and ethnic unrest are inextricably tied to their geographical context. To understand modern Iraq, it is not enough to know who is in power and how the political system functions. We also need to know something about the country's ethnic groups and their settlement patterns, the different physical environments and resources within the country, and its ties to surrounding countries and trading partners.

Those matters are sometimes addressed by practitioners of other disciplines, of course, but they are rarely central to the analysis. Instead, generalizations are often made at the level of the state, and little attention is given to spatial patterns and practices that play out on local levels or across international boundaries. Such preoccupations help to explain why many scholars were caught off guard by the explosion of ethnic unrest in Eastern Europe following the fall of the Iron Curtain.

Similarly, comprehending the dynamics of El Niño requires more than knowledge of the behavior of ocean and air currents; it is also important to understand how those currents are situated with respect to land masses and how they relate to other climatic patterns, some of which have been altered by the burning of fossil fuels and other human activities. And any attempt to understand the nature and extent of humans' impact on the environment requires consideration of the relationship between human and physical contributions to environmental change. The factories and cars in a city produce smog, but surrounding mountains may trap it, increasing air pollution significantly.

TODAY, academics in fields including history, economics, and conservation biology are turning to geographers for help with some of their concerns. Paul Krugman, a noted economist at the Massachusetts Institute of Technology, for example, has turned conventional wisdom on its head by pointing out the role of historically rooted regional inequities in how international trade is structured.

Geographers work on issues ranging from climate change to ethnic conflict to urban sprawl. What unites their work is its focus on the shifting organization and character of the earth's surface. Geographers examine changing patterns of vegetation to study global warming; they analyze where ethnic groups live in Bosnia to help understand the pros and cons of competing administrative solutions to the civil war there; they map AIDS cases in Africa to learn how to reduce the spread of the disease.

Geography is reclaiming attention because it addresses such questions in their relevant spatial and environmental contexts. A growing number of scholars in other disciplines are realizing that it is a mistake to treat all places as if they were essentially the same (think of the assumptions in most economic models), or to undertake research on the environment that does not include consideration of the relationships between human and physical processes in particular regions.

Still, the challenges to the discipline are great. Only a small number of primary- and secondary-school teachers have enough training in geography to offer students an exciting introduction to the subject. At the college level, many geography departments are small; they are absent altogether at some high-profile universities.

Perhaps the greatest challenge is to overcome the public's view of geography as a simple exercise in place-name recognition. Much of geography's power lies in the insights it sheds on the nature and meaning of the evolving spatial arrangements and landscapes that make up our world. The importance of those insights should not be underestimated at a time of changing political boundaries, accelerated human alteration of the environment, and rapidly shifting patterns of human interaction.

Alexander B. Murphy is a professor and head of the geography department at the University of Oregon, and a vice-president of the American Geographical Society.

The Four Traditions of Geography

William D. Pattison

Late Summer, 1990

To Readers of the *Journal of Geography:*

I am honored to be introducing, for a return to the pages of the *Journal* after more than 25 years, "The Four Traditions of Geography," an article which circulated widely, in this country and others, long after its initial appearance—in reprint, in xerographic copy, and in translation. A second round of life at a level of general interest even approaching that of the first may be too much to expect, but I want you to know in any event that I presented the paper in the beginning as my gift to the geographic community, not as a personal property, and that I re-offer it now in the same spirit.

In my judgment, the article continues to deserve serious attention—perhaps especially so, let me add, among persons aware of the specific problem it was intended to resolve. The background for the paper was my experience as first director of the High School Geography Project (1961–63)—not all of that experience but only the part that found me listening, during numerous conference sessions and associated interviews, to academic geographers as they responded to the project's invitation to locate "basic ideas" representative of them all. I came away with the conclusion that I had been witnessing not a search for consensus but rather a blind struggle for supremacy among honest persons of contrary intellectual commitment. In their dialogue, two or more different terms had been used, often unknowingly, with a single reference, and no less disturbingly, a single term had been used, again often unknowingly, with two or more different references. The article was my attempt to stabilize the discourse. I was proposing a basic nomenclature (with explicitly associated ideas) that would, I trusted, permit the de-velopment of mutual comprehension **and** confront all parties concerned with the pluralism inherent in geographic thought.

This intention alone could not have justified my turning to the NCGE as a forum, of course. The fact is that from the onset of my discomfiting realization I had looked forward to larger consequences of a kind consistent with NCGE goals. As finally formulated, my wish was that the article would serve "to greatly expedite the task of maintaining an alliance between professional geography and pedagogical geography and at the same time to promote communication with laymen" (see my fourth paragraph). I must tell you that I have doubts, in 1990, about the acceptability of my word choice, in saying "professional," "pedagogical," and "layman" in this context, but the message otherwise is as expressive of my hope now as it was then.

I can report to you that twice since its appearance in the *Journal,* my interpretation has received more or less official acceptance—both times, as it happens, at the expense of the earth science tradition. The first occasion was Edward Taaffe's delivery of his presidential address at the 1973 meeting of the Association of American Geographers (see *Annals AAG,* March 1974, pp. 1–16). Taaffe's working-through of aspects of an interrelations among the spatial, area studies, and man-land traditions is by far the most thoughtful and thorough of any of which I am aware. Rather than fault him for omission of the fourth tradition, I compliment him on the grace with which he set it aside in conformity to a meta-epistemology of the American university which decrees the integrity of the social sciences as a consortium in their own right. He was sacrificing such holistic claims as geography might be able to muster for a freedom to argue the case for geography as a social science.

The second occasion was the publication in 1984 of *Guidelines for Geographic Education: Elementary and Secondary Schools,* authored by a committee jointly representing the AAG and the NCGE. Thanks to a recently published letter (see *Journal of Geography,* March-April 1990, pp. 85–86), we know that, of five themes commended to teachers in this source,

The committee lifted the human environmental interaction theme directly from Pattison. The themes of place and location are based on Pattison's spatial or geometric geography, and the theme of region comes from Pattison's area studies or regional geography.

Having thus drawn on my spatial, area studies, and man-land traditions for four of the five themes, the committee could have found the remaining theme, movement, there too—in the spatial tradition (see my sixth paragraph). However that may be, they did not avail themselves of the earth science tradition, their reasons being readily surmised. Peculiar to the elementary and secondary schools is a curriculum category framed as much by theory of citizenship as by theory of knowledge: the social studies. With admiration, I see already in the committee members' adoption of the theme idea a strategy for assimilation of their program to the established repertoire of social studies practice. I see in their exclusion of the earth science tradition an intelligent respect for social studies' purpose.

Here's to the future of education in geography: may it prosper as never before.

W. D. P., 1990

From *Journal of Geography,* September/October 1990, pp. 202–206. © 1990 by the National Council for Geographic Education. Reprinted by permission.

Reprinted from the Journal of Geography, 1964, pp. 211–216.

In 1905, one year after professional geography in this country achieved full social identity through the founding of the Association of American Geographers, William Morris Davis responded to a familiar suspicion that geography is simply an undisciplined "omnium-gatherum" by describing an approach that as he saw it imparts a "geographical quality" to some knowledge and accounts for the absence of the quality elsewhere.[1] Davis spoke as president of the AAG. He set an example that was followed by more than one president of that organization. An enduring official concern led the AAG to publish, in 1939 and in 1959, monographs exclusively devoted to a critical review of definitions and their implications.[2]

Every one of the well-known definitions of geography advanced since the founding of the AAG has had its measure of success. Tending to displace one another by turns, each definition has said something true of geography.[3] But from the vantage point of 1964, one can see that each one has also failed. All of them adopted in one way or another a monistic view, a singleness of preference, certain to omit if not to alienate numerous professionals who were in good conscience continuing to participate creatively in the broad geographic enterprise.

The thesis of the present paper is that the work of American geographers, although not conforming to the restrictions implied by any one of these definitions, has exhibited a broad consistency, and that this essential unity has been attributable to a small number of distinct but affiliated traditions, operant as binders in the minds of members of the profession. These traditions are all of great age and have passed into American geography as parts of a general legacy of Western thought. They are shared today by geographers of other nations.

There are four traditions whose identification provides an alternative to the competing monistic definitions that have been the geographer's lot. The resulting pluralistic basis for judgment promises, by full accommodation of what geographers do and by plain-spoken representation thereof, to greatly expedite the task of maintaining an alliance between professional geography and pedagogical geography and at the same time to promote communication with laymen. The following discussion treats the traditions in this order: (1) a spatial tradition, (2) an area studies tradition, (3) a man-land tradition and (4) an earth science tradition.

Spatial Tradition

Entrenched in Western thought is a belief in the importance of spatial analysis, of the act of separating from the happenings of experience such aspects as distance, form, direction and position. It was not until the 17th century that philosophers concentrated attention on these aspects by asking whether or not they were properties of things-in-themselves. Later, when the 18th century writings of Immanuel Kant had become generally circulated, the notion of space as a category including all of these aspects came into widespread use. However, it is evident that particular spatial questions were the subject of highly organized answering attempts long before the time of any of these cogitations. To confirm this point, one need only be reminded of the compilation of elaborate records concerning the location of things in ancient Greece. These were records of sailing distances, of coastlines and of landmarks that grew until they formed the raw material for the great *Geographia* of Claudius Ptolemy in the 2nd century A.D.

A review of American professional geography from the time of its formal organization shows that the spatial tradition of thought had made a deep penetration from the very beginning. For Davis, for Henry Gannett and for most if not all of the 44 other men of the original AAG, the determination and display of spatial aspects of reality through mapping were of undoubted importance, whether contemporary definitions of geography happened to acknowledge this fact or not. One can go further and, by probing beneath the art of mapping, recognize in the behavior of geographers of that time an active interest in the true essentials of the spatial tradition—*geometry* and *movement*. One can trace a basic favoring of movement as a subject of study from the turn-of-the-century work of Emory R. Johnson, writing as professor of transportation at the University of Pennsylvania, through the highly influential theoretical and substantive work of Edward L. Ullman during the past 20 years and thence to an article by a younger geographer on railroad freight traffic in the U.S. and Canada in the *Annals* of the AAG for September 1963.[4]

One can trace a deep attachment to geometry, or positioning-and-layout, from articles on boundaries and population densities in early 20th century volumes of the *Bulletin of the American Geographical Society,* through a controversial pronouncement by Joseph Schaefer in 1953 that granted geographical legitimacy only to studies of spatial patterns[5] and so onward to a recent *Annals* report on electronic scanning of cropland patterns in Pennsylvania.[6]

One might inquire, is discussion of the spatial tradition, after the manner of the remarks just made, likely to bring people within geography closer to an understanding of one another and people outside geography closer to an understanding of geographers? There seem to be at least two reasons for being hopeful. First, an appreciation of this tradition allows one to see a bond of fellowship uniting the elementary school teacher, who attempts the most rudimentary instruction in directions and mapping, with the contemporary research geographer, who dedicates himself to an exploration of central-place theory. One cannot only open the eyes of many teachers to the potentialities of their own instruction, through proper exposition of the spatial tradition, but one can also "hang a bell" on research quantifiers in geography, who are often thought to have wandered so far in their intellectual adventures as to have become lost from the rest. Looking outside geography, one may anticipate benefits from the readiness of countless persons to associate the name "geography" with maps. Latent within this readiness is a willingness to recognize as geography, too, what maps are about—and that is the geometry of and the movement of what is mapped.

Area Studies Tradition

The area studies tradition, like the spatial tradition, is quite strikingly represented in classical antiquity by a practitioner to whose surviving work we can point. He is Strabo, celebrated for his *Geography* which is a massive production addressed to the statesmen of Augustan Rome and intended to sum up and regularize knowledge not of the location of places and associated cartographic facts, as in the somewhat later case of Ptolemy, but of the nature of places, their character and their differentiation. Strabo exhibits interesting attributes of the area-studies tradition that can hardly be overemphasized. They are a pronounced tendency toward subscription primarily to literary standards, an almost omnivorous appetite for information and a self-conscious companionship with history.

It is an extreme good fortune to have in the ranks of modern American geography the scholar Richard Hartshorne, who has pondered the meaning of the area-studies tradition with a legal acuteness that few persons would challenge. In his *Nature of Geography,* his 1939 monograph already cited,[7] he scrutinizes exhaustively the implications of the "interesting attributes" identified in connection with Strabo, even though his concern is with quite other and much later authors, largely German. The major literary problem of unities or wholes he considers from every angle. The Gargantuan appetite for miscellaneous information he accepts and rationalizes. The companionship between area studies and history he clarifies by appraising the so-called idiographic content of both and by affirming the tie of both to what he and Sauer have called "naively given reality."

The area-studies tradition (otherwise known as the chorographic tradition) tended to be excluded from early American professional geography. Today it is beset by certain champions of the spatial tradition who would have one believe that somehow the area-studies way of organizing knowledge is only a subdepartment of spatialism. Still, area-studies as a method of presentation lives and prospers in its own right. One can turn today for reassurance on this score to practically any issue of the *Geographical Review*, just as earlier readers could turn at the opening of the century to that magazine's forerunner.

What is gained by singling out this tradition? It helps toward restoring the faith of many teachers who, being accustomed to administering learning in the area-studies style, have begun to wonder if by doing so they really were keeping in touch with professional geography. (Their doubts are owed all too much to the obscuring effect of technical words attributable to the very professionals who have been intent, ironically, upon protecting that tradition.) Among persons outside the classroom the geographer stands to gain greatly in intelligibility. The title "area-studies" itself carries an understood message in the United States today wherever there is contact with the usages of the academic community. The purpose of characterizing a place, be it neighborhood or nation-state, is readily grasped. Furthermore, recognition of the right of a geographer to be unspecialized may be expected to be forthcoming from people generally, if application for such recognition is made on the merits of this tradition, explicitly.

Man-Land Tradition

That geographers are much given to exploring man-land questions is especially evident to anyone who examines geographic output, not only in this country but also abroad. O. H. K. Spate, taking an international view, has felt justified by his observations in nominating as the most significant ancient precursor of today's geography neither Ptolemy nor Strabo nor writers typified in their outlook by the geographies of either of these two men, but rather Hippocrates, Greek physician of the 5th century B.C. who left to posterity an extended essay, *On Airs, Waters and Places*.[8] In this work made up of reflections on human health and conditions of external nature, the questions asked are such as to confine thought almost altogether to presumed influence passing from the latter to the former, questions largely about the effects of winds, drinking water and seasonal changes upon man. Understandable though this uni-directional concern may have been for Hippocrates as medical commentator, and defensible as may be the attraction that this same approach held for students of the

condition of man for many, many centuries thereafter, one can only regret that this narrowed version of the man-land tradition, combining all too easily with social Darwinism of the late 19th century, practically overpowered American professional geography in the first generation of its history.[9] The premises of this version governed scores of studies by American geographers in interpreting the rise and fall of nations, the strategy of battles and the construction of public improvements. Eventually this special bias, known as environmentalism, came to be confused with the whole of the man-land tradition in the minds of many people. One can see now, looking back to the years after the ascendancy of environmentalism, that although the spatial tradition was asserting itself with varying degrees of forwardness, and that although the area-studies tradition was also making itself felt, perhaps the most interesting chapters in the story of American professional geography were being written by academicians who were reacting against environmentalism while deliberately remaining within the broad man-land tradition. The rise of culture historians during the last 30 years has meant the dropping of a curtain of culture between land and man, through which it is asserted all influence must pass. Furthermore work of both culture historians and other geographers has exhibited a reversal of the direction of the effects in Hippocrates, man appearing as an independent agent, and the land as a sufferer from action. This trend as presented in published research has reached a high point in the collection of papers titled *Man's Role in Changing the Face of the Earth*. Finally, books and articles can be called to mind that have addressed themselves to the most difficult task of all, a balanced tracing out of interaction between man and environment. Some chapters in the book mentioned above undertake just this. In fact the separateness of this approach is discerned only with difficulty in many places; however, its significance as a general research design that rises above environmentalism, while refusing to abandon the man-land tradition, cannot be mistaken.

The NCGE seems to have associated itself with the man-land tradition, from the time of founding to the present day, more than with any other tradition, although all four of the traditions are amply represented in its official magazine, *The Journal of Geography* and in the proceedings of its annual meetings. This apparent preference on the part of the NCGE members *for defining geography in terms of the man-land tradition* is strong evidence of the appeal that man-land ideas, separately stated, have for persons whose main job is teaching. It should be noted, too, that this inclination reflects a proven acceptance by the general public of learning that centers on resource use and conservation.

Earth Science Tradition

The earth science tradition, embracing study of the earth, the waters of the earth, the atmosphere surrounding the earth and the association between earth and sun, confronts one with a paradox. On the one hand one is assured by professional geographers that their participation in this tradition has declined precipitously in the course of the past few decades, while on the other one knows that college departments of geography across the nation rely substantially, for justification of their role in general education, upon curricular content springing directly from this tradition. From all the reasons that combine to account for this state of affairs, one may, by selecting only two, go far toward achieving an understanding of this tradition. First, there is the fact that American college geography, growing out of departments of geology in many crucial instances, was at one time greatly overweighted in favor of earth science, thus rendering the field unusually liable to a sense of loss as better balance came into being. (This one-time disproportion found reciprocate support for many years in the narrowed, environmentalistic interpretation of the man-land tradition.) Second, here alone in earth science does one encounter subject matter in the normal sense of the term as one reviews geographic traditions. The spatial tradition abstracts certain aspects of reality; area studies is distinguished by a point of view; the man-land tradition dwells upon relationships; but earth science is identifiable through concrete objects. Historians, sociologists and other academicians tend not only to accept but also to ask for help from this part of geography. They readily appreciate earth science as something physically associated with their subjects of study, yet generally beyond their competence to treat. From this appreciation comes strength for geography-as-earth-science in the curriculum.

Only by granting full stature to the earth science tradition can one make sense out of the oft-repeated addage, "Geography is the mother of sciences." This is the tradition that emerged in ancient Greece, most clearly in the work of Aristotle, as a wide-ranging study of natural processes in and near the surface of the earth. This is the tradition that was rejuvenated by Varenius in the 17th century as "Geographia Generalis." This is the tradition that has been subjected to subdivision as the development of science has approached the present day, yielding mineralogy, paleontology, glaciology, meterology and other specialized fields of learning.

Readers who are acquainted with American junior high schools may want to make a challenge at this point, being aware that a current revival of earth sciences is being sponsored in those schools by the field of geology. Belatedly, geography has joined in support of this revival.[10] It may be said that in this connection and in others, American professional geography may have faltered in its adherence to the earth science tradition but not given it up.

In describing geography, there would appear to be some advantages attached to isolating this final tradition. Separation improves the geographer's chances of successfully explaining to educators why geography has extreme difficulty in accommodating itself to social studies programs. Again, separate attention allows one to make understanding contact with members of the American public for whom surrounding nature is known as the geographic environment. And finally, specific reference to the geographer's earth science tradition brings into the open the basis of what is, almost without a doubt, morally the most significant concept in the entire geographic heritage, that of the earth as a unity, the single common habitat of man.

An Overview

The four traditions though distinct in logic are joined in action. One can say of geography that it pursues concurrently all four of them. Taking the traditions in varying combinations, the geographer can explain the conventional divisions of the field. Human or cultural geography turns out to consist of the first three traditions applied to human societies; physical geography, it becomes evi-

dent, is the fourth tradition prosecuted under constraints from the first and second traditions. Going further, one can uncover the meanings of "systematic geography," "regional geography," "urban geography," "industrial geography," etc.

It is to be hoped that through a widened willingness to conceive of and discuss the field in terms of these traditions, geography will be better able to secure the inner unity and outer intelligibility to which reference was made at the opening of this paper, and that thereby the effectiveness of geography's contribution to American education and to the general American welfare will be appreciably increased.

Notes

1. William Morris Davis, "An Inductive Study of the Content of Geography," *Bulletin of the American Geographical Society,* Vol. 38, No. 1 (1906), 71.
2. Richard Hartshorne, *The Nature of Geography,* Association of American Geographers (1939), and idem., *Perspective on the Nature of Geography,* Association of American Geographers (1959).
3. The essentials of several of these definitions appear in Barry N. Floyd, "Putting Geography in Its Place," *The Journal of Geography,* Vol. 62, No. 3 (March, 1963). 117–120.
4. William H. Wallace, "Freight Traffic Functions of Anglo-American Railroads," *Annals of the Association of American Geographers,* Vol. 53, No. 3 (September, 1963), 312–331.
5. Fred K. Schaefer, "Exceptionalism in Geography: A Methodological Examination," *Annals of the Association of American Geographers,* Vol. 43, No. 3 (September, 1953), 226–249.
6. James P. Latham, "Methodology for an Instrumental Geographic Analysis," *Annals of the Association of American Geographers,* Vol. 53, No. 2 (June, 1963), 194–209.
7. Hartshorne's 1959 monograph, *Perspective on the Nature of Geography,* was also cited earlier. In this later work, he responds to dissents from geographers whose preferred primary commitment lies outside the area studies tradition.
8. O. H. K. Spate, "Quantity and Quality in Geography," *Annals of the Association of American Geographers,* Vol. 50, No. 4 (December, 1960), 379.
9. Evidence of this dominance may be found in Davis's 1905 declaration: "Any statement is of geographical quality if it contains . . . some relation between an element of inorganic control and one of organic response" (Davis, *loc. cit.*).
10. Geography is represented on both the Steering Committee and Advisory Board of the Earth Science Curriculum Project, potentially the most influential organization acting on behalf of earth science in the schools.

The American Geographies

Americans are fast becoming strangers in a strange land, where one roiling river, one scarred patch of desert, is as good as another. America the beautiful exists—a select few still know it intimately—but many of us are settling for a homogenized national geography.

Barry Lopez

Barry Lopez has written The Rediscovery of North America *(Vintage), and his most recent book is* Field Notes *(Knopf).*

It has become commonplace to observe that Americans know little of the geography of their country, that they are innocent of it as a landscape of rivers, mountains, and towns. They do not know, supposedly, the location of the Delaware Water Gap, the Olympic Mountains, or the Piedmont Plateau; and, the indictment continues, they have little conception of the way the individual components of this landscape are imperiled, from a human perspective, by modern farming practices or industrial pollution.

I do not know how true this is, but it is easy to believe that it is truer than most of us would wish. A recent Gallup Organization and National Geographic Society survey found Americans woefully ignorant of world geography. Three out of four couldn't locate the Persian Gulf. The implication was that we knew no more about our own homeland, and that this ignorance undermined the integrity of our political processes and the efficiency of our business enterprises.

As Americans, we profess a sincere and fierce love for the American landscape, for our rolling prairies, freeflowing rivers, and "purple mountains' majesty"; but it is hard to imagine, actually, where this particular landscape is. It is not just that a nostalgic landscape has passed away—Mark Twain's Mississippi is now dammed from Minnesota to Missouri and the prairies have all been sold and fenced. It is that it has always been a romantic's landscape. In the attenuated form in which it is presented on television today, in magazine articles and in calendar photographs, the essential wildness of the American landscape is reduced to attractive scenery. We look out on a familiar, memorized landscape that portends adventure and promises enrichment. There are no distracting people in it and few artifacts of human life. The animals are all beautiful, diligent, one might even say well-

---

To truly understand geography requires not only time but a kind of local expertise, an intimacy with place few of us ever develop.

---

behaved. Nature's unruliness, the power of rivers and skies to intimidate, and any evidence of disastrous human land management practices are all but invisible. It is, in short, a magnificent garden, a colonial vision of paradise imposed on a real place that is, at best, only selectively known.

The real American landscape is a face of almost incomprehensible depth and complexity. If one were to sit for a few days, for example, among the ponderosa pine forests and black lava fields of the Cascade Mountains in western Oregon, inhaling the pines' sweet balm on an evening breeze from some point on the barren rock, and then were to step off to the Olympic Peninsula in Washington, to those rain forests with sphagnum moss floors soft as fleece underfoot and Douglas firs too big around for five people to hug, and then head south to walk the ephemeral creeks and sun-blistered playas of the Mojave Desert in southern California, one would be reeling under the sensations. The contrast is not only one of plants and soils, a different array say, of brilliantly colored beetles. The shock to the senses comes from a different shape to the silence, a difference in the very quality of light, in the weight of the air. And this relatively short journey down the West Coast would still leave the traveler with all that lay to the east to explore—the anomalous sand hills of Nebraska, the heat and frog voices of Okefenokee Swamp, the fetch of Chesapeake Bay, the hardwood copses and black bears of the Ozark Mountains.

No one of these places, of course, can be entirely fathomed, biologically or aesthetically. They are mysteries upon which we impose names. Enchantments. We tick the names off glibly but lovingly. We mean no disrespect. Our genuine desire, though we might be skeptical about the time it would take and uncertain of its practical value to us, is to actually know these places. As deeply ingrained in the American psyche as the desire to conquer and control the land is the desire to sojourn in it, to sail up and down Pamlico Sound, to paddle a

Originally from *Orion* magazine, Autumn 1989, pp. 52–61. © 1989 by Barry Holstun Lopez. Reprinted by permission of Sterling Lord Literistic, Inc.

canoe through Minnesota's boundary waters, to walk on the desert of the Great Salt Lake, to camp in the stony hardwood valleys of Vermont.

To do this well, to really come to an understanding of a specific American geography, requires not only time but a kind of local expertise, an intimacy with place few of us ever develop. There is no way around the former requirement: If you want to know you must take the time. It is not in books. A specific geographical understanding, however, can be sought out and borrowed. It resides with men and women more or less sworn to a place, who abide there, who have a feel for the soil and history, for the turn of leaves and night sounds. Often they are glad to take the outlander in tow.

These local geniuses of American landscape, in my experience, are people in whom geography thrives. They are the antithesis of geographical ignorance. Rarely known outside their own communities, they often seem, at the first encounter, unremarkable and anonymous. They may not be able to recall the name of a particular wildflower—or they may have given it a name known only to them. They might have forgotten the precise circumstances of a local historical event. Or they can't say for certain when the last of the Canada geese passed through in the fall, or can't differentiate between two kinds of trout in the same creek. Like all of us, they have fallen prey to the fallacies of memory and are burdened with ignorance; but they are nearly flawless in the respect they bear these places they love. Their knowledge is intimate rather than encyclopedic, human but not necessarily scholarly. It rings with the concrete details of experience.

America, I believe, teems with such people. The paradox here, between a faulty grasp of geographical knowledge for which Americans are indicted and the intimate, apparently contradictory familiarity of a group of largely anonymous people, is not solely a matter of confused scale. (The local landscape is easier to know than a national geography.) And it is not simply ironic. The paradox is dark. To be succinct: The politics and advertising that seek a national audience must project a national geography; to be broadly useful that geography must, inevitably, be generalized and it is often romantic. It is therefore frequently misleading and

imprecise. The same holds true with the entertainment industry, but here the problem might be clearer. The same films, magazines, and television features that honor an imaginary American landscape also tout the worth of the anonymous men and women who interpret it. Their affinity for the land is lauded, their local allegiance admired. But the rigor of their local geographies, taken as a whole, contradicts a patriotic, national vision of unspoiled, untroubled land. These men and women are ultimately forgotten, along with the details of the landscapes they speak for, in the face of more pressing national matters. It is the chilling nature of modern society to find an ignorance of geography, local or national, as excusable as an ignorance of hand tools; and to find the commitment of people to their home places only momentarily entertaining. And finally naive.

If one were to pass time among Basawara people in the Kalahari Desert, or with Kreen-Akrora in the Amazon Basin, or with Pitjantjatjara Aborigines in Australia, the most salient impression they might leave is of an absolutely stunning knowledge of their local geography—geology, hydrology, biology, and weather. In short, the extensive particulars of their intercourse with it.

In 40,000 years of human history, it has only been in the last few hundred years or so that a people could afford to ignore their local geographies as completely as we do and still survive. Technological innovations from refrigerated trucks to artificial fertilizers, from sophisticated cost accounting to mass air transportation, have utterly changed concepts of season, distance, soil productivity, and the real cost of drawing sustenance from the land. It is now possible for a resident of Boston to bite into a fresh strawberry in the dead of winter; for someone in San Francisco to travel to Atlanta in a few hours with no worry of how formidable might be crossing of the Great Basin Desert or the Mississippi River; for an absentee farmer to gain a tax advantage from a farm that leaches poisons into its water table and on which crops are left to rot. The Pitjantjatjara might shake their heads in bewilderment and bemusement, not because they are primitive or ignorant people, not because they have no sense of irony or are incapable of marveling, but because they have not (many would say not yet) realized a world in which such manipulation of the land—sur-

mounting the imperatives of distance it imposes, for example, or turning the large-scale destruction of forests and arable land in wealth—is desirable or plausible.

In the years I have traveled through America, in cars and on horseback, on foot and by raft, I have repeatedly been brought to a sudden state of awe by some gracile or savage movement of animal, some odd wrapping of tree's foliage by the wind, an unimpeded run of dew-laden prairie stretching to a horizon flat as a coin where a pin-dot sun pales the dawn sky pink. I know these things are beyond intellection, that they are the vivid edges of a world that includes but also transcends the human world. In memory, when I dwell on these things, I know that in a truly national literature there should be odes to the Triassic reds of the Colorado Plateau, to the sharp and ghostly light of the Florida Keys, to the aeolian soils of southern Minnesota, and the Palouse in Washington, though the modern mind abjures the literary potential of such subjects. (If the sand and flood water farmers of Arizona and New Mexico were to take the black loams of Louisiana in their hands they would be flabbergasted, and that is the beginning of literature.) I know there should be eloquent evocations of the cobbled beaches of Maine, the plutonic walls of the Sierra Nevada, the orange canyons of the Kaibab Plateau. I have no doubt, in fact, that there are. They are as numerous and diverse as the eyes and fingers that ponder the country—it is that only a handful of them are known. The great majority are to be found in drawers and boxes, in the letters and private journals of millions of workaday people who have regarded their encounters with the land as an engagement bordering on the spiritual, as being fundamentally linked to their state of health.

One cannot acknowledge the extent and the history of this kind of testimony without being forced to the realization that something strange, if not dangerous, is afoot. Year by year, the number of people with firsthand experience in the land dwindles. Rural populations continue to shift to the cities. The family farm is in a state of demise, and government and industry continue to apply pressure on the native peoples of North America to sever their ties with the land. In the wake of this loss of personal and local knowledge from which a real geography is derived, the knowledge on which a country must ultimately stand, has [be]come something hard to define but I think sinister and unsettling—the packaging and marketing of land as a form of entertainment. An incipient industry, capitalizing on the nostalgia Americans feel for the imagined virgin landscapes of their fathers, and on a desire for adventure, now offers people a convenient though sometimes incomplete or even spurious geography as an inducement to purchase a unique experience. But the line between authentic experience and a superficial exposure to the elements of experience is blurred. And the real landscape, in all its complexity, is distorted even further in the public imagination. No longer innately mysterious and dignified, a ground from which experience grows, it becomes a curiously generic backdrop on which experience is imposed.

In theme parks the profound, subtle, and protracted experience of running a river is reduced to a loud, quick, safe equivalence, a pleasant distraction. People only able to venture into the countryside on annual vacations are, increasingly, schooled in the belief that wild land will, and should, provide thrills and exceptional scenery on a timely basis. If it does not, something is wrong, either with the land itself or possibly with the company outfitting the trip.

People in America, then, face a convoluted situation. The land itself, vast and differentiated, defies the notion of a national geography. If applied at all it must be applied lightly and it must grow out of the concrete detail of local geographies. Yet Americans are daily presented with, and have become accustomed to talking about, a homogenized national geography. One that seems to operate independently of the land, a collection of objects rather than a continuous bolt of fabric. It appears in advertisements, as a background in movies, and in patriotic calendars. The suggestion is that there can be national geography because the constituent parts are interchangeable and can be treated as commodities. In day-to-day affairs, in other words, one place serves as well as another to convey one's point. On reflection, this is an appalling condescension and a terrible imprecision, the very antithesis of knowledge. The idea that either the Green River in Utah or the Salmon River in Idaho will do, or that the valleys of Kentucky and West Virginia are virtually interchangeable, is not just misleading. For people still dependent on the soil for their sustenance, or for people whose memories tie them to those places, it betrays a numbing casualness, utilitarian, expedient, and commercial frame of mind. It heralds a society in which it is no longer necessary for human beings to know where they live, except as those places are described and fixed by numbers. The truly difficult and lifelong task of discovering where one lives is finally disdained.

If a society forgets or no longer cares where it lives, then anyone with the political power and the will to do so can manipulate the landscape to conform to certain social ideals or nostalgic visions. People may hardly notice that anything has happened, or assume that whatever happens—a mountain stripped of timber and eroding into its creeks—is for the common good. The more superficial a society's knowledge of the real dimensions of the land it occupies becomes, the more vulnerable the land is to exploitation, to manipulation for short-term gain. The land, virtually powerless before political and commercial entities, finds itself finally with no defenders. It finds itself bereft of intimates with indispensable, concrete knowledge. (Oddly, or perhaps not oddly, while American society continues to value local knowledge as a quaint part of its heritage, it continues to cut such people off from any real political power. This is as true for small farmers and illiterate cowboys as it is for American Indians, native Hawaiians, and Eskimos.)

The intense pressure of imagery in America, and the manipulation of images necessary to a society with specific goals, means the land will inevitably be treated like a commodity; and voices that tend to contradict the proffered image will, one way or another, be silenced or discredited by those in power. This is not new to America; the promulgation in America of a false or imposed geography has been the case from the beginning. All local geographies, as they were defined by hundreds of separate, independent native traditions, were denied in the beginning in favor of an imported and unifying vision of America's natural history. The country, the landscape itself, was eventually defined according to dictates of Progress like Manifest Destiny, and laws like the Homestead Act which reflected a poor understanding of the physical lay of the land.

When I was growing up in southern California, I formed the rudiments of a local geography—eucalyptus trees, February rains, Santa Ana winds. I lost much of it when my family moved to New York City, a move typical of the modern, peripatetic style of American life, responding to the exigencies of divorce and employment. As a boy I felt a hunger to know the American landscape that was extreme; when I was finally able to travel on my own, I did so. Eventually I visited most of the United States, living for brief periods of time in Arizona, Indiana, Alabama, Georgia, Wyoming, New Jersey, and Montana before settling 20 years ago in western Oregon.

The astonishing level of my ignorance confronted me everywhere I went. I knew early on that the country could not be held together in a few phrases, that its geography was magnificent and incomprehensible, that a man or woman could devote a lifetime to its elucidation and still feel in the end that he had but sailed many thousands of miles over the surface of the ocean. So I came into the habit of traversing landscapes I wanted to know with local tutors and reading what had previously been written about, and in, those places. I came to value exceedingly novels and essays and works of nonfiction that connected human enterprise to real and specific places, and I grew to be mildly distrustful of work that occurred in no particular place, work so cerebral and detached as to be refutable only in an argument of ideas.

These sojourns in various corners of the country infused me, somewhat to my surprise on thinking about it, with a great sense of hope. Whatever despair I had come to feel at a waning sense of the real land and the emergence of false geographies—elements of the land being manipulated, for example, to create erroneous but useful patterns in advertising—was dispelled by the depth of a single person's local knowledge, by the serenity that seemed to come with that intelligence. Any harm that might be done by people who cared nothing for the land, to whom it was not innately worthy but only something ultimately for sale, I thought, would one day have to meet this kind of integrity, people with the same dignity and transcendence as the land they occupied. So when I traveled, when I rolled my sleeping bag out on the shores of the Beaufort Sea, or in the high pastures of the Absaroka

Range in Wyoming, or at the bottom of the Grand Canyon, I absorbed those particular testaments to life, the indigenous color and songbird song, the smell of sun-bleached rock, damp earth, and wild honey, with some crude appreciation of the singular magnificence of each of those places. And the reassurance I felt expanded in the knowledge that there were, and would likely always be, people speaking out whenever they felt the dignity of the Earth imperiled in those places.

The promulgation of false geographies, which threaten the fundamental notion of what it means to live somewhere, is a current with a stable and perhaps growing countercurrent. People living in New York City are familiar with the stone basements, the cratonic geology, of that island and have a feeling for birds migrating through in the fall, their sequence and number. They do not find the city alien but human, its attenuated natural history merely different from that of rural Georgia or Kansas. I find the countermeasure, too, among Eskimos who cannot read but who might engage you for days on the subtleties of sea-ice topography. And among men and women who, though they have followed in the footsteps of their parents, have come to the conclusion that they cannot farm or fish or log in the way their ancestors did; the finite boundaries to this sort of wealth have appeared in their lifetime. Or among young men and women who have taken several decades of book-learned agronomy, zoology, silviculture and horticulture, ecology, ethnobotany, and fluvial geomorphology and turned it into a new kind of local knowledge, who have taken up residence in a place and sought, both because of and in spite of their education, to develop a deep intimacy with it. Or they have gone to work, idealistically, for the National Park Service or the fish and wildlife services or for a private institution like the Nature Conservancy. They are people to whom the land is more than politics and economics. These are people for whom the land is alive. It feeds them, directly, and that is how and why they learn its geography.

In the end, then, if one begins among the blue crabs of Chesapeake Bay and wanders for several years, down through the Smoky Mountains and back to the bluegrass hills, along the drainages of the Ohio and into the hill country of Missouri, where in summer a chorus of cicadas might drown out human conversation, then up the Missouri itself, reading on the way the entries of Meriwether Lewis and William Clark and musing on the demise of the plains grizzly and the sturgeon, crosses west into the drainage of the Platte and spends the evenings with Gene Weltfish's *The Lost Universe,* her book about the Pawnee who once thrived there, then drops south to the Palo Duro Canyon and the irrigated farms of the Llano Estacado in Texas, turns west across the Sangre de Cristo, southernmost of the Rocky Mountain ranges, and moves north and west up onto the slickrock mesas of Utah, those browns and oranges, the ocherous hues reverberating in the deep canyons, then goes north, swinging west to the insular ranges that sit like battleships in the pelagic space of Nevada, camps at the steaming edge of the sulfur springs in the Black Rock desert, where alkaline pans are glazed with a ferocious light, a heat to melt iron, then crosses the northern Sierra Nevada, waist-deep in summer snow in the passes, to descend to the valley of the Sacramento, and rises through groves of the elephantine redwoods in the Coast Range, to arrive at Cape Mendocino, before Balboa's Pacific, cormorants and gulls, gray whales headed north for Unimak Pass in the Aleutians, the winds crashing down on you, facing the ocean over the blue ocean that gives the scene its true vastness, making this crossing, having been so often astonished at the line and the color of the land, the ingenious lives of its plants and animals, the varieties of its darknesses, the intensity of the stars overhead, you would be ashamed to discover, then, in yourself, any capacity to focus on ravages in the land that left you unsettled. You would have seen so much, breathtaking, startling, and outsize, that you might not be able for a long time to break the spell, the sense, especially finishing your journey in the West, that the land had not been as rearranged or quite as compromised as you had first imagined.

After you had slept some nights on the beach, however, with that finite line of the ocean before you and the land stretching out behind you, the wind first battering then cradling you, you would be compelled by memory, obligated by your own involvement, to speak of what left you troubled. To find the rivers dammed and shrunken, the soil washed away, the land fenced, a tracery of pipes and wires and roads laid down everywhere and animals, cutting the eye off repeatedly and confining it—you had expected this. It troubles you no more than your despair over the ruthlessness, the insensitivity, the impetuousness of modern life. What underlies this obvious change, however, is a less noticeable pattern of disruption: acidic lakes, the skies empty of birds, fouled beaches, the poisonous slags of industry, the sun burning like a molten coin in ruined air.

It is a tenet of certain ideologies that man is responsible for all that is ugly, that everything nature creates is beautiful. Nature's darkness goes partly unreported, of course, and human brilliance is often perversely ignored. What is true is that man has a power, literally beyond his comprehension, to destroy. The lethality of some of what he manufactures, the incompetence with which he stores it or seeks to dispose of it, the cavalier way in which he employs in his daily living substances that threaten his health, the leniency of the courts in these matters (as though products as well as people enjoyed the protection of the Fifth Amendment), and the treatment of open land, rivers, and the atmosphere as if, in some medieval way they could still be regarded as disposal sinks of infinite capacity, would make you wonder, standing face to in the wind at Cape Mendocino, if we weren't bent on an errant of madness.

The geographies of North America, the myriad small landscapes that make up the national fabric, are threatened—by ignorance of what makes them unique, by utilitarian attitudes, by failure to include them in the moral universe, and by brutal disregard. A testament of minor voices can clear away an ignorance of any place, can inform us of its special qualities; but no voice, by merely telling a story, can cause the poisonous wastes that saturate some parts of the land to decompose, to evaporate. This responsibility falls ultimately to the national community, a vague and fragile entity to be sure, but one that, in America, can be ferocious in exerting its will.

Geography, the formal way in which we grapple with this areal mystery, is finally knowledge that calls up something in the land we recognize and respond to. It gives us a sense of place and a sense of community. Both are indispensable to a state of well-being, an individual's and a country's.

One afternoon on the Siuslaw River in the Coast Range of Oregon, in January, I hooked a steelhead, a sea-run trout, that told me, through the muscles of my hands and arms and shoulders, something of the nature of the thing I was calling "the Siuslaw River." Years ago I had stood under a pecan tree in Upson Country, Georgia, idly eating the nuts, when slowly it occurred to me that these nuts would taste different from pecans growing somewhere up in South Carolina. I didn't need a sharp sense of taste to know this, only to pay attention at a level no one had ever told me was necessary. One November dawn, long before the sun rose, I began a vigil at the Dumont Dunes in the Mojave Desert in California, which I kept until a few minutes after the sun broke the horizon. During that time I named to myself the colors by which the sky changed and by which the sand itself flowed like a rising tide through grays and silvers and blues into yellows, pinks, washed duns, and fallow beiges.

It is through the power of observation, the gifts of eye and ear, of tongue and nose and finger, that a place first rises up in our mind; afterward, it is memory that carries the place, that allows it to grow in depth and complexity. For as long as our records go back we have held these two things dear, landscape and memory. Each infuses us with a different kind of life. The one feeds us, figuratively and literally. The other protects us from lies and tyranny. To keep landscapes intact and the memory of them, our history in them, alive, seems as imperative a task in modern time as finding the extent to which individual expression can be accommodated, before it threatens to destroy the fabric of society.

If I were now to visit another country, I would ask my local companion, before I saw any museum or library, any factory or fabled town, to walk me in the country of his or her youth, to tell me the names of things and how, traditionally, they have been fitted together in a community. I would ask for the stories, the voice of memory over the land. I would ask about the history of storms there, the age of the trees, the winter color of the hills. Only then would I ask to see the museum. I would want first the sense of a real place, to know that I was not inhabiting an idea. I would want to know the lay of the land first, the real geography, and take some measure of the love of it in my companion before [having] stood before the painting or read works of scholarship. I would want to have something real and remembered against which I might hope to measure their truth.

World Prisms

The Future of Sovereign States and International Order

*A*s this century closes, the contradictory organizational energies associated with globalization and fragmentation are mounting concerted attacks on the primacy of the sovereign, territorial state as the sole building block of world order. To begin with, transnational social forces seem on the verge of forming some kind of global civil society over the course of the next several decades, providing a foundation for the project of "global democracy." Also significant is the resurgence of religion, and closely linked, the rise of civilizational consciousness.

BY RICHARD FALK

At the same time, the resilience of the state and its twin, the ideology of nationalism, strongly suggest that we have yet to experience the definitive waning of the state system, which is the form of world order that has dominated political imagination and history books for several centuries. And let us not overlook, in this preliminary examination of the world order, the potential of regionalism, which is often underestimated. Our dualistic mental habits lead many of us to think only of "states" and "the world," which involves comparing the most familiar part of our experience to an imagined whole while excluding from consideration all other possibilities.

Several salient issues warrant attention. Given the potent dynamic of economic globalization, how can market forces become effectively regulated in the future? The mobility of capital and the relative immobility of labor will challenge governments to balance their interest in promoting trade and profits against their concern with the well-being of their citizenry. At the same time, economic globalization and the information revolution, with their accompanying compression of time and space, could encourage an emerging political globalization; the processes and institutional and ethical consequences associated with this transformation

RICHARD FALK is Albert G. Milbank Professor of International Law and Practice at Princeton University.

have many ramifications for the future of a globalized international community. Within this changing global milieu, world cities are becoming political actors that form their own networks of transnational relationships that are producing a new layer of world order.

Other lines of inquiry point to the uncertain nature of international institutions in today's world. Is there a crucial role for regional institutions as a halfway house between utopian globalism and outmoded statism? Also at issue is whether the current eclipse of the United Nations is merely a temporary phenomenon associated with the high incidence of civic violence, or rather a more durable development that reflects the peculiarities of the current phase of peace and security issues dominated by civil wars and ethnic strife. Are the main political actors, especially states, adapting their roles in response to new challenges and realities, or are they being superseded and outflanked by alternative problem-solving and institutional frameworks? What is the impact of the particular style and substance of global leadership provided by the United States, and is this likely to change due to internal developments, lessons learned, and external challenges?

Building Pressures

These trends will exert pressure on existing institutions, creating receptivity and resistance to various proposals for institutional

innovation, as well as encouraging a variety of regressive withdrawals from interstate cooperation. Perhaps the most ominous of these trends relates to the pressure of an expected population increase of more than two billion over the next two decades, more than 90 percent of which will be concentrated in developing countries. Such a pattern will impact many political arrangements and will challenge the adequacy of food and freshwater supplies in many settings that are already economically disadvantaged. Comparably serious concerns relate to the impacts of greenhouse gas emissions on climate change, deforestation and desertification, the extinction of species, ozone depletion, and the further spread, development, and retention of weapons of mass destruction. Such trends suggest that existing managerial and adaptive capacities will be badly overstretched, thresholds of ecological and social balance will likely be crossed, and tensions and conflicts will abound. The increased incidence of conflicts will prompt contradictory forms of reliance on hyper-modern and primitive types of coercion—that is, on super-smart weaponry of great technical sophistication and on terrorist violence. Both types of violence deeply challenge basic assumptions about legal, moral, and political limits on conflict.

Against such an ominous background, it is important to identify several potentially promising developments that could mitigate, if not overcome, these mounting dangers. None of these developments is free of ambiguity, and the overall prospect of the future is embedded in the fluidity of the historical present.

The Eclipse of the UN

Ever since its founding in 1945, many of the hopes for a more peaceful and benign world have reflected confidence in the United Nations. The United Nations was created as a response to World War II in a climate of resolve to erase the memories of failure associated with the League of Nations. The new organization was promoted as embodying a new approach to world order based on the commitment of leading states to establish and enact "collective security." With the onset of the Cold War, this undertaking by leading states to ground global security upon the collective mechanisms of the United Nations became untenable, and balance-of-power politics was retrofitted to serve the security needs of bipolar conflict in the nuclear age. Although the United Nations managed to accomplish many useful tasks on behalf of the peoples of the world, including a variety of peace-keeping ventures at the edges of world politics, it was sidelined in relation to its central peace-and-security mission by the effects of East-West gridlock. When Mikhail Gorbachev came to power in Moscow in the mid-1980s, a flurry of UN activity ensued, enabling the organization to play a leading role in the resolution of a series of regional conflicts, including those in Afghanistan, El Salvador, Angola, Cambodia, Iran, and Iraq. This suggested a bright future for the organization when and if the superpowers tempered their opposition to each other.

After the surprisingly abrupt end of the Cold War in 1989, there emerged a strong expectation that a golden age for the United Nations lay just beyond the horizon. The Gulf War of 1991 confirmed this optimistic view for many observers of the global scene, demonstrating for the first time that the Security Council could organize an effective collective response to aggression, precisely along the lines envisaged by the drafters of the UN Charter so long ago. It should be noted, however, that this reinvigorated United Nations also aroused concern in some quarters, as some argued that the Security Council was becoming a geopolitical instrument in the hands of the United States and its allies. With the restoration of Kuwaiti sovereignty, the debate about the benefits and burdens of UN potency reached its climax, and was soon over-shadowed in the mid-1990s by a revived sense of futility, initially in reaction to Somalia, but most definitively with respect to ethnic cleansing in Bosnia and genocide in Rwanda. It then became evident that the United Nations had

File Photo

Alternatives to the United Nations as a global security organization may be needed.

been used as a convenient mobilizing arena in the specific setting of strategic interests at risk in the Gulf Crisis. In contrast, where strategic interests were trivial or not present—as was the case in Somalia, Rwanda, and Bosnia—the political will of the Permanent Five was far too weak to support effective forms of humanitarian intervention under UN auspices.

Arguably, the United Nation's military role had been carefully confined, reflecting considerations of prudence and legal doctrine, to addressing warfare between states. When established, the organization was endowed with neither the competence nor the capability to respond to matters of civil strife. Nevertheless, neither governments nor the media made such a distinction in the early 1990s. Unable to meet these new demands, the United Nations was seriously discredited and once again sidelined in matters of peace and security. The new approach to global security at the end of the 1990s seems to be a mixture of old-fashioned geopolitics—that is, unilateralism by the leading military power—and coalition diplomacy under the aegis of NATO, as seen in this year's Kosovo intervention. If collective action on a world community scale has been seen as a world-order goal since World War I, then it would seem that the century is ending on a regressive note; a retreat to traditional methods of maintaining international order that depend on the strong-arm tactics of dominant states and their allies with little or no attention to international law.

The future of the United Nations as the basis of global security is not currently bright, but such an outlook could change rapidly. Memories are short, and if there are setbacks associated with US-NATO initiatives, a revived reliance on an augmented UN Security Council could materialize

quickly. Even a change in the form and substance of US political leadership in the White House and Congress could rather abruptly confer upon the United Nations a new opportunity to play an important global security role. From the present vantage point, however, this outcome is not too likely. If the main security challenges continue to arise from intrastate violence, it seems unlikely that the United Nations, as presently constituted, could respond in a consistently useful manner. Also, the US approach to global leadership, given such recently hyped challenges as biowar and international terrorism, seems likely to avoid being constrained by the collective procedures of the United Nations.

Prospects for Reform

For these reasons, it seems appropriate not to expect too much from the United Nations in the peace-and-security area over the course of the next decade. At the same time, the United Nations remains a vital actor in relation to a wide range of global concerns: environment, food, health, humanitarian relief, human rights, refugees, lawmaking, and development.

It is also helpful to keep in mind that the United Nations was established by states to serve states—that it was international in conception and operation, and not supranational or intranational. As such, it has been difficult for the United Nations to accommodate the various strands of transnational activity that together compose the phenomenon of "globalization." Neither the private sector nor global civil society can gain meaningful access to the main arenas of the United Nations, an exclusion that has resulted in high-profile activity engaging grassroots par-

European regionalism remains the most dramatic challenge to the state system. The historical irony is evident: the region that invented the modern state and its diplomacy is taking the lead in establishing a form of world order that is sufficiently different to qualify as a sequel.

ticipation beyond the normal arenas of the United Nations. If the United Nations were constitutionally recreated in the year 2000, it would not likely resemble the organization set up in 1945. The awareness of such a discrepancy between what was done then and what is needed now is made even more serious due to the inability to reform the organization in even minimal respects. It has not been possible to agree on how to alter the composition of the Security Council to take account of the enormous changes in the makeup of international society over the course of the last half-century. It is evident to even casual observers that to retain permanent seats on the Security Council for both Great Britain and France, while denying such a presence to India and Brazil, or Japan and Germany, is to mock the current distribution of influence in world politics. The legitimacy of the United Nations, and its credibility as an actor, depends on its own structures of authority more or less mirroring the structure of relations among states and regions.

It is far too soon to write off the United Nations as a viable global security organization, but there are few reasons to believe that it will play a central role in peace, security, and development activity over the course of the next decade or so. Fortunately, there are more promising alternatives.

Learning from Europe

The most daring world-order experiment of this century had undoubtedly been the European regional movement. It has encroached on the sovereign rights of territorial states far more than anything attempted by the United Nations, even though regional organizations are formally subordinated to the UN Security Council with respect to the use of force. The European Union has used its authority structures, its institutional implementing procedures, and the integrative political will of government leaders to put a variety of supranational moves into operation: the supremacy of European Community Law; the external accountability of governments in relation to human-rights claims; the minimalization of obstacles to intra-regional trade, investment, customs, immigration, and travel controls; the existence of a common currency and central bank; and a directly

elected regional parliament. The cumulative effect of these developments over the course of more than 50 years has been to bring into existence a distinctive European entity that is far from being either a single European superstate or merely a collaborative framework for a collection of states in Europe. Europe continues to evolve. Its final shape will not be known for several more decades.

The European Union is something quite new and different; it is likely to keep changing. If it is perceived outside of Europe as a success in these latest supranationalizing moves, it will provide Asia, Africa, and Latin America, or portions thereof, with positive models that they will be tempted to imitate and adapt to their regional conditions. The regional model is so promising that by 2025 and 2030 it might become natural to speak of "a world of regions" as an overarching successor metaphor to "a world of states."

At the same time, there are uncertainties associated with these European develop-

ments. The present turmoil in the Balkans could spread beyond the former Yugoslavia—and even if it does not, it could stimulate some drastic rethinking concerning the security implications of a regional approach, especially if linked, as in the Cold War, to US leadership. The NATO operation in Kosovo, besides suggesting the abandonment of the United Nations in the setting of global security, may discourage any further transfers of sovereignty by European states or contrarily provide momentum for the creation of an all-European security system. The Kosovo ordeal may also be regarded outside Europe as an indirect reaffirmation of the state as the basic ordering framework for most peoples in the world. It is also possible that Europeans will reevaluate NATO itself, since it is increasingly perceived as an alliance arrangement that belongs anachronistically to an era of old geopolitics—it is neither regional in orientation, nor collective in operation, nor responsive to the historical

File Photo

The effectiveness of the United Nations as an international actor remains in doubt.

File Photo

Cities are becoming transnational actors with their own worldviews, interests, and networks.

realities of Europe after the Cold War. If this type of revisionist thinking takes hold, a pooling of European security resources and at least a partial disengagement from dependence on the United States will follow, clarifying for a time the character of European regionalism.

However the Balkan crisis is eventually resolved, European regionalism remains the most fundamental challenge to the state system. The historical irony is that Europe, as the region that invented the modern state and its diplomacy, is taking the lead in establishing a form of world order that may soon turn out to be sufficiently different as to qualify as a sequel. Non-European forms of regionalism have also been moving ahead in this period, and need to be brought much more explicitly into our thinking about order, security, development, and justice as aspects of an emergent system of global governance.

The Re-empowered State

As I explained in greater detail in my work *Predatory Globalization: A Critique,* the effects of neo-liberal globalization have been to disempower the state with respect to solving both internal social and cultural problems. This generalization pertains to such external challenges as those arising from environmental deterioration, scarcity of renewable resources, and protection of the global commons for future generations. The idea of capital mobility on a regional and global scale in response to market factors offers the main insight into the disempowerment of the state, converting governments into capital facilitators and limiting the space for responsible political debate and party rivalry. Other important sources of this phe-

nomenon of disempowerment have resulted from the weakening of organized labor as a source of societal pressure and the revolutionary impacts of information technology, the computer, and the internet, all tendencies that alter the role and identity of the state and weaken its disposition toward innovation and problem-solving.

Re-empowerment of the state implies reversing these trends, especially finding ways to offset the presiding influence of global market forces on the outlook and behavior of governing elites. Such a rebalancing of contending social force could result from the further mobilization of civil society in relation to such shared objectives as environmental quality, human rights, demilitarization, labor and welfare policy, and regulation of global capital flows. The Asian financial crisis which began in 1997, combined with the troubles in a series of other important countries such as Japan, Russia, and Brazil, has already caused a partial ideological retreat from extreme versions of neo-liberal globalization. Further retreat could be prompted by instability generated in part by a spreading sense that the benefits of economic globalization are being unfairly distributed, with a steady increase in income and wealth disparities (a pattern clearly documented in the annual volumes of the *Human Development Report*).

Another reason for re-empowerment might arise from a growing anxiety, including among business elites, of backlash threats ranging from extremist religions, micro-nationalisms, and neo-fascist political movements. It may come to be appreciated by economic elites that the global economy needs to be steered with a greater attentiveness to social concerns and to the global public good, as well as in relation to effi-

ciency of returns on capital. Otherwise, a new round of dangerous and costly revolutionary struggles looms menacingly on the horizon. A further supportive tendency is the appearance of "a new internationalism" in the form of various coalitions among many "normal" governments (those without extra-regional global claims) and large numbers of transnational civic associations. This type of coalition was integral to the campaigns to ban anti-personnel landmines, to establish an international criminal court, and to call into question the legality and possession of nuclear weaponry. This set of initiatives put many governments in an activist mode and in opposition to the policy positions taken by the geopolitical leadership of the world, mainly the US government.

There is another somewhat silent re-empowerment taking place in the form of an extension of the authority of the state to manage the world's ocean's and outer space. The 1982 Law of the Sea Treaty, for example, validated an enormous expansion of coastal authority in the form of a 200-mile Exclusive Economic Zone. The magnitude of this re-empowerment can be appreciated when it is realized that 95 percent of the beneficial use of the ocean lies within these coastal waters.

The re-empowerment of the state is not meant to negate the benefits or reality of economic globalization. It is instead a matter of making the state more of a regulative mechanism in relation to market forces, and less of a facilitative force. Out of such re-empowerment might yet emerge an informal global social contract that could help provide the world economy with the sort of political and social stability that will be needed if "sustainable development" is to become a credible future reality.

Coordinating World Cities

It is not possible to do more than highlight this formidable frontier for ordering relations among peoples in ways that elude traditional and familiar frameworks. With an increasing proportion of wealth, culture, people, and innovation concentrated in or appropriated by world cities, these entities are more and more self-consciously becoming transnational actors with their own agendas, worldviews, and networks. The economic and political success of such city-states as Singapore and Hong Kong is also suggestive of the possibility that world order need not be premised on territorial dominion in the future. Whether states will successfully contain this disempowering trend or appropriate it for their own goals remains largely uncertain, as is the impact of deepening European-style regionalism. China's relationship to Hong Kong will be a test of whether the city as a political actor can withstand the challenges to its autonomy being posed by Beijing. In any event, the role and future of city-states in the next century is definitely an idea worth including on any chart depicting emergent forms of world order.

Creeping Functionalism

Among the most evident world-order trends is the impulse of governmental and other bureaucracies to coordinate specialist activities across state boundaries by way of consultation, periodic meetings, and informal codes of conduct. Through banking, shipping, and insurance specialists, an enormous proliferation of ad hoc functional arrangements are being negotiated and implemented in a wide variety of international arenas. Some interpreters of the global scene have identified the proliferation of such undertakings as the wave of the future, as a disaggregating response of sovereign states to the complexity of a highly interconnected world. Anne-Marie Slaughter articulates this position and gives it a positive spin in the 75th Anniversary Issue of *Foreign Affairs*. In a basic sense, this extension of functional modes of coordination to address a dazzling array of technical and commercial issues represents a series of practical adjustments to the growing complexity of international life. The role of the state is being tailored to work more on behalf of common economic interests associated with globalization and to be less preoccupied with the promotion of exclusively national economic interests.

In some circles, this devolution and dispersion of authority, with more direct rulemaking participation by private-sector representatives, along with the overall erosion of public-private sector distinctions, is sometimes referred to as "the new medievalism." Such a terminology deliberately recalls the world order of feudal Europe with its overlapping patterns of authority and the importance of both local and universal institutional actors. It was the territorial consolidation of this confusing reality that gave rise initially to the absolute state ruled by a monarch, and later to a constitutional government legitimized by the consent of the governed. The postmodern state is in the process of formation, and is as varied in character and orientation as are the circumstances of differing cultures, stages of development, degrees of integration, and respect for human rights that exist around the world.

A New World Order?

It is evident that in the institutional ferment of the moment, several trends and counter-trends make it impossible to depict the future shape of the world order with any confidence. Central to this future is the uncertain degree to which the sovereign state can adapt its behavior and role to a series of deterritorializing forces associated with markets, transnational social forces, cyberspace, demographic and environmental pressures, and urbanism. Also critical to the future is the fate of the European Union and the way in which it is reflected in non-European opinion. Seemingly less crucial, but still of interest, is whether the United Nations can find ways to retrieve its reputation as relevant to peace and security while continuing to engineer a myriad of useful activities beyond the gaze of the media. At issue, finally, is the sort of global leadership provided by the United States, and the nature of leadership alternatives, if any exist. Crossing the millennial threshold is likely to clarify the mix of these developments, but probably not in a definitive enough pattern to be worthy of being labeled "a new world order," at least for several decades.

Human Domination of Earth's Ecosystems

Peter M. Vitousek, Harold A. Mooney, Jane Lubchenco, Jerry M. Melillo

Human alternation of Earth is substantial and growing. Between one-third and one-half of the land surface has been transformed by human action; the carbon dioxide concentration in the atmosphere has increased by nearly 30 percent since the beginning of the Industrial Revolution; more atmospheric nitrogen is fixed by humanity than by all natural terrestrial sources combined; more than half of all accessible surface fresh water is put to use by humanity; and about one-quarter of the bird species on Earth have been driven to extinction. By these and other standards, it is clear that we live on a human-dominated planet.

All organisms modify their environment, and humans are no exception. As the human population has grown and the power of technology has expanded, the scope and nature of this modification has changed drastically. Until recently, the term "human-dominated ecosystems" would have elicited images of agricultural fields, pastures, or urban landscapes; now it applies with greater or lesser force to all of Earth. Many ecosystems are dominated directly by humanity, and no ecosystem on Earth's surface is free of pervasive human influence.

This article provides an overview of human effects on Earth's ecosystems. It is not intended as a litany of environmental disasters, though some disastrous situations are described; nor is it intended either to downplay or to celebrate environmental successes, of which there have been many. Rather, we explore how large humanity looms as a presence on the globe—how, even on the grandest scale, most aspects of the structure and functioning of Earth's ecosystems cannot be understood without accounting for the strong, often dominant influence of humanity.

We view human alterations to the Earth system as operating through the interacting processes summarized in Fig. 1. The growth

P. M. Vitousek and H. A. Mooney are in the Department of Biological Sciences, Stanford University, Stanford, CA 94305, USA. J. Lubchenco is in the Department of Zoology, Oregon State University, Corvallis, OR 97331, USA. J. M. Melillo is at the U.S. Office of Science and Technology Policy, Old Executive Office Building, Room 443, Washington, DC 20502, USA.

of the human population, and growth in the resource base used by humanity, is maintained by a suite of human enterprises such as agriculture, industry, fishing, and international commerce. These enterprises transform the land surface (through cropping, forestry, and urbanization), alter the major biogeochemical cycles, and add or remove species and genetically distinct populations in most of Earth's ecosystems. Many of these changes are substantial and reasonably well quantified; all are ongoing. These relatively well-documented changes in turn entrain further alterations to the functioning of the Earth system, most notably by driving global climatic change (1) and causing irreversible losses of biological diversity (2).

Fig. 1 A conceptual model illustrating humanity's direct and indirect effects on the Earth system [modified from (56)].

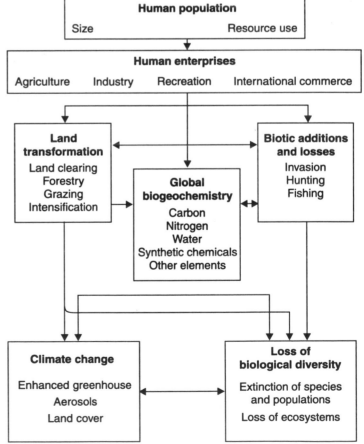

Land Transformation

The use of land to yield goods and services represents the most substantial human alteration of the Earth system. Human use of land alters the structure and functioning of ecosystems, and it alters how ecosystems interact with the atmosphere, with aquatic systems, and with surrounding land. Moreover, land transformation interacts strongly with most other components of global environmental change.

The measurement of land transformation on a global scale is challenging; changes can be measured more or less straightforwardly at a given site, but it is difficult to aggregate these changes regionally and globally. In contrast to analyses of human alteration of the global carbon cycle, we cannot install instruments on a tropical mountain to collect evidence of land transformation. Remote sensing is a most useful technique, but only recently has there been a serious scientific effort to use high-resolution civilian satellite imagery to evaluate even the more visible forms of land transformation, such as deforestation, on continental to global scales (3).

Land transformation encompasses a wide variety of activities that vary substantially in their intensity and consequences. At one extreme, 10 to 15% of Earth's land surface is occupied by row-crop agriculture or by urban-industrial areas, and another 6 to 8% has been converted to pastureland (4); these systems are wholly changed by human activity. At the other extreme, every terrestrial ecosystem is affected by increased atmospheric carbon dioxide (CO2), and most ecosystems have a history of hunting and other low-intensity resource extraction. Between these extremes lie grassland and semiarid ecosystems that are grazed (and sometimes degraded) by domestic animals, and forests and woodlands from which wood products have been harvested; together, these represent the majority of Earth's vegetated surface.

The variety of human effects on land makes any attempt to summarize land transformations globally a matter of semantics as well as substantial uncertainty. Estimates of the fraction of land transformed or degraded by humanity (or its corollary, the fraction of the land's biological production that is used or dominated) fall in the range of 39 to 50% (5) (Fig. 2). These numbers have large uncertainties, but the fact that they are large is not at all uncertain. Moreover, if anything these estimates understate the global impact of land transformation, in that land that has not been transformed often has been divided into fragments by human alteration of the surrounding areas. This fragmentation affects the species composition and functioning of otherwise little modified ecosystems (6).

Overall, land transformation represents the primary driving force in the loss of biological diversity worldwide. Moreover, the effects of land transformation extend far beyond the boundaries of transformed lands. Land transformation can affect climate directly at local and even regional scales. It contributes ~20% to current anthropogenic CO_2 emissions, and more substantially to the increasing concentrations of the greenhouse gases methane and nitrous oxide; fires associated with it alter the reactive chemistry of the troposphere, bringing elevated carbon monoxide concentrations and episodes of urban-like photochemical air pollution to remote tropical areas of Africa and South America; and it causes runoff of sediment and nutrients that drive substantial changes in stream, lake, estuarine, and coral reef ecosystems (7–10).

The central importance of land transformation is well recognized within the community of researchers concerned with global environmental change. Several research programs are focused on aspects of it (9, 11); recent and substantial progress toward understanding these aspects has been made (3), and much more progress can be anticipated. Understanding land transformation is a difficult challenge; it requires integrating the social, economic, and cultural causes of land transformation with evaluations of its biophysical nature and consequences. This interdisciplinary approach is essential to predicting the course, and to any hope of affecting the consequences, of human-caused land transformation.

Oceans

Human alterations of marine ecosystems are more difficult to quantify than those of terrestrial ecosystems, but several kinds of information suggest that they are substantial. The human population is concentrated near coasts—about 60% within 100 km—and the oceans' productive coastal margins have been affected strongly by humanity. Coastal wetlands that mediate interactions between land and sea have been altered over large areas; for example, approximately 50% of mangrove ecosystems globally have been transformed or destroyed by human activity (12). Moreover, a recent analysis suggested that although humans use about 8% of the primary production of the oceans, that fraction grows to more than 25% for upwelling areas and to 35% for temperate continental shelf systems (13).

Many of the fisheries that capture marine productivity are focused on top predators, whose removal can alter marine ecosystems out of proportion to their abundance. Moreover, many such fisheries have proved to be unsustainable, at least at our present level of knowledge and control. As of 1995, 22% of recognized marine fisheries were overexploited or already depleted, and 44% more were at their limit of exploitation (14) (Figs. 2 and 3). The consequences of fisheries are not restricted to their target organisms; commercial marine fisheries around the world discard 27 million tons of nontarget animals annually, a quantity nearly one-third as large as total landings (15). Moreover, the dredges and trawls used in some fisheries damage habitats substantially as they are dragged along the sea floor.

A recent increase in the frequency, extent, and duration of harmful algal blooms in coastal areas (16) suggests that human activity has affected the base as well as the top of marine food chains. Harmful algal blooms

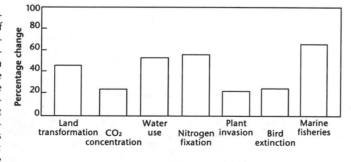

Fig. 2. Human dominance or alteration of several major components of the Earth system, expressed as (from left to right) percentage of the land surface transformed (5); percentage of the current atmospheric CO_2 concentration that results from human action (17); percentage of accessible surface fresh water used (20); percentage of terrestrial N fixation that is human-caused (28); percentage of plant species in Canada that humanity has introduced from elsewhere (48); percentage of bird species on Earth that have become extinct in the past two millennia, almost all of them as a consequence of human activity (42); and percentage of major marine fisheries that are fully exploited, overexploited, or depleted (14).

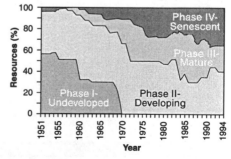

Fig. 3. Percentage of major world marine fish resources in different phases of development, 1951 to 1994 [from (57)]. Undeveloped = a low and relatively constant level of catches; developing = rapidly increasing catches; mature = a high and plateauing level of catches; senescent = catches declining from higher levels.

are sudden increases in the abundance of marine phytoplankton that produce harmful structures or chemicals. Some but not all of these phytoplankton are strongly pigmented (red or brown tides). Algal blooms usually are correlated with changes in temperature, nutrients, or salinity; nutrients in coastal waters, in particular, are much modified by human activity. Algal blooms can cause extensive fish kills through toxins and by causing anoxia; they also lead to paralytic shellfish poisoning and amnesic shellfish poisoning in humans. Although the existence of harmful algal blooms has long been recognized, they have spread widely in the past two decades (*16*).

Alterations of the Biogeo-chemical Cycles

Carbon. Life on Earth is based on carbon, and the CO_2 in the atmosphere is the primary resource for photosynthesis. Humanity adds CO_2 to the atmosphere by mining and burning fossil fuels, the residue of life from the distant past, and by converting forests and grasslands to agricultural and other low-biomass ecosystems. The net result of both activities is that organic carbon from rocks, organisms, and soils is released into the atmosphere as CO_2.

The modern increase in CO_2 represents the clearest and best documented signal of human alteration of the Earth system. Thanks to the foresight of Roger Revelle, Charles Keeling, and others who initiated careful and systematic measurements of atmospheric CO_2 in 1957 and sustained them through budget crises and changes in scientific fashions, we have observed the concentration of CO_2 as it has increased steadily from 315 ppm to 362 ppm. Analysis of air bubbles extracted from the Antarctic and Greenland ice caps extends the record back much further; the CO_2 concentration was more or less stable near 280 ppm for thousands of years until about 1800, and has increased exponentially since then (*17*).

There is no doubt that this increase has been driven by human activity, today primarily by fossil fuel combustion. The sources of CO_2 can be traced isotopically; before the period of extensive nuclear testing in the atmosphere, carbon depleted in ^{14}C was a specific tracer of CO_2 derived from fossil fuel combustion, whereas carbon depleted in ^{13}C characterized CO_2 from both fossil fuels and land transformation. Direct measurements in the atmosphere, and analyses of carbon isotopes in tree rings, show that both ^{13}C and ^{14}C in CO_2 were diluted in the atmosphere relative to ^{12}C as the CO_2 concentration in the atmosphere increased.

Fossil fuel combustion now adds 5.5 ± 0.5 billion metric tons of CO_2-C to the atmosphere annually, mostly in economically developed regions of the temperate zone (*18*) (Fig. 4). The annual accumulation of CO_2-C has averaged 3.2 ± 0.2 billion metric tons recently (*17*). The other major terms in the atmospheric carbon balance are net ocean-atmosphere flux, net release of carbon during land transformation, and net storage in terrestrial biomass and soil organic matter. All of these terms are smaller and less certain than fossil fuel combustion or annual atmospheric accumulation; they represent rich areas of current research, analysis, and sometimes contention.

The human-caused increase in atmospheric CO_2 already represents nearly a 30% change relative to the pre-industrial era (Fig. 2), and CO_2 will continue to increase for the foreseeable future. Increased CO_2 represents the most important human enhancement to the greenhouse effect; the consensus of the climate research community is that it probably already affects climate detectably and will drive substantial climate change in the next century (*1*). The direct effects of increased CO_2 on plants and ecosystems may be even more important. The growth of most plants is enhanced by elevated CO_2, but to very different extents; the tissue chemistry

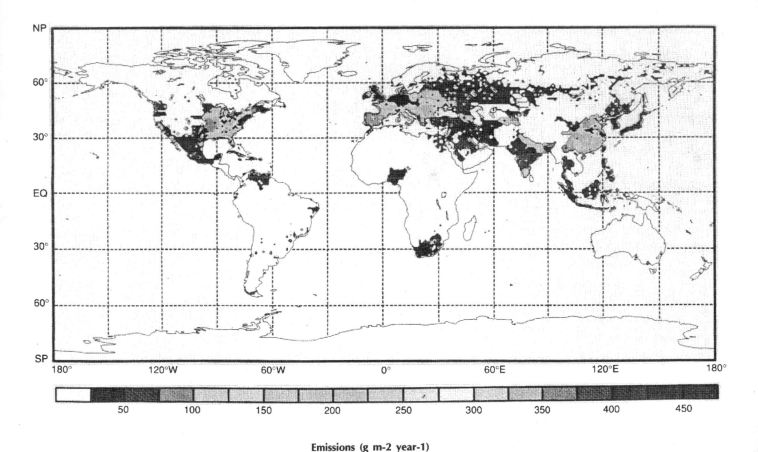

Emissions (g m-2 year-1)

Fig. 4 Geographical distribution of fossil fuel sources of CO_2 as of 1990. The global mean is 12.2 g m^{-2} year^{-1}; most emissions occur in economically developed regions of the north temperate zone. EQ, equator; NP, North Pole; SP, South Pole. [Prepared by A.S. Denning, from information in (*18*)]

of plants that respond to CO_2 is altered in ways that decrease food quality for animals and microbes; and the water use efficiency of plants and ecosystems generally is increased. The fact that increased CO_2 affects species differentially means that it is likely to drive substantial changes in the species composition and dynamics of all terrestrial ecosystems (19).

Water. Water is essential to all life. Its movement by gravity, and through evaporation and condensation, contributes to driving Earth's biogeochemical cycles and to controlling its climate. Very little of the water on Earth is directly usable by humans; most is either saline or frozen. Globally, humanity now uses more than half of the runoff water that is fresh and reasonably accessible, with about 70% of this use in agriculture (20) (Fig. 2). To meet increasing demands for the limited supply of fresh water, humanity has extensively altered river systems through diversions and impoundments. In the United States only 2% of the rivers run unimpeded, and by the end of this century the flow of about two-thirds of all of Earth's rivers will be regulated (21). At present, as much as 6% of Earth's river runoff is evaporated as a consequence of human manipulations (22). Major rivers, including the Colorado, the Nile, and the Ganges, are used so extensively that little water reaches the sea. Massive inland water bodies, including the Aral Sea and Lake Chad, have been greatly reduced in extent by water diversions for agriculture. Reduction in the volume of the Aral Sea resulted in the demise of native fishes and the loss of other biota; the loss of a major fishery; exposure of the salt-laden sea bottom, thereby providing a major source of windblown dust; the production of a drier and more continental local climate and a decrease in water quality in the general region; and an increase in human diseases (23).

Impounding and impeding the flow of rivers provides reservoirs of water that can be used for energy generation as well as for agriculture. Waterways also are managed for transport, for flood control, and for the dilution of chemical wastes. Together, these activities have altered Earth's freshwater ecosystems profoundly, to a greater extent than terrestrial ecosystems have been altered. The construction of dams affects biotic habitats indirectly as well; the damming of the Danube River, for example, has altered the silica chemistry of the entire Black Sea. The large number of operational dams (36,000) in the world, in conjunction with the many that are planned, ensure that humanity's effects on aquatic biological systems will continue (24). Where surface water is sparse or over-exploited, humans use groundwater—and in many areas the groundwater that is drawn upon is nonrenewable, or fossil, water (25). For example, three-quarters of the water supply of Saudi Arabia currently comes from fossil water (26).

Alterations to the hydrological cycle can affect regional climate. Irrigation increases atmospheric humidity in semiarid areas, often increasing precipitation and thunderstorm frequency (27). In contrast, land transformation from forest to agriculture or pasture increases albedo and decreases surface roughness; simulations suggest that the net effect of this transformation is to increase temperature and decrease precipitation regionally (7, 26).

Conflicts arising from the global use of water will be exacerbated in the years ahead, with a growing human population and with the stresses that global changes will impose on water quality and availability. Of all of the environmental security issues facing nations, an adequate supply of clean water will be the most important.

Nitrogen. Nitrogen (N) is unique among the major elements required for life, in that its cycle includes a vast atmospheric reservoir (N_2) that must be fixed (combined with carbon, hydrogen, or oxygen) before it can be used by most organisms. The supply of this fixed N controls (at least in part) the productivity, carbon storage, and species composition of many ecosystems. Before the extensive human alteration of the N cycle, 90 to 130 million metric tons of N (Tg N) were fixed biologically on land each year; rates of biological fixation in marine systems are less certain, but perhaps as much was fixed there (28).

Human activity has altered the global cycle of N substantially by fixing N_2—deliberately for fertilizer and inadvertently during fossil fuel combustion. Industrial fixation of N fertilizer increased from <10 Tg/year in 1950 to 80 Tg/year in 1990; after a brief dip caused by economic dislocations in the former Soviet Union, it is expected to increase to >135 Tg/year by 2030 (29). Cultivation of soybeans, alfalfa, and other legume crops that fix N symbiotically enhances fixation by another ~40 Tg/year, and fossil fuel combustion puts >20 Tg/year of reactive N into the atmosphere globally—some by fixing N_2, more from the mobilization of N in the fuel. Overall, human activity adds at least as much fixed N to terrestrial ecosystems as do all natural sources combined (Fig. 2), and it mobilizes >50 Tg/year more during land transformation (28, 30).

Alteration of the N cycle has multiple consequences. In the atmosphere, these include (i) an increasing concentration of the greenhouse gas nitrous oxide globally; (ii) substantial increases in fluxes of reactive N gases (two-thirds or more of both nitric oxide and ammonia emissions globally are human-caused); and (iii) a substantial contribution to acid rain and to the photochemical smog that afflicts urban and agricultural areas throughout the world (31). Reactive N that is emitted to the atmosphere is deposited downwind, where it can influence the dynamics of recipient ecosystems. In regions where fixed N was in short supply, added N generally increases productivity and C storage within ecosystems, and ultimately increases losses of N and cations from soils, in a set of processes termed "N saturation" (32). Where added N increases the productivity of ecosystems, usually it also decreases their biological diversity (33).

Human-fixed N also can move from agriculture, from sewage systems, and from N-saturated terrestrial systems to streams, rivers, groundwater, and ultimately the oceans. Fluxes of N through streams and rivers have increased markedly as human alteration of the N cycle has accelerated; river nitrate is highly correlated with the human population of river basins and with the sum of human-caused N inputs to those basins (8). Increases in river N drive the eutrophication of most estuaries, causing blooms of nuisance and even toxic algae, and threatening the sustainability of marine fisheries (16, 34).

Other cycles. The cycles of carbon, water, and nitrogen are not alone in being altered by human activity. Humanity is also the largest source of oxidized sulfur gases in the atmosphere; these affect regional air quality, biogeochemistry, and climate. Moreover, mining and mobilization of phosphorus and of many metals exceed their natural fluxes; some of the metals that are concentrated and mobilized are highly toxic (including lead, cadmium, and mercury) (35). Beyond any doubt, humanity is a major biogeochemical force on Earth.

Synthetic organic chemicals. Synthetic organic chemicals have brought humanity many beneficial services. However, many are toxic to humans and other species, and some are hazardous in concentrations as low as 1 part per billion. Many chemicals persist in the environment for decades; some are both toxic and persistent. Long-lived organochlorine compounds provide the clearest examples of environmental consequences of persistent compounds. Insecticides such as DDT and its relatives, and industrial compounds like polychlorinated biphenyls (PCBs), were used widely in North America in the 1950s and 1960s. They were transported globally, accumulated in organisms, and magnified in concentration through food chains; they devastated populations of some predators (notably falcons and eagles) and entered parts of the human food supply in concentrations higher than was prudent. Domestic use of these compounds was phased out in the 1970s in the United States and Canada, and their concentrations declined thereafter. However, PCBs in particular remain readily detectable in many organisms, sometimes approaching thresholds of public health concern (36). They will continue to circulate through organisms for many decades.

Synthetic chemicals need not be toxic to cause environmental problems. The fact that the persistent and volatile chlorofluorocarbons (CFCs) are wholly nontoxic contributed to their widespread use as refrigerants and even aerosol propellants. The subsequent discovery that CFCs drive the breakdown of stratospheric ozone, and especially the later discovery of the Antarctic ozone hole and their role in it, represent great surprises in global environmental science (37). Moreover, the response of the international political system to those discoveries is the best extant illustration that global environmental change can be dealt with effectively (38).

Particular compounds that pose serious health and environmental threats can be and often have been phased out (although PCB production is growing in Asia). Nonetheless, each year the chemical industry produces more than 100 million tons of organic chemicals representing some 70,000 different compounds, with about 1000 new ones being added annually (39). Only a small fraction of the many chemicals produced and released into the environment are tested adequately for health hazards or environmental impact (40).

Biotic Changes

Human modification of Earth's biological resources—its species and genetically distinct populations—is substantial and growing. Extinction is a natural process, but the current rate of loss of genetic variability, of populations, and of species is far above background rates; it is ongoing; and it represents a wholly irreversible global change. At the same time, human transport of species around Earth is homogenizing Earth's biota, introducing many species into new areas where they can disrupt both natural and human systems.

Losses. Rates of extinction are difficult to determine globally, in part because the majority of species on Earth have not yet been identified. Nevertheless, recent calculations suggest that rates of species extinction are now on the order of 100 to 1000 times those before humanity's dominance of Earth (41). For particular well-known groups, rates of loss are even greater; as many as one-quarter of Earth's bird species have been driven to extinction by human activities over the past two millennia, particularly on oceanic islands (42) (Fig. 2). At present, 11% of the remaining birds, 18% of the mammals, 5% of fish, and 8% of plant species on Earth are threatened with extinction (43). There has been a disproportionate loss of large mammal species because of hunting; these species played a dominant role in many ecosystems, and their loss has resulted in a fundamental change in the dynamics of those systems (44), one that could lead to further extinctions. The largest organisms in marine systems have been affected similarly, by fishing and whaling. Land transformation is the single most important cause of extinction, and current rates of land transformation eventually will drive many more species to

extinction, although with a time lag that masks the true dimensions of the crisis (45). Moreover, the effects of other components of global environmental change—of altered carbon and nitrogen cycles, and of anthropogenic climate change—are just beginning.

As high as they are, these losses of species understate the magnitude of loss of genetic variation. The loss to land transformation of locally adapted populations within species, and of genetic material within populations, is a human-caused change that reduces the resilience of species and ecosystems while precluding human use of the library of natural products and genetic material that they represent (46).

Although conservation efforts focused on individual endangered species have yielded some successes, they are expensive—and the protection or restoration of whole ecosystems often represents the most effective way to sustain genetic, population, and species diversity. Moreover, ecosystems themselves may play important roles in both natural and human-dominated landscapes. For example, mangrove ecosystems protect coastal areas from erosion and provide nurseries for offshore fisheries, but they are threatened by land transformation in many areas.

Invasions. In addition to extinction, humanity has caused a rearrangement of Earth's biotic systems, through the mixing of floras and faunas that had long been isolated geographically. The magnitude of transport of species, termed "biological invasion," is enormous (47); invading species are present almost everywhere. On many islands, more than half of the plant species are nonindigenous, and in many continental areas the figure is 20% or more (48) (Fig. 2).

As with extinction, biological invasion occurs naturally—and as with extinction, human activity has accelerated its rate by orders of magnitude. Land transformation interacts strongly with biological invasion, in that human-altered ecosystems generally provide the primary foci for invasions, while in some cases land transformation itself is driven by biological invasions (49). International commerce is also a primary cause of the breakdown of biogeographic barriers; trade in live organisms is massive and global, and many other organisms are inadvertently taken along for the ride. In freshwater systems, the combination of upstream land transformation, altered hydrology, and numerous deliberate and accidental species introductions has led to particularly widespread invasion, in continental as well as island ecosystems (50).

In some regions, invasions are becoming more frequent. For example, in the San Francisco Bay of California, an average of one new species has been established every 36 weeks since 1850, every 24 weeks since 1970, and every 12 weeks for the last decade (51). Some introduced species quickly become invasive over large areas (for example, the Asian clam in the San Francisco Bay), whereas

others become widespread only after a lag of decades, or even over a century (52).

Many biological invasions are effectively irreversible; once replicating biological material is released into the environment and becomes successful there, calling it back is difficult and expensive at best. Moreover, some species introductions have consequences. Some degrade human health and that of other species; after all, most infectious diseases are invaders over most of their range. Others have caused economic losses amounting to billions of dollars; the recent invasion of North America by the zebra mussel is a well-publicized example. Some disrupt ecosystem processes, altering the structure and functioning of whole ecosystems. Finally, some invasions drive losses in the biological diversity of native species and populations; after land transformation, they are the next most important cause of extinction (53).

Conclusions

The global consequences of human activity are not something to face in the future—as Fig. 2 illustrates, they are with us now. All of these changes are ongoing, and in many cases accelerating; many of them were entrained long before their importance was recognized. Moreover, all of these seemingly disparate phenomena trace to a single cause—the growing scale of the human enterprise. The rates, scales, kinds, and combinations of changes occurring now are fundamentally different from those at any other time in history; we are changing Earth more rapidly than we are understanding it. We live on a human-dominated planet—and the momentum of human population growth, together with the imperative for further economic development in most of the world, ensures that our dominance will increase.

The papers in this special section summarize our knowledge of and provide specific policy recommendations concerning major human-dominated ecosystems. In addition, we suggest that the rate and extent of human alteration of Earth should affect how we think about Earth. It is clear that we control much of Earth, and that our activities affect the rest. In a very real sense, the world is in our hands—and how we handle it will determine its composition and dynamics, and our fate.

Recognition of the global consequences of the human enterprise suggests three complementary directions. First, we can work to reduce the rate at which we alter the Earth system. Humans and human-dominated systems may be able to adapt to slower change, and ecosystems and the species they support may cope more effectively with the changes we impose, if those changes are slow. Our footprint on the planet (54) might then be stabilized at a point where enough space and resources remain to sustain most of the other species on Earth, for their sake and our own. Reducing the rate of growth in human effects on Earth involves slowing human population growth and using resources as efficiently as is practical. Often it is the waste products and by-products of human activity that drive global environmental change.

Second, we can accelerate our efforts to understand Earth's ecosystems and how they interact with the numerous components of human-caused global change. Ecological research is inherently complex and demanding: It requires measurement and monitoring of populations and ecosystems; experimental studies to elucidate the regulation of ecological processes; the development, testing, and validation of regional and global models; and integration with a broad range of biological, earth, atmospheric, and marine sciences. The challenge of understanding a human-dominated planet further requires that the human dimensions of global change—the social, economic, cultural, and other drivers of human actions—be included within our analyses.

Finally, humanity's dominance of Earth means that we cannot escape responsibility for managing the planet. Our activities are causing rapid, novel, and substantial changes to Earth's ecosystems. Maintaining populations, species, and ecosystems in the face of those changes, and maintaining the flow of goods and services they provide humanity (55), will require active management for the foreseeable future. There is no clearer illustration of the extent of human dominance of Earth than the fact that maintaining the diversity of "wild" species and the functioning of "wild" ecosystems will require increasing human involvement.

REFERENCES AND NOTES

1. Intergovernmental Panel on Climate Change, *Climate Change 1995* (Cambridge Univ. Press, Cambridge, 1996), pp. 9–49.
2. United Nations Environment Program, *Global Biodiversity Assessment,* V. H. Heywood, Ed. (Cambridge Univ. Press, Cambridge, 1995).
3. D. Skole and C. J. Tucker, *Science* **260**, 1905 (1993).
4. J. S. Olson, J. A. Watts, L. J. Allison, *Carbon in Live Vegetation of Major World Ecosystems* (Office of Energy Research, U.S. Department of Energy, Washington, DC, 1983).
5. P. M. Vitousek, P. R. Ehrlich, A. H. Ehrlich, P. A. Matson, *Bioscience* **36**, 368 (1986); R. W. Kates, B. L. Turner, W. C. Clark, in (35), pp. 1–17; G. C. Daily, *Science* **269**, 350 (1995).
6. D. A. Saunders, R. J. Hobbs, C. R. Margules, *Conserv. Biol.* **5**, 18 (1991).
7. J. Shukla, C. Nobre, P. Sellers, *Science* **247**, 1322 (1990).
8. R. W. Howarth *et al., Biogeochemistry* **35**, 75 (1996).
9. W. B. Meyer and B. L. Turner II, *Changes in Land Use and Land Cover: A Global Perspec-* *tive* (Cambridge Univ. Press, Cambridge, 1994).
10. S. R. Carpenter, S. G. Fisher, N. B. Grimm, J. F. Kitchell, *Annu. Rev. Ecol. Syst.* **23**, 119 (1992); S. V. Smith and R. W. Buddemeier, *ibid.,* p. 89; J. M. Melillo, I. C. Prentice, G. D. Farquhar, E.-D. Schulze, O. E. Sala, in (1), pp. 449–481.
11. R. Leemans and G. Zuidema, *Trends Ecol. Evol.* **10**, 76 (1995).
12. World Resources Institute, *World Resources 1996– 1997* (Oxford Univ. Press, New York, 1996).
13. D. Pauly and V. Christensen, *Nature* **374**, 257 (1995).
14. Food and Agricultural Organization (FAO), *FAO Fisheries Tech. Pap. 335* (1994).
15. D. L. Alverson, M. H. Freeberg. S. A. Murawski, J. G. Pope, *FAO Fisheries Tech. Pap. 339* (1994).
16. G. M. Hallegraeff, *Phycologia* **32**, 79 (1993).
17. D. S. Schimel *et al.,* in *Climate Change 1994: Radiative Forcing of Climate Change,* J. T. Houghton *et al.,* Eds. (Cambridge Univ. Press, Cambridge, 1995), pp. 39–71.
18. R. J. Andres, G. Marland, I. Y. Fung, E. Matthews, *Global Biogeochem. Cycles* **10**, 419 (1996).
19. G. W. Koch and H. A. Mooney, *Carbon Dioxide and Terrestrial Ecosystems* (Academic Press, San Diego, CA, 1996); C. Körner and F. A. Bazzaz, *Carbon Dioxide, Populations, and Communities* (Academic Press, San Diego, CA, 1996).
20. S. L. Postel, G. C. Daily, P. R. Ehrlich, *Science* **271**, 785 (1996).
21. J. N. Abramovitz, *Imperiled Waters, Impoverished Future: The Decline of Freshwater Ecosystems* (Worldwatch Institute, Washington, DC, 1996).
22. M. I. L'vovich and G. F. White, in (35), pp. 235–252; M. Dynesius and C. Nilsson, *Science* **266**, 753 (1994).
23. P. Micklin, *Science* **241**, 1170 (1988); V. Kotlyakov, *Environment* **33**, 4 (1991).
24. C. Humborg, V. Ittekkot, A. Cociasu, B. Bodungen, *Nature* **386**, 385 (1997).
25. P. H. Gleick, Ed., *Water in Crisis* (Oxford Univ. Press, New York, 1993).
26. V. Gornitz, C. Rosenzweig, D. Hillel, *Global Planet. Change* **14**, 147 (1997).
27. P. C. Milly and K. A. Dunne, *J. Clim.* **7**, 506 (1994).
28. J. N. Galloway, W. H. Schlesinger, H. Levy II, A. Michaels, J. L. Schnoor, *Global Biogeochem. Cycles* **9**, 235 (1995).
29. J. N. Galloway, H. Levy II, P. S. Kasibhatla, *Ambio* **23**, 120 (1994).
30. V. Smil, in (35), pp. 423–436.
31. P. M. Vitousek *et al., Ecol. Appl.,* in press.
32. J. D. Aber, J. M. Melillo, K. J. Nadelhoffer, J. Pastor, R. D. Boone, *ibid.* **1**, 303 (1991).
33. D. Tilman, *Ecol. Monogr.* **57**, 189 (1987).
34. S. W. Nixon et al., *Biogeochemistry* **35**, 141 (1996).
35. B. L. Turner II *et al.,* Eds., *The Earth As Transformed by Human Action* (Cambridge Univ. Press, Cambridge, 1990).
36. C. A. Stow, S. R. Carpenter, C. P. Madenjian, L. A. Eby, L. J. Jackson, *Bioscience* **45**, 752 (1995).
37. F. S. Rowland, *Am. Sci.* **77**, 36 (1989); S. Solomon, *Nature* **347**, 347 (1990).
38. M. K. Tolba *et al.,* Eds., *The World Environment 1972–1992* (Chapman & Hall, London, 1992).
39. S. Postel, *Defusing the Toxics Threat: Controlling Pesticides and Industrial Waste* (Worldwatch Institute, Washington, DC, 1987).
40. United Nations Environment Program (UNEP). *Saving Our Planet—Challenges and Hopes* (UNEP, Nairobi, 1992).
41. J. H. Lawton and R. M. May, Eds., *Extinction Rates* (Oxford Univ. Press, Oxford, 1995); S. L. Pimm, G. J. Russell, J. L. Gittleman, T. Brooks, *Science* **269**, 347 (1995).
42. S. L. Olson, in *Conservation for the Twenty-First Century,* D. Western and M. C. Pearl, Eds. (Oxford Univ. Press, Oxford, 1989), p. 50; D. W. Steadman, *Science* **267**, 1123 (1995).
43. R. Barbault and S. Sastrapradja. in (2), pp. 193–274.
44. R. Dirzo and A. Miranda, in *Plant-Animal Interactions,* P. W. Price, T. M. Lewinsohn, W. Fernandes, W. W. Benson, Eds. (Wiley Interscience, New York, 1991), p. 273.
45. D. Tilman, R. M. May, C. Lehman, M. A. Nowak, *Nature* **371**, 65 (1994).
46. H. A. Mooney, J. Lubchenco, R. Dirzo, O. E. Sala, in (2). pp. 279–325.
47. C. Elton, *The Ecology of Invasions by Animals and Plants* (Methuen, London, 1958); J. A. Drake *et al.,* Eds., *Biological Invasions. A Global Perspective* (Wiley, Chichester, UK, 1989).
48. M. Rejmanek and J. Randall, *Madrono* **41**, 161 (1994).
49. C. M. D'Antonio and P. M. Vitousek, *Annu. Rev. Ecol. Syst.* **23**, 63 (1992).
50. D. M. Lodge, *Trends Ecol. Evol.* **8**, 133 (1993).
51. A. N. Cohen and J. T. Carlton, *Biological Study: Nonindigenous Aquatic Species in a United States Estuary. A Case Study of the Biological Invasions of the San Francisco Bay and Delta* (U.S. Fish and Wildlife Service, Washington, DC, 1995).
52. I. Kowarik, in *Plant Invasions—General Aspects and Special Problems,* P. Pysek, K. Prach, M. Rejmánek, M. Wade, Eds. (SPB Academic, Amsterdam, 1995), p. 15.
53. P. M. Vitousek, C. M. D'Antonio, L. L. Loope, R. Westbrooks, *Am. Sci.* **84**, 468 (1996).
54. W. E. Rees and M. Wackernagel, in *Investing in Natural Capital: The Ecological Economics Approach to Sustainability,* A. M. Jansson, M. Hammer, C. Folke, R. Costanza, Eds. (Island, Washington, DC, 1994).
55. G. C. Daily, Ed., *Nature's Services* (Island, Washington, DC, 1997).
56. J. Lubchenco *et al., Ecology,* **72**, 371 (1991), P. M. Vitousek, *ibid.* **75**, 1861 (1994).
57. S. M. Garcia and R. Grainger. *FAO Fisheries Tech. Pap.* 359 (1996).
58. We thank G. C. Daily, C. B. Field, S. Hobbie, D. Gordon, P.A. Matson, and R. L. Naylor for constructive comments on this paper, A. S. Denning and S. M. Garcia for assistance with illustrations, and C. Nakashima and B. Lilley for preparing text and figures for publication.

The Role of Science in Policy

THE CLIMATE CHANGE DEBATE IN THE UNITED STATES

by Eugene B. Skolnikoff

Climate change is on the international policy agenda primarily because of warnings from scientists. Their forecasts of a potentially dangerous increase in the average global temperature, fortuitously assisted by unusual weather events, have prompted governments to enter into perhaps the most complicated—and most significant—set of negotiations ever attempted. Key questions—the rapidity of global climate change, its effects on the natural systems on which humans depend, and the options available to lessen or adapt to such change—have energized the scientific and related communities in analyses that are deeply dependent on scientific evidence and research.

At both the national and international levels, the policy debate over climate change is unfolding rapidly. But it is also becoming increasingly mired in controversy, and nowhere more so than in the United States. This raises a crucial question: Why is it that this country—the undisputed leader of the world in science and technology—is finding it so difficult to agree on policies to address an ecological threat that, if it materializes, could have catastrophic consequences for itself and the rest of the world?

The perhaps surprising answer is that in the U.S. policy process, climate change is not now a scientific issue.[1] Although much of the controversy appears to revolve around scientific principles, political and economic forces actually dominate. In a sense, this is not surprising: In dealing with possible climate change, policymakers, stakeholders, and the public have to confront competing economic interests, significant political change, and such difficult issues as intergenerational equity, international competition, national sovereignty, and the role (and competence) of international institutions.[2] What are the primary factors that determine policy outcomes on this complex subject? Detailing them vividly demonstrates how scientific knowledge interacts with the formulation of policy on a significant issue in the United States.

The Policy Setting

Governments first formally addressed the issue of global warming in the Framework Convention on Climate Change (FCCC), which was negotiated at the Earth Summit in Rio de Janeiro in 1992 and subsequently ratified by 175 countries (including the United States). This agreement called for voluntary reductions in emissions of carbon dioxide and other greenhouse gases from the arbitrary base year of 1990, but there has been little response to these commitments. As a result, a process was set in motion to develop mandatory reductions. This culminated in Kyoto, Japan, in 1997, when representatives from more than 160 countries negotiated what has been called the Kyoto protocol. Under this protocol, all Annex I countries (i.e., the members of the Organisation for Economic Cooperation and Development (OECD) plus those of the former Soviet bloc) would face mandatory reductions in greenhouse gas emissions, while other countries would be encouraged to reduce their emissions but not actually required to do so. The United States would have to reduce its average emissions for the period 2008–12 by 7 percent relative to their 1990 level. The protocol targets six greenhouse gases[3] for reduction, making allowance for the creation of "sinks" (such as newly planted forests) to absorb them. It also provides various "flexibility mechanisms" (emissions trading, joint implementation, and a "Clean Development Mechanism") for reducing the overall cost of emissions reductions.[4]

The protocol's entry into force requires that 55 countries ratify—not just sign—it and that the emissions of these countries represent at least 55 percent of the total for all Annex I countries. As of March 1999, 84 countries had signed the agreement; however, only 7 had ratified it, which is well below the number required for entry into force. Although the United States was instrumental (at the last moment) in bringing about agreement at Kyoto, it did not sign the protocol for many months. The Clinton administration has indicated that it currently

From *Environment*, June 1999, pp. 16-20, 42-45. Reprinted with permission of the Helen Dwight Reid Educational Foundation. Published by Heldref Publications, 1319 Eighteenth St., NW, Washington, DC 20036-1802. © 1999.

has no intention of submitting the agreement to the Senate, where the prospects for ratification are quite dim. Many key issues remain unresolved, including the sinks, the various trading ideas, the Clean Development Mechanism, noncompliance procedures, and financial assistance to developing countries. These issues were discussed at a preparatory meeting in Bonn in June 1998 and at the Fourth Conference of the Parties to the FCCC in Buenos Aires in November 1998. No formal agreements have yet been reached, however, and the issues will require further consideration at future Conferences of the Parties. (An analysis of the outcome of the Buenos Aires meeting and subsequent discussions is forthcoming in *Environment*.)

The Role of Scientific Evidence

Of the many factors that can affect the role scientific evidence plays in questions of public policy, six appear to be most important in the case of climate change: the uncertainty of the scientific evidence; the structure of government; debatable economic assessments; the international framework; the media; and partisan politics. Although they are all significant, the uncertainty of the evidence on this issue permeates all the others.

Uncertainty

At its core, the climate change issue hinges on scientific evidence and forecasts. To date, there has been no conclusive demonstration that global warming is occurring. Rather, the entire subject is on the world's agenda because scientists have forecast that such warming will occur if the greenhouse gases produced by humans continue to accumulate as they have since the beginning of the Industrial Revolution. Concerns grew when a series of hot summers in the 1980s and 1990s appeared to the public to confirm these forecasts, and continuing assessments by the Intergovernmental Panel on Climate Change (IPCC) have strengthened the general perception that the phenomenon is real. In fact, the last IPCC assessment in 1995 cited the increase in the Earth's mean surface temperature and the changes in the patterns of atmospheric temperatures to justify its assertion (in its summary statement) that "the balance of evidence . . . suggests that there is a discernible human influence on global climate."[5]

But the evidence on climate change is not clear-cut. There is considerable uncertainty both about the basic conclusion of a demonstrable anthropogenic

"fingerprint" and, at least as important, about the scale and timing of any warming that might take place. Forecasting the scale and timing of climate change is crucial to estimating its effects and assessing the resulting costs and benefits—and thus to identifying the interests that would be affected and designing measures to reduce emissions.

Such uncertainty is always a serious problem in the formulation of public policy. It gives full play to those who oppose taking corrective action, allowing them to question the legitimacy of the forecast risks and to argue that regulation may be harmful if the risks are overestimated.

> ## Scientific evidence has a long row to hoe to have a decisive impact on policy.

It also leaves the door open to alternative scientific analyses (in fact, it stimulates such analyses) by those who perceive that their interests are threatened, thus increasing the perception that the science is uncertain.

In the case of climate change, moreover, the uncertainty is not limited to the evidence on warming, as there are even greater doubts about the ecological, physical, and economic consequences of a significant change in climate. The costs of the measures to mitigate warming are equally contentious because they are affected by different assumptions about technological change, the temporal sequencing of mitigation policies, and the basic policy framework (e.g., which countries will participate in efforts to reduce emissions and whether or not emissions trading will be allowed).

It will be a decade or more before these uncertainties are substantially reduced. In fact, for a while they may actually grow as evidence accumulates that at times seems to support one view and at other times another; as computer simulations take a larger and larger number of variables into consideration; and as the participants in the debates become more articulate in defending their positions.[6] For the present, at least, the United States will find it very difficult to reach agreement on a climate change policy. Major industries have taken strong

positions against ratifying the Kyoto protocol, labor unions have expressed reservations, and scientific "skeptics" have challenged the IPCC's basic position.[7]

Another area of uncertainty that is of central importance to this issue is the role technological change can play in reducing greenhouse emissions through increases in the efficiency of energy production and use, the development of noncarbon energy sources, and reductions in the cost of adaptation. The assumptions that analysts make about these matters are critical to their forecasts of the extent of climate change and the costs of responding to the problem. The difficulty, however, is that advances in knowledge cannot be "known" in advance. Moreover, technological change depends on policies that are explicitly designed to support research and development in both the public and private sectors. The Clinton administration has made this avenue its first response to the Kyoto commitments, proposing a 5-year, $6.3 billion program of R&D and incentives for improving the efficiency of energy production and use.[8]

The Structure of Government

The structure of government in the United States makes it harder for this country to reach closure on an issue with such major implications and levels of uncertainty than it is for any other industrial democracy. With a fundamental division of power between the executive and legislative branches and an adversarial approach to resolving policy differences, the government necessarily finds itself in deep conflict over any issue that touches major interests and ideologies. To compound matters, almost every agency in the executive branch has some legitimate interest in the climate issue, while most congressional committees are (or will be) involved in the debate—each with turf to defend or expand and each with a limited vision of the national interest. Moreover, as a result of the fragmented committee structure in Congress and weak party discipline, interest groups have easy access to the levers of power.

In this setting, scientific evidence has a long row to hoe to have a decisive impact on policy. Although that evidence may be crucial in placing an issue on the political agenda, or in influencing how that issue evolves as new knowledge is acquired, at any given time its role in the actual determination of policy is usually far less important than that of the political, economic, and other interests involved. Or, if the level of uncertainty is high enough, science may become the principal lever that all sides use to justify positions reached primarily on other grounds.

The problem is magnified when the issue has high visibility and the economic stakes are large, as is the case with climate change. Those who stand to lose from efforts to reduce emissions find it more acceptable to question the science than to defend their interests directly. Challenging the science is also more effective because most of the public cannot judge the attacks critically and thus can be easily misled or confused. As a result, disagreements among scientists are amplified and the science itself appears more uncertain—to both the public and Congress—than would be the case with a less prominent issue or one with fewer consequences.

Scientific analysis is likely to play a larger role in the executive branch than in Congress because the former has a formal structure for conducting analyses and determining policy choices. In addition, the White House has its own science adviser to evaluate scientific assessments and present conclusions in the highest policy councils. It would be a mistake, however, to assume that science plays the dominant role in determining an administration's position on a complex issue. Administrations do have many other factors to consider, as well as other influences on them, including pressure from industry, concerns over the state of the economy and the public's reaction to their positions, tradeoffs with other policy goals, relations with other nations, and, not least, the need to sell a particular policy to Congress when there are many other items on the agenda. Finally, there are the partisan factors of a party's electoral prospects and personal electoral ambitions as well.

Adding substantially to the difficulties that science faces in the political arena is the fact that the benefits of present expenditures may not be realized until far in the future. No politician likes to be in the position of advocating such expenditures when there are more immediate needs to be addressed—and especially when the case for such expenditures can be challenged as "not proven."[9]

Congress is in an even more politicized position because it is structurally more exposed to the interests of influential stakeholders. Moreover, Congress has no adequate analytical capability of its own to assess the validity and implications of scientific evidence.[10] Committees must rely on staff work and hearings to acquire and assess the knowledge produced by the executive branch and other interested parties. In such a situation, it is all too easy for individual members or their staffs (the latter of whom are often influential as a

consequence of the organization and constitutional role of Congress) to judge the validity of evidence as their own politics or ideology dictate. Even highly convincing scientific cases are often overridden when important interests or influential constituents will be adversely affected.

Congress's handling of the global warming issue illustrates this only too well. The threat of higher prices for fossil fuels or regulatory measures that would force greater efficiency in energy use has led to hearings in which the selection of witnesses is heavily biased toward those who disavow any scientific basis for concern. In fact, efforts by the Clinton administration to promote mild policies that would make sense even

> ## Congress has no adequate analytical capability of its own to assess the validity and implications of scientific evidence.

without the threat of global warming (e.g., emissions trading and R&D to improve the efficiency of energy production and use) have been attacked as "end-runs" around the Kyoto protocol ratification process.[11]

Of course, those who would benefit from lower emissions of carbon or higher energy efficiency are also able to influence the policy process.[12] But in a Congress dominated since 1995 by a Republican Party with a strong (even radically) conservative wing, the influence of environmentalists has been quite modest. Some manufacturing companies and trade organizations have also lobbied in favor of policy actions to limit greenhouse gas emissions,[13] but their influence, though symbolically important, has been marginal so far and will remain so as long as the uncertainty about the science remains high.

The range of policy options is further constrained by the attitudes of U.S. vot-

ers, particularly their antipathy toward additional taxes. This is serious because measures to limit or reduce greenhouse gas emissions may well have to include some form of tax on fossil fuels. Even if such taxes are obscured by calling them fees or some other neutral term or offset by reductions in other taxes, they can easily be attacked in a political atmosphere in which any tax "increase" is unacceptable. Moreover, because those who would be harmed by a carbon tax are likely to be clearly focused and politically powerful while those who would benefit are widely dispersed (and the benefits themselves fairly distant), any tax proposal is doubly in danger.

The separation of powers between the executive and the legislative branches, coupled with the bicameral structure of Congress and the decentralization of authority among numerous committees, further complicates the negotiations necessary to reach agreement on a consequential issue like global warming. Moreover, the tradeoffs implicit in such negotiations may be quite different from those encountered at the international level. At Kyoto, for instance, the bargaining was over emissions commitments and flexibility mechanisms; in the United States, the debate will be over the specific measures that are necessary and their economic implications.

Given all this, it may seem surprising that the Clinton administration agreed to a cut of 7 percent in U.S. emissions by 2008–12. It did so primarily because it (and particularly Vice President Al Gore) had made a prior commitment that could not be disregarded without political cost—particularly when the international negotiating process had developed so much momentum. By the end of the Kyoto negotiations, the administration must have calculated that it would pay a higher political price at home if it scuttled the negotiations than it would by acceding to some commitment. In addition, the Kyoto protocol provides a number of possible loopholes (e.g., additional greenhouse gases that may be included, the "purchase" of unused emission allowances from Russia, and credit for creating sinks) that may facilitate compliance when the time comes. The administration may also have been willing to take some political risks because policy action well in advance of 2008 did not appear urgent on political grounds (even though a delay will in fact make the commitment almost impossible to meet). A decade may be a short time as far as climate change is concerned, but it is an eternity in politics when there are three presidential elections along the way.

Economic Impacts

As suggested above, the economic implications of global warming and the measures that might be taken to prevent it play a central role in the politics of the issue. The key questions are the costs (to both the United States and the rest of the world) of climate change and the comparable costs of controlling or reducing greenhouse gas emissions;[14] the effects on international competitiveness if the commitments made by nations are uneven (as they will be because major developing countries have not accepted binding commitments); and the effects on employment and growth in specific industries or sectors.

Unfortunately, the economic dimension of global warming is more speculative than the scientific dimension and even less amenable to convincing analysis. Even the element that is most directly tied to science, i.e., the advances in alternative energy technologies that can be expected, cannot be determined in advance. The canonical figure usually used by economists, that technological change will lead to improvements in productivity (indicated by primary energy consumption per unit of GDP) of 1 percent per year, is only an assumption made for modeling purposes. And in any case, increases in productivity depend on the extent to which resources are committed to R&D.

There have been a number of attempts to analyze the economic costs of global warming and to design policies to minimize them.[15] These analyses are necessarily based on a variety of assumptions and estimates that by their very nature are uncertain. They have also tended to support policies that are desirable from an analytic standpoint but questionable politically. For example, most analyses suggest that an efficient emissions trading system would minimize the cost of reducing greenhouse emissions over the next century. But the likelihood of creating even a marginally satisfactory trading system—let alone an optimal one prior to 2008—is slim indeed. The conditions that would have to be met (e.g., agreement on national caps on emissions, an effective emissions monitoring capability, and an initial allocation of permits that would appear to reward existing patterns of consumption, among others) would not only be difficult to negotiate but in many cases politically inflammatory as well.

Even with a strong economy and low fuel prices, neither the Clinton administration nor Congress (nor the public, for that matter) wishes to adopt policies that might dampen growth when there is no evidence of an imminent ecological crisis. As far as the politics of the issue is concerned, it is irrelevant whether mitigation measures would entail only minor economic costs in the long run (and possibly benefit some sectors); whether the policies would apportion the burden equitably; and whether additional tax revenues would be used to offset other taxes. Because such measures appear to endanger the present economic prosperity (or can easily be made to), it is difficult for a politician to press for them without convincing evidence of imminent danger.

Further bedeviling the issue is the fact that fossil fuel prices have fallen steadily in recent years due to a worldwide glut of oil and (after adjustment for inflation) are now about what they were at the time of the oil shocks in the 1970s.

> The way the media presents an issue is at least as important to public attitudes as scientific evidence and economic analyses.

Low prices, of course, simply encourage consumption of these fuels, which are the major source of anthropogenically produced carbon dioxide. Higher fossil fuel prices, which would be achieved by the imposition of a carbon tax, would both reduce consumption and create incentives for improving energy efficiency throughout the economy. But in the current political climate, such a policy is politically unthinkable whatever its merit.

The United States' Inward Focus

Climate change is a quintessentially global problem, and in many ways the international response to it has been astonishing. In a relatively short time, the nations of the world have created an organizational structure to deal with this problem, launched a massive scientific assessment effort, and negotiated binding emissions reduction targets and timetables. A whirlwind of further study, meetings, and negotiations is now under way. At the same time, however, several of the international aspects complicate the debate within the United States.

One of the most contentious issues is the role of developing countries. It is clear that some of the larger ones, especially China, Brazil, India, and Indonesia, will become major emitters of greenhouse gases as their economies grow. Yet the Kyoto protocol specifically exempts all developing countries from binding emissions reductions. That exemption was agreed to in 1995 when the Berlin mandate (the negotiating process that led to the Kyoto protocol) was adopted. It's purpose was to place most of the responsibility for reducing emissions on the richer countries that had created the problem in the first place. However, without commitments from the developing countries, opponents of the Kyoto protocol can easily argue that the agreement means little and would unfairly penalize U.S. companies and workers. The Senate—which will have to ratify the protocol—has already passed a resolution (by a vote of 95 to 0) stating that the president should not submit the treaty for ratification unless it "also mandates new specific scheduled commitments to limit or reduce greenhouse gas emissions for Developing Country Parties within the same compliance period."[16]

Perhaps the most disturbing hindrance to international action on climate change is the reluctance of the United States to participate in any effort in which the United Nations (UN) and other international bodies will play a central role. In recent years, there has been a growing climate of xenophobia in Congress, which is partly reflected in the electorate and which is challenging the role of the nation in world affairs and particularly the work of the UN and organizations such as the World Bank and the International Monetary Fund. Although the mindless fears of UN "black helicopters" are an extreme example, a vocal portion of the public is turning away from international efforts unless they are U.S. led and rejecting policies that they perceive as in any way infringing U.S. sovereignty. In this context, any agreement negotiated under the auspices of the UN that would affect the U.S. economy is immediately suspect. The scientific evidence is of little moment in this situation, especially as one cannot claim that without an agreement ecological disaster is certain.

These international issues can overshadow the science because the Kyoto protocol may in fact be a flawed approach to the threat of global warming. The agreement sets a target the United States will almost certainly be unable to meet, especially given the increased emissions resulting from the country's robust economic growth since 1990. By focusing attention on near-term targets, the protocol detracts from the essential task of creating the institutions and policies that will be necessary to meet a century-long goal that includes determining an acceptable final concentration of carbon dioxide in the atmosphere, establishing a trading system that will mini-

> **One cannot overstate the importance of maintaining a credible scientific basis for policy measures and their acceptance by the public.**

mize costs, mounting a sensible R&D program, and finding ways to enlist developing countries.[17] These and other tasks will take time and experimentation to bring to fruition; failure to meet the rather arbitrary and costly goal of reducing emissions in 10 years may well undermine the long-term effort to curb emissions.

The United States agreed to the negotiations that led to the Kyoto protocol even though the ground rules were clearly deficient.[18] Now there is the danger that the whole international process will be tainted by the unrealistic actions of its supporters as well as its opponents. If the United States refuses to ratify the protocol—or cannot fulfill the commitments that it made under it—the resulting disillusionment could severely impede the development of the international structure that may well prove to be essential in the next century.

The Media

The way the media presents an issue is at least as important to public attitudes as scientific evidence and economic analyses. The media prefers issues that are either controversial or apocalyptic, and global warming can fit both criteria. Hence, the largely empty debate between the small band of climate "skeptics" (who are certain that climate change is not a threat) and most of the scientific community receives substantial press. This implies that there is something of a standoff between the two sides, a considerable misreading of the actual situation. In the same vein, unusual weather events tend to receive a lot of attention, the implication being that global warming is beginning or, if temperatures are abnormally low, that the theory is not valid. In both cases, the implied conclusions are an artifact of the way the media handles the issue rather than a true reflection of the scientific evidence and debate. This is hardly surprising, as the evidence is fuzzy and most reporters are not able to evaluate it critically.

In this situation, the public cannot help but be confused,[19] and all the more so when the information presented by the media is used to support the differing positions of different groups. Without more clear-cut scientific evidence, this is simply unavoidable. Unfortunately, the science of climate change will not be sufficiently certain to short-circuit these divisions for many years into the future.

The other side of that coin, however, is that severe climatic events may lead to public acceptance of the reality of global warming whether or not those events are actually related to such warming. For example, the very destructive El Niño events of 1997–98 had such an effect even though El Niños long predate global warming. The succession of 100 °F days that occurred in Texas in the summer of 1998 may be taken as another indicator that global warming is real. In any case, it will probably take a catastrophic ecosystem event that can readily be linked to global warming to lead to public support for policies to reduce greenhouse emissions. Otherwise, the debate in the United States is likely to turn not on the science but on the myriad other issues raised by the subject.

Partisan Politics

Finally, partisan politics is of central importance to the way in which science influences climate policy. The Republicans in Congress tend to see global warming as a Democratic issue, even though it was first placed on the agenda by the Bush administration. In particular, they associate it with Vice President Al Gore, who wrote a well-known book on the subject when he was a senator and who—not irrelevantly—may be the Democratic candidate for president in 2000.[20] This situation is ready made for a partisan conflict in which the Republicans emphasize the social and economic costs of the Kyoto protocol while the Democrats play those costs down. The uncertainty of the scientific evidence makes it easy for both parties to take their respective tacks; in fact, it has led the Clinton administration to exaggerate the impending danger to make its case for early action.[21]

Though much of the debate appears to turn on the scientific evidence, this is largely a convenient cover for the pursuit of political goals. In fact, the issue of climate policy is likely to be pressed by the more conservative elements of the Republican Party because it offers many opportunities to exploit public opposition to new taxes and the export of jobs as well as its desire for smaller government and a minimal role for the United Nations.

The Clinton administration will find it very difficult to persuade the Senate to ratify the Kyoto protocol. The task will not be made any easier by the disproportionate number of senators from states rich in natural resources. In fact, the administration will probably not even submit the treaty until after the presidential election of 2000, even though the congressional elections of 1998 may actually have improved the prospects for ratification. (Although the party lineup in the Senate remained unchanged, two prominent Republican opponents—Lauch Faircloth (N.C.) and Alphonse D'Amato (N.Y.)—were defeated.) Thus, the treaty will remain prominently on the agenda well into the future, masquerading as a scientific issue though in fact an integral part of long-standing political controversies.

Conclusion

Global warming is an issue with potentially enormous environmental, political, and economic consequences that was put on the national and international agendas by scientists. It has stimulated an intense international process of institution building, interaction, and negotiations on the part of governments, their citizens, and the United Nations. Yet in the United States, it has become entwined in internal political and economic debates made possible by the degree of uncertainty surrounding the science. As long as that uncertainty persists, other factors will bedevil efforts to agree on a policy direction.

The uncertainty does not have to be removed entirely to permit a new political consensus to be attained, however. Continuing research on the forces at work, the indicators of climate change, and the available policy options is essential and it should gradually lead to greater knowledge about the issue and the development of real policy choices. However, in a situation that is so dominated by extraneous pressures, it is especially important that the scientific and engineering communities (including economists and other social scientists) maintain their professional integrity and objectivity.

Maintaining objectivity—and the perception of it—is often not as easy as it sounds. Many scientists are tempted to intervene in the policy arena because of their personal views about climate change. Although it may be appropriate for some to do this, one cannot overstate the importance of maintaining a credible scientific basis for policy measures and their acceptance by the public. At a time when many argue that the scientific community should play a greater role in this and other policy matters, it is critical that the scientific community remain objective and not slant its results according to personal prejudices. This applies not only to individual scientists but also to endeavors such as those being carried out by IPCC. The somewhat sloppy procedure for drafting the "Summary for Policymakers" in the second IPCC assessment, in which the summary appeared to go beyond the report without adequate review, allowed accusations of bias to be made. This must not be allowed to recur;[22] the stakes are simply too large.

In the end, a growing scientific consensus will be only one factor in eventually achieving a political consensus on climate policy. A crisis that can plausibly be linked to global warming, such as the El Niño events of 1997–98, the heat wave and floods in Texas, and the damaging hurricanes of the 1998 season, will probably be equally important. Though none of the events that have occurred so far has been devastating enough to shift public opinion, a more dramatic one, such as a major shift in the climate of northern Europe, could do so, especially if the media were to link it forcefully to climate change. In time, a succession of less spectacular events, such as hot summers accompanied by more numerous and intense storms, might also convince the public that the issue cannot be ignored even if addressing it will be costly.

Whether the body politic comes to understand the climate change issue through the gradual emergence of a scientific consensus or through environmental crises, the research must continue and it must be adequately communicated to both the public and policymakers—designing and implementing the policy measures that may be necessary depend on it. This implies not only continuing study of the fundamental phenomenon of climate change but also of the effects of that change, ways to reduce emissions at minimum cost, and (because there will be a substantial accumulation of greenhouse gases in the next century whatever actions are taken) the options that are available to adapt to climate change. It will greatly ease the political difficulties of taking action if there are policy options that will reduce the costs both generally and to the major stakeholders.

The menu for the scientific and technological communities is large, even if at present political factors dominate the issue. Eventually, however, the work of these communities will provide the necessary underpinnings for policy decisions. But it is important not to assume that current research and analysis will automatically determine policy. They will enrich the debate, to be sure, but that debate will hinge on a different calculus for some time to come. Disillusionment with this situation is not useful; realistic assessment of the role of knowledge is.

Eugene B. Skolnikoff is an emeritus professor of political science at the Massachusetts Institute of Technology (MIT) in Cambridge, Massachusetts. This article was written in conjunction with MIT's Joint Program on the Science and Policy of Global Change. The author may be contacted at the Massachusetts Institute of Technology, 77 Massachusetts Avenue, Cambridge, MA 02139-4307 (telephone: 617-253-3140; e-mail: ebskol @mit.edu).

NOTES

1. In this article, the term *science* refers only to the physical sciences, not to economics and other social sciences.

2. See E.B. Skolnikoff, "The Policy Gridlock on Global Warming," *Foreign Policy* 79 (1990): 77.

3. Carbon dioxide, methane, nitrous oxide, hydrofluorocarbons, perfluorocarbons, and sulfur hexafluoride.

4. See L. D. D. Harvey and E. J. Bush, "Joint Implementation: An Effective Strategy for Combating Global Warming?," *Environment*, October 1997, 14; and J. Lanchbery, "Expectations for the Climate Talks in Buenos Aires," *Environment*, October 1998, 16.

5. J. T. Houghton et al., eds., *Climate Change 1995: The Science of Climate Change,* Contribution of Working Group I to the Second Assessment Report of the Intergovernmental Panel on Climate Change (New York: Cambridge University Press, 1996), 5. A review of this report by William C. Clark and Jill Jäger appears in the November 1997 issue of *Environment*.

6. Even if the uncertainties were less pronounced, the issue would still raise the difficult ethical questions of intra- and intergenerational equity, which are highly charged politically.

7. For a typical example of industry's reactions to the Kyoto protocol, see J. M. Broder, "Auto Makers See Nothing but Trouble in a Warmer World," *New York Times,* 16 October 1997, 19. Some companies, however, have responded more positively. See, for example, the full-page ad entitled "It's Time to Step Up to the Plate on Climate Change" that the Business Environmental Leadership Council ran in the national weekly edition of the *Washington Post* on 9 November 1998 (page S6). The council has 18 members, including such major corporations as BP America, DuPont, and Boeing. One attempt to discredit the IPCC's conclusions was rather astonishing in its brazenness: A former president of the U.S. National Academy of Sciences distributed an article appearing to be a reprint from the Academy's highly respected, peer-reviewed journal that debunked the scientific basis of global warming. In fact, the article had never been peer-reviewed or even submitted for publication. The current president of the Academy severely criticized this move.

8. Council of Economic Advisors, *The Kyoto Protocol and the President's Policies to Address Climate Change: Administration Economic Analysis* (Washington, D.C., 1998), 7.

9. Benefit-cost analysis compounds the problem of justifying current costs by means of future benefits because it discounts values that occur in the future. For more details, see S. Farrow and M. Toman, "Using Benefit-Cost Analysis to Improve Environmental Regulations," *Environment,* March 1999, 12.

10. Congress's most authoritative research arm, the Office of Technology Assessment, was abolished in 1995, and the other support agencies (the Congressional Research Service, the General Accounting Office, and the Congressional Budget Office) are not well suited to provide in-depth analysis on scientific issues.

11. U.S. House of Representatives, Government Reform and Oversight Committee, Subcommittee on National Economic Growth, Natural Resources, and Regulatory Affairs, "McIntosh to Monitor Clinton Regulatory 'End-Run' on Kyoto," press release, 2 March 1998.

12. For them, the scientific evidence is already compelling enough to justify a precautionary approach involving strong measures to reduce greenhouse gas emissions. Several major European countries have taken that position, at least rhetorically. See E. B. Skolnikoff, *Same Science, Differing Policies: The Saga of Global Climate Change*, MIT Joint Program on the Science and Policy of Global Change, Report No. 22 (Cambridge, Mass., 1997).

13. Business Environmental Leadership Council, note 7 above.

14. See H. D. Jacoby, R. G. Prinn, and R. Schmalensee, *The Road from Kyoto*, MIT Joint Program on the Science and Policy of Global Change, Report No. 32 (Cambridge, Mass., 1998), 3.

15. There is a large and growing literature on the economics of climate change. One of the best analyses is A. S. Manne and R. G. Richels, *Buying Greenhouse Insurance: The Economic Costs of CO_2, Emissions Limits* (Cambridge, Mass.: MIT Press, 1992). See also W. D. Nordhaus, *Managing the Global Commons: The Economics of Climate Change* (Cambridge, Mass.: MIT Press, 1994); and H. D. Jacoby et al. "CO_2 Emissions Limits: Economic Adjustments and the Distribution of Burdens," *Energy Journal* 18, no. 3 (1997).

16. *Expressing the Sense of the Senate regarding the Conditions for the United States Becoming a Signatory to Any International Agreement on Greenhouse Gas Emissions under the United Nations,* 105th Cong., 1st sess., S. R. 98.

17. For a balanced view of the problems that the Kyoto approach may lead to, see H. D. Jacoby, R. G. Prinn, and R. Schmalensee, "Kyoto's Unfinished Business," *Foreign Affairs*, July/August 1998, 54.

18. The countries in the European Union pushed hard for the process, taking positions determined more by internal politics and economics than by realistic environmental concerns. In doing so, they engaged in a good deal of posturing, knowing that the United States would save them from having to actually achieve the substantial emissions reductions that they first proposed. In the long run, however, this may well set back rather than advance their avowed goals. See Skolnikoff, note 12 above.

19. See W. Kempton, "How the Public Views Climate Change," *Environment,* November 1997, 12.

20. A. Gore, *Earth in the Balance: Ecology and the Human Spirit* (Boston, Mass.: Houghton Mifflin, 1992).

21. As one congressional staffer noted, "The reality is that anything with Gore's name on it is dead on arrival up here." C. Macilwain, "Gore Calls for Action on Climate Change as Congress Stalls," *Nature* 394 (23 July 1998): 305. In the same article, Gore was quoted as saying (in reference to the heat wave in Texas and forest fires in Florida) that "[t]he evidence of global warming keeps piling up."

22. See E. Masood and A. Ochert, "UN Climate Change Report Turns Up the Heat," *Nature* 378 (9 November 1995): 119.

California fumes over oil

Offshore-drilling decision gives new life to an old battle

BY BETSY STREISAND

Against the glittering backdrop of the Pacific Ocean, activists late for their day jobs clumsily try to arrange themselves to spell "GET OIL OUT!"

If they seem out of practice, they are. Environmentalists in California haven't faced a major battle over offshore oil drilling in almost a decade. (The "monogrammed" white HAZMAT jumpsuits the protesters wore in Santa Barbara last week were left over from the early 1970s.) But they could be seeing a lot of action now.

Federal officials are expected this month to grant oil companies permission to develop 40 sea-bottom tracts off California's tourist-rich central coast, which includes Santa Barbara and San Luis Obispo. Oil Companies for several years had been prevented from developing the leases,which were sold before the latest moratorium on new leases off the California coast took effect in 1990. If the oil industry prevails, renewed exploration could begin later this year, and new oil platforms could start dotting the seascape as early as 2005. Ultimately, the leases could account for 1 billion barrels of crude oil—as much as has been taken from California waters in the entire 20th century.

But fighting offshore oil drilling is practically a leisure-time activity in California. Many residents recall the massive platform blowout off the Santa Barbara coast in 1969 that created an 800-square-mile oil slick, devastating the coast and its marine life for years. Many smaller spills have followed.

The California Coastal Commission vowed during a Santa Barbara meeting last week to keep the leases from being developed. "The oil wars are back," says Susan Jordan, a League for Coastal Protection board member. "California isn't going to let this happen." (Although the commission can't ban drilling, it can easily muck up the works with legal and regulatory challenges.) Democratic Gov. Gray Davis and other big-name officials, including Sens. Barbara Boxer and Dianne Feinstein, have pledged to fight any new drilling. And Davis is also exploring whether the leases could be repurchased by the federal government. That approach worked in Florida and Alaska, but it requires congressional approval and is very expensive.

Environmentalists, politicians, and most Californians argue that the oil, whose quality is so low it would be used mainly for asphalt, isn't worth it. California's coastal economy, from fishing to tourism, is valued at $17 billion a year—if, that is, the shoreline is free of oil. "Oil cleanup is an oxymoron wrapped in a political illusion," says Mark Massara of the Sierra Club. "You can't clean it up, and we're not willing to risk the coast to pave the road."

The oil industry, which paid some $1.2 billion for the leases in the '80s, argues that it is sitting on a stack of valid contracts and that it operates far more safely—under volumes of new regulations—and much less obtrusively than in the past. "I don't know what everyone is so worked up about, "says J. Lisle Reed, regional director of the U.S. Minerals Management Service, which has the final word on offshore oil development. Only six new platforms—at most—are planned, for instance, since horizontal and long-reach drilling have reduced the "platform cites" of the past. Reed also points out that there are months, perhaps years, of paperwork to plow through before any new oil rigs appear. And, according to MMS, in the past 30 years only 820 barrels of oil (out of nearly 1 billion) have spilled into the offshore waters of the state, the equivalent of one week's worth of natural oil seepage.

Leaky line. Yet Reed agrees that cleanup is an iffy proposition. "We're in the business of preventing spills, not cleaning them up," he says. A leak discovered last week coming from a pipeline off the coast of Orange County and estimated by MMS at about a half barrel created a sheen 2 miles long and more than 100 yards wide.

Armed with statistics like those and well aware that California is an electoral mecca, activists intend to make the leases an environmental litmus test for presidential candidates in 2000. "Don't even bother campaigning in this state if you're not prepared to discuss how to stop these leases," warns Jordan. That may be less of a problem for Vice President Gore than for George W. Bush, whose home state of Texas is an offshore driller's paradise. The federal moratorium does not apply to Texas or Louisiana. In fact, oil production has been surging in the Gulf of Mexico. "How do you tell George Bush it's OK for Texas but not California?" asks Reed. Based on the statewide disdain for offshore drilling, Californians will likely find a way.

With Rebecca Sinderbrand

Lead in the Inner Cities

*Policies to reduce children's exposure to lead may be overlooking
a major source of lead in the environment*

Howard W. Mielke

In the middle of the 1970s, U.S. health officials identified what some called a "silent epidemic." They were referring to childhood lead poisoning, a problem that is easily overlooked and underappreciated. Of all of the metal-poisoning episodes to date, none has come close in sheer numbers. Toxicology textbooks mention cadmium poisoning in Japan in the 1950s and methyl mercury poisoning in both Japan in the 1950s and Iraq in 1972. Some hundreds of deaths were attributed to each of these events. But most textbooks fail to mention lead poisoning, in spite of the fact that since the 1920s millions of American children have been quietly poisoned by lead, and thousands of deaths are attributed to this over the long term.

Although childhood lead exposure has diminished over the past 20 or so years, the problem has by no means been solved. Rather,

Howard Mielke is a professor at the College of Pharmacy of Xavier University in New Orleans. He served as a Peace Corps Volunteer in Malawi, Africa, before obtaining his M.S. in biology and his Ph.D. in geography at the University of Michigan. He began his urban lead research in 1971 while teaching at the University of California, Los Angeles, and continued these studies at the University of Maryland, Baltimore County, and later at Macalester College in Minnesota. Mielke serves as the program director, and is a principal investigator of a multimedia study of metals in the urban and rural environment, for the Substance Specific Applied Research Program as part of a cooperative agreement with the Minority Health Professions Foundation/Agency for Toxic Substances and Disease Registry. He is a corresponding member of the Working Group on Geoscience and Health, Commission on Geological Sciences for Environmental Planning of the International Union of Geological Sciences. Address: Xavier University of Louisiana, College of Pharmacy, 7325 Palmetto Street, New Orleans, LA 70125. Internet: hmielke@mail. xula .edu.

the demographics have shifted. Some groups, mainly minority and poor children living in the inner city, suffer from high rates of lead poisoning. Over 50 percent (some studies place this figure at around 70 percent) of children living in the inner cities of New Orleans and Philadelphia have blood lead levels above the current guideline of 10 micrograms per deciliter (ug/dl). In contrast, in the concrete jungle of Manhattan, where very little of the soil is exposed and almost all apartments and housing contain lead-based paints, between 5 and 7 percent of children under the age of 6 have been reported to have blood-lead levels of 10 ug/dl or higher. Interestingly, just across the river in Brooklyn, where yards containing soil are common, the percentage of affected children is several times higher.

Exposure remains such a problem that early in the 1990s, the U.S. Centers for Disease Control and Prevention (CDC) in Atlanta called lead poisoning "one of the most common pediatric health problems in the United States today," but added that the problem was "entirely preventable." Effective prevention, however, assumes an accurate identification of the environmental reservoirs of lead.

Current policies to reduce lead exposure are based on the assumption that the greatest lead hazard comes from lead-based paints. Poorly maintained paints decay and release lead on their surfaces in the form of dust. In addition, lead tastes sweet, and young children may be tempted to eat leaded paint chips as though they were candy. The health consequences of this can be severe. Lead is a neurotoxin that can be especially dangerous to the developing nervous systems of infants and young children. To deal with this problem, most lead paints have been removed from the market, some lead paints are being stripped

From the *American Scientist*, January/February 1999, pp. 62–73. © 1999 by Sigma Xi, The Scientific Research Society, Inc. Reprinted by permission.

Figure 1. Lead additives were used in gasoline for over 50 years until they were banned in 1986. Although lead is no longer allowed in most gasoline products, the legacy of its use remains embedded in soils along roadways and in U.S. cities. Where there is a long history of traffic congestion, such as in the inner cities of many urban areas, lead accumulations are especially high in the soils. Children who live in these areas and play in these soils are particularly vulnerable to lead poisoning. Some simple and relatively inexpensive methods can be used, says the author, to reduce children's exposure to this environmental toxin. (Except where noted, photographs are courtesy of the author.)

off of walls, and parents are instructed to guard their children from eating paint flakes.

But such policies deal with only part of the lead hazard. Work done in my laboratory demonstrates that there exist additional sources of lead in the environment that pose as great, if not greater, threats to children. To be sure, lead paint is a major contributor of the lead in the environment. Lead was used in residential paint between 1884 and 1978 and remains on the walls of many older buildings. But paint is neither the most abundant nor the most accessible source of lead. The common problem is lead dust. For most children the issue is whether there is a source of lead dust in the environment. Research in my laboratory and others shows that in predictable lo-

cations of many cities, the soil is a giant reservoir of tiny particles of lead. This means that many children face their greatest risk for exposure in the yards around their houses and, to a lesser extent, in the open spaces such as public playgrounds in which they play. My colleagues and I believe that an accurate and complete appreciation of the distribution of lead in the environment can help shape policies that more effectively protect the health of children.

Sources of Lead

Lead is versatile and formulated into many products. Some products, such as common lead-acid batteries used in cars, trucks, boats,

Mildred Mead/Chicago Historical Society/PNI

Figure 2. Bare soils in play areas, especially those near busy roads, constitute one of the most common vectors by which children become poisoned. Children play in these soils, put their hands in their mouths, or touch objects that do, and ingest the lead. Children are most vulnerable to the toxic effects of lead, which include damage to the nervous system, learning impairments and behavioral problems.

motorcycles and the like, are sealed and if appropriately recycled, should not be the cause of poisoning of ordinary citizens. Other products allow lead to be released into the pathway of human exposure. Lead solder was used to seal seams in the canning industry until it was voluntarily withdrawn, first from baby food containers and then all canning facilities in the early 1980s. The same canning solder is used in other countries, and imported canned food continues to be tainted with lead. Some brightly colored ceramic plates and cups as well as leaded crystal may, in the presence of acidic foods (tomatoes, pineapple, wine etc.), release lead and contaminate the food. Lead-based paint was banned for household use in 1978, but lead is still an ingredient of "specialty" paints. Leaded gasoline was banned in 1986, although lead additives are still in use in racing fuels (up to 6 grams per gallon), boat fuels, farm tractors and personal watercraft fuels despite the fact that they are not required in any of these applications. Alternative octane boosters are available.

Lead acetate, or "sugar of lead," is water-soluble and one of the most bioavailable forms of lead. It is an ingredient in some hair-coloring cosmetic products. The Food and Drug Administration allows up to 6,000 parts per million of lead acetate in cosmetics. Several brands of slow-acting hair-coloring cosmetics are used daily by a sizable number of people with graying hair in many homes both

Joe Willis/Southern Stock/PNI

in the U.S. and abroad. When users pour the cosmetic into their bare hands to rub into their hair, they become conveyers of a very toxic substance. Some lead acetate may be spilled on the sink, and it is indistinguishable from drops of water. On the hands it can be easily transferred to other items such as toothbrushes, faucet handles, combs and dental floss. It can be absorbed through the skin and shows up in sweat and saliva, but not in blood, as does ingested lead. Many plastics and vinyl products contain lead as a stabilizer or coloring agent. Products become a hazard if they deteriorate into fine dust particles or otherwise directly release lead onto hands (or paws), from which it is transferred into the mouths of unsuspecting creatures, including people.

Lead in paint and gasoline together accounts for most of the lead now in the human environment. In terms of raw tonnage, the amount of lead in gasoline over only the 57 years of its use from 1929 to 1986 roughly equals all of the lead in paints in 94 years of lead-paint production, from 1884 to 1978. The peak use of lead-based paint came during the 1920s when the U.S. economy was largely agrarian and rural. Most lead paints still exist as a thin mass on the walls and structures of older buildings. Deteriorated or sanded and scraped paint contributes to lead dust accumulation in the soil.

In contrast, the use of leaded gasoline peaked early in the 1970s, a time during which the U.S. economy had become industrial and urban and reliant on automobiles for transportation. About 75 percent of gasoline lead was emitted from exhaust pipes in the form of a fine lead dust (the remaining 25 percent of the lead ended up in the oil or was trapped on the internal sur-

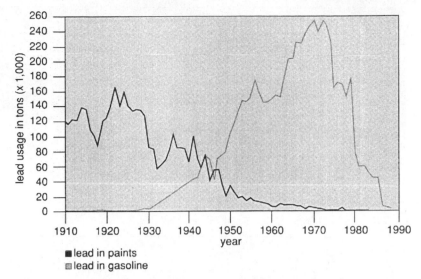

Figure 3. Lead use in paint peaked in the U.S. in the mid-1920s and then gradually died out, just as the use of leaded gasoline was on the rise. The use of leaded gasoline reached its peak in the 1970s and then declined until Congress banned it in 1986. There is a common perception that lead-based paints alone account for the amount of lead in the environment. In reality, both sources of lead contribute to the problem.

faces of the engine and exhaust system). It is estimated that the use of 5.9 million metric tons of lead in gasoline left a residue of 4 to 5 million metric tons in the environment. From these facts, we expected to find that lead would be disproportionately concentrated along roadways with the highest traffic flows—those running through cities.

Figure 4. Car exhaust spewed lead particles into the air when leaded gasolines were used. Soils in areas with a long history of traffic congestion, such as the inner cities, are heavily contaminated with lead. Lead resists movement and does not decompose, so it remains a permanent feature of the environment until people take measures to deal with it.

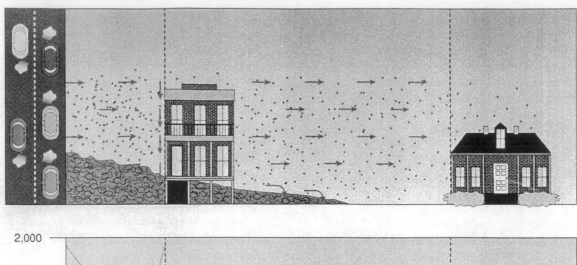

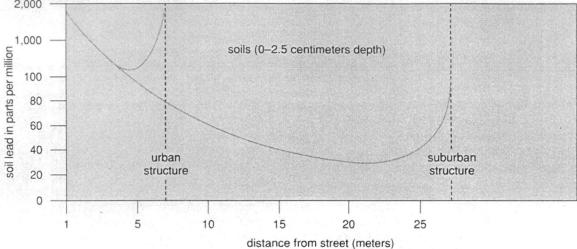

Figure 5. Two different-sized lead particles are spewed into the air when leaded gasoline is used. The trajectory of a typical particle is indicated by the arrows (*top*). Larger, heavier particles settle near the street. Smaller particles are carried by the wind until they meet a barrier, such as a tree or a house, to which they stick. Eventually, they are washed into the soil by rain. Lighter particles can be carried a greater distance and are eventually scavenged by precipitation and then fall to the ground. Assuming unpainted homes, the graph (*bottom*) generalizes the quantity of lead in the soil around an inner-city home situated about 7 meters from a heavily traveled road versus around a home more than 25 meters from the road.

Health Consequences

Lead poisoning was known in colonial times and has roots in antiquity. Clinical symptoms of lead poisoning were observed in the work force of trades that handled lead. Benjamin Franklin noticed in 1786 that some of his lead type–setters at the print shop suffered from a debilitating weakness of the hands called dropsy. He observed that those workers who suffered from dropsy failed to wash their hands before eating their sandwiches, and attributed the disease to their slovenly and unclean habits. When he learned of a 60-year-old report noting similar health problems, he commented, "and you will observe with concern how long a useful truth may be known and exist before it is generally received and practiced on."

The clinical problems of lead poisoning were described by the Greeks and were so prevalent during Roman times that some attribute the fall of Rome to lead poisoning of the aristocracy. It was considered a scourge (Saturnine poisoning) in Europe in the Middle Ages, when people frequently sweetened sour wine with lead acetate.

In other words, lead poisoning is not new, but because of the industrial way of life the poisoning has shifted to children, and the number of its victims has skyrocketed. Consider the magnitude of the man-made products that use and release lead into global environmental circulation. Jerome Nriagu of the University of Michigan School of Public Health estimates the total amount of lead produced from mines to be about 260 million metric tons in the past 3,000 years. And all of this mining has left its imprint in the geological record. Lead is found trapped in datable layers of the glaciers of Greenland, peat-bog cores of Switzerland and sediment cores of the Mississippi Gulf Coast. Although this geochemical perspective provides a framework for the history of the global quantity of mining and environmental emissions over three millennia, it does not assist us in appreciating the nature and degree of the local insult that lead has had on our urbanized society. The impact—particularly on children was not fully appreciated until the past few decades.

Re-evaluations

Until the 1970s in the United States, the health guidelines for lead exposures were essentially the same for children as for adults. In 1979 guidelines for both the amount of lead and the differences in vulnerability between children and adults were revised when Herbert Needleman at Harvard Medical School reported that children's cognitive abilities were affected by much lower lead exposures than previously recognized.

At about the same time that Needleman was doing his work, my colleague Rufus Chaney and I were conducting a study at the U.S. Department of Agriculture in Beltsville, Maryland, of the lead content of garden soils of Baltimore, Maryland. While we were performing the field studies, supervisors and other people frequently reminded me that the major source of lead in soil is lead-based paint. When we pored over the results of our analytical determinations, we came up with a problem of statistics. The distribution of the numerical results did not follow a normal curve. The statisticians I initially worked with wanted to truncate the results so that they conformed to the normal distribution.

In essence, the highest numbers were being removed from the database to satisfy the requirements of the parametric statistical models being used at that time. My brother, Paul Mielke, is a professor of statistics at Colorado State University, and I called on him for help. Paul and Ken Berry, also at Colorado State University, had recently completed their work on a non-parametric statistical test, Multi-Response Permutation Procedures, which provided a method for testing the kind of data that I had.

Paul had an appropriate nonparametric statistical model but no data base to analyze; I had the data base with no model to evaluate it. Serendipity played a role in our initial realization about the extraordinary accumulation of lead in urban environments. The research results were surprising. The pattern of lead did not match what we expected to find if paint alone was the major source of lead in Baltimore soils.

Our data did not support the lead-based paint hypothesis. Our observation was that in the inner city, where the soil-lead concentrations are highest, Baltimore had mostly unpainted brick buildings, since its inner city underwent a huge fire in 1904. On reconstruction, the new building codes required the use of fireproof (hence, brick) construction materials.

The sites of old housing constructed with painted wood siding are located in outlying parts of the city, where the lead content of garden soils is lowest. The highest lead-containing garden soils appeared to be associated with the inner-city location, not the presence of painted wood structures. The geographic pattern was extremely strong. Because of this simple observation, I sought an alternative source of lead.

Leaded gasoline fit the requirements for that alternative. We learned that gaso-

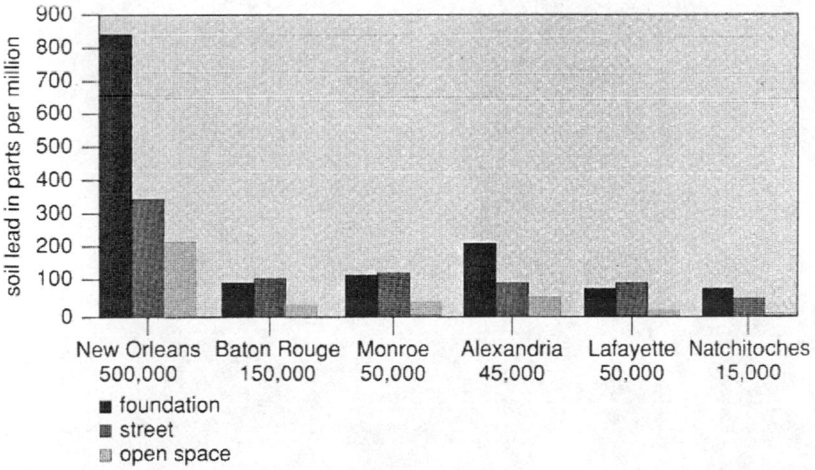

Figure 6. Soil-lead concentrations are greatest in the soils of large cities, such as New Orleans, and lower in the soils of smaller cities and towns, such as Baton Rouge, Monroe, Alexandria, Lafayette and Natchitoches. Soil surrounding the foundations of houses and buildings is more heavily contaminated than are soils around the street or in open areas. (The 1990 population of each city is given below the city's name.)

line was a huge source of lead, and we predicted that it was exhausted in a pattern that corresponded to the flow and congestion of traffic. The environment of the city was undergoing lead accumulations because of the daily commuter traffic flows that strongly characterize the industrial way of urban life. We suspected that the lead particles released with gasoline exhaust could travel through the air until they hit a barrier, such as the side of a house or apartment building. We also imagined that these lead particles would be washed down the sides of the buildings and into the soil anytime it rained. Based on this scenario, we would expect the greatest accumulations of lead to be found in the soil surrounding the foundations of buildings. We also thought it was possible that lead accumulated in soil was a major contributor to the childhood lead problem of Baltimore.

Our findings were published in 1983. Did the automobile act as a toxic-substance delivery system in all large cities? Critics of our publication argued that Baltimore is an unusual city because of all of the heavy industries there. They asserted that industrial emissions, and not leaded gasoline, accounted for the soil-lead accumulation problem in Baltimore. In 1979 I moved to Macalester College in St. Paul, Minnesota. I received a small grant to study St. Paul soils. The family of one of my students, Handy Wade of Boston, funded the purchase of analytical equipment to determine the lead content of soil samples. By 1984, the results were available.

We found that St. Paul and Minneapolis, where, in contrast to Baltimore, there is no heavy industry, had the same inner-city soil-lead accumulation as first observed in Baltimore. Even the quantities were similar. We also studied small towns and large cities and found that city size was a more important factor than age of city.

The soil-lead concentration in old communities of large cities is 10 to 100 times greater than comparable neighborhoods of smaller cities. In addition, soil-lead concentrations diminish with distance from a city center.

For example, in Baltimore, the highest garden-soil contamination is so tightly clustered around the city center that the probability that this distribution could be due to chance is 1 in 10^{23}. The concentration of lead in the soil of the Twin Cities (Minneapolis and St. Paul) is 10 times greater than that in adjacent suburbs that have old housing, where lead-based paint concentrations are as high. In 1988, I joined the faculty of Xavier University in New Orleans. I repeated the soil-lead studies conducted in Minnesota and obtained similar results. Soil-lead concentrations were higher in congested high-traffic, inner-city regions of New Orleans, as compared with its older suburbs and with smaller towns. Although the age of the houses in these communities is similar, the traffic flows vary widely. In New Orleans, with a population of about 500,000 people, the daily traffic count at major intersections in the inner-city averaged about 100,000 vehicles in 1970. In contrast, in a small town, such as Thibodaux, with its population

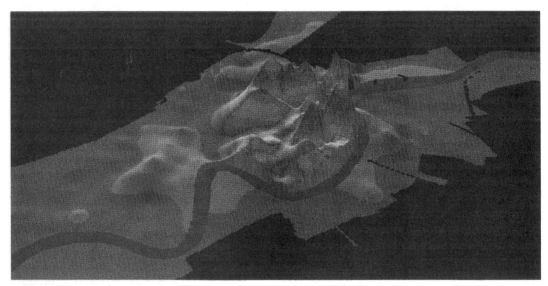

Figure 7. Three-dimensional surface plot shows that New Orleans soil-lead concentrations peak in the inner city, where traffic is heaviest, and then decrease exponentially toward the outlying parts of the city. (Image by James Pepersack, courtesy of the author.)

of about 14,000, the busiest intersections averaged only about 10,000 vehicles a day in 1970. These data strongly suggested to us that proximity to a high-traffic route is a better predictor of soil-lead concentrations than is the age of the buildings in the area or the amount of lead-based paint in the buildings. We further speculated that the soil of larger cities is more concentrated in lead than is the soil of smaller cities because of the greater volume of traffic in larger cities and because each vehicle remains inside the larger city longer than it would inside a smaller city.

In New Orleans, we tested our hypothesis that soil-lead accumulations do indeed follow traffic volumes with a comparison of large versus small cities. To do this, my colleagues and I calculated the quantities of lead emitted within one-half mile of major intersections of the inner-city regions of New Orleans and Thibodaux using the average daily vehicle traffic records during the peak lead-use years. Both of these communities have older housing with lead-based paint located around the community core, but their traffic flows differ by a factor of 10.

Next, we estimated the lead emissions in these two communities, basing our calculations on available records of vehicular use of gasoline during the years in which lead gasoline was in use, which we applied to the traffic volumes in the two communities. From our calculations, we determined that during the peak years, 5.15 metric tons of lead was emitted annually in New Orleans, as opposed to 0.45 metric ton in Thibodaux—less than one-tenth of the New Orleans concentrations. The

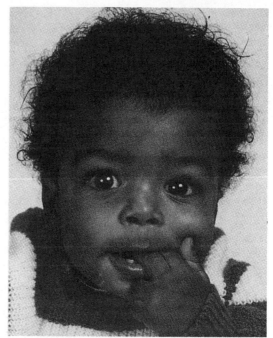

Erika Stone (Photo Researchers, Inc.)

Figure 8. Young children frequently put objects and hands in their mouths, by which they ingest lead-contaminated dust and soil.

distinctive pattern of lead that remains in cities has created "urban metal islands." A consequence of the accumulation of lead in urban communities is that a widespread environmental health hazard is now imposed on a large segment of society.

The fact that the lead is not distributed uniformly throughout the city provides strong support for our hypothesis. Within the cities we studied in Maryland, Minnesota and Louisiana, the greatest amount of lead can be found in the central part of the city where the

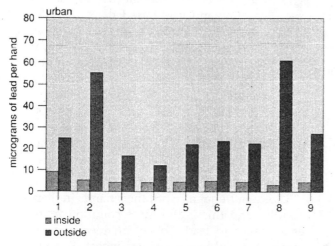

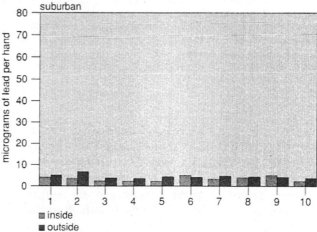

Figure 9. Measuring the lead on children's hands reveals that children pick up more lead when they play outdoors than indoors. This graph reflects data collected at private day-care centers at residential dwellings. The children of inner-city New Orleans (*left graph*) playing outdoors encounter as much as ten times the lead on their hands as children who play outdoors in non-inner city and suburban neighborhoods (*right graph*) around New Orleans.

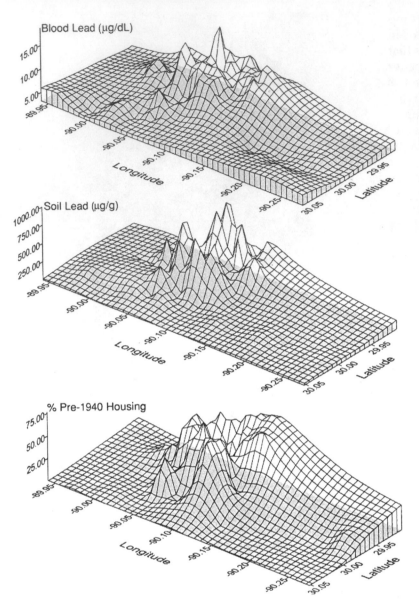

Figure 10. Three-dimensional surface plots, matched by census tracts, relate blood-lead levels (*a*), soil-lead quantities (*b*) and percentage of houses built before 1940 in New Orleans. Peaks for soil-lead and blood-lead levels are visibly similar. The percent of pre-1940 housing, which corresponds to the period of peak use of lead-based paint, does not match as closely to blood-lead levels as do soil-lead levels. Soil lead was found to be more predictive of children's exposure than was the age of housing. (Image by Christopher Gonzales, courtesy of the author.)

highway networks concentrate traffic. In effect, urban highways can be considered inadvertent dispersal systems of lead in densely populated areas surrounding the city center. Some people may argue that since lead has been removed from gasoline, it no longer poses a threat to children's health. But the clays and organic matter in soil weakly bind lead, and our research findings give us reason to believe that lead remains in the soil of the inner cities for a long time. Our results refute the commonly held notion that lead concen-

trations are highest in *all* areas with older housing.

Children at Risk

Lead is a neurotoxin that can be especially harmful to the developing brains and nervous systems of young people. High doses can be fatal. Chronic exposures to as little as 10 micrograms of lead per deciliter of blood can permanently impair the brain's functioning, thus limiting a child's intellectual and social development. It can also result in behavioral problems, which are disruptive to other children in the classroom.

The survey conducted in the 1970s showed that nine of eleven children in the United States had lead concentrations of 10 ug/dL or higher in their blood. By that time, lead-based paint production had been curtailed. Since that time, lead has been removed by mandate from gasoline and eliminated from food processing, so that a survey conducted late in the 1980s indicated that only one out of eleven children was exposed to dangerously high levels of lead. Surveys show, however, that the children affected now are disproportionately African-Americans living in the inner cities of major U.S. cities.

The Environmental Protection Agency (EPA) has estimated that more than 12 million children living within U.S. urban environments are exposed to risk from 10 million metric tons of lead residues resulting from the use of leaded gasoline and lead-based paint. A child can safely tolerate an intake of only 6 micrograms of lead per day. The mass of lead in America's urban environments that is potentially available to children is about 19 orders of magnitude (10^{19}) greater than safe levels. Clearly, there is an unacceptable amount of lead *potentially* available to children. In the laboratory we have attempted to measure how much lead is *actually* available to the child.

Children frequently put their hands in their mouths, and this is the most common route of exposure to lead. To get an idea of how much lead children routinely ingest, Latonia Viverette, then on staff at Xavier, and I measured the amount of lead on children's hands in New Orleans.

Figure 11. Sandblasting and other traditional sanding methods for removing lead-based paint from buildings release find lead-dust particles into the environment. Measures to limit lead contamination of soil include using modern sanders that capture lead dust and alternative methods to remove or encapsulate paint.

We visited several day-care centers in various parts of the city. Children's hands were wiped after they had played indoors only and then a second time after they had played outside. Our general conclusion was that children have several times more lead on their hands after playing outdoors than they do after playing indoors. The amount of lead varies directly with amount of lead measured in the outdoor soil. For private day-care centers located in ordinary homes with soil in the yards, the amount of lead picked up by children varied directly with the amount of lead in the yard soil. The amount of lead in the soil also related to the part of the city in which the day-care center was located. In the inner city, the children picked up the most lead, whereas in day-care facilities in the outer city, children were least exposed. We also observed that soil was absent from public Head Start day-care centers. The soil in outside play areas was completely covered with rubberized matting or other playground covering and did not contain bare soil. Even in the inner city, children at public day-care centers did not pick up appreciable amounts of lead on their hands.

In collaboration with Dianne Dugas and her staff at the Department of Environmental Epidemiology and Toxicology in the Louisiana Office of Public Health, we then looked at the age of homes and soil lead to determine how they related to children's blood-lead levels. Correlating the age of housing with blood-lead levels yields mixed results. In 96 percent of the areas where we found low levels of lead in children's blood, a majority of the houses were newer, having been built after 1940, when the use of lead-based paint had been reduced. By contrast, in areas where children had high levels of lead in their blood, 51 percent of the houses in these areas were built after 1940, and 49 percent were built before. Newer housing is a good predictor of low blood-lead levels, whereas old housing is a poor predictor for the group of greatest interest, children with high blood-lead levels.

The strongest correlation linked soil-lead concentrations and blood-lead concentrations, as we had expected. The association between soil lead and blood lead was 12 orders of magnitude (10^{12}) stronger than the association between the age of housing and blood lead levels. Soil-lead concentrations are more predictive of childhood lead exposure than is the age of housing. In New Orleans, housing tracts with low levels of lead in the soil are very strongly associated with low blood-lead concentrations, whereas areas high in soil lead are likewise associated with high blood-lead concentrations.

Figure 12. Artificial surfaces, such as this wooden deck, can cover over already-contaminated soils, thus limiting a child's potential exposure to toxic soils. The child in this household recovered rapidly from lead poisoning after this deck was installed. Other relatively inexpensive landscaping measures can likewise safeguard children from exposure to environmental lead.

Prevention

During a routine check-up in 1983, my daughter Beverly, then about two, was found to have a blood-lead level of about 10 to 15 ug/dL. I became haunted by the fact that my daughter and all other children were being poisoned, a finding that was reinforced by the 1978—83 national survey results. Given what I knew, I spent several days observing her activities and taking soil and paint samples. Her licensed daycare home was located several hundred yards from a major freeway. The state's licensing inspection focused on the condition of the inside facilities. The sandbox where the children played daily contained a mixture of soil and sand, was located next to the house and was surrounded by soil that contained over 500 parts per million of lead. I replaced the sand and laid down indoor-outdoor carpet on the bare soil around the sandbox. A month after I made these changes, Beverly's blood-lead levels dropped by half. My daughter's exposure riveted my attention to our nation's lead problem.

This experience and the results from our work in Minnesota prompted the formation of the Minnesota Lead Coalition. Patrick Reagan of the Midwest Environmental Education and Research Association became a major co-worker with the coalition. The lead coalition lobbied the Minnesota legislature to ban leaded gasoline. The effort failed because, unknown to us, additives to gasoline, as stated in the Federal Air Pollution Regulations, are the exclusive jurisdiction of the federal government. The only way that lead could be banned from gasoline was by EPA regulation or congressional action. Senator David Durenberger was on the subcommittee that conducted hearings to consider new lead regulations.

The Senate hearings took place in June of 1984, and I was invited to make a presentation on behalf of the Minnesota Lead Coalition. Our statement played a prominent role in the deliberations. Senator Durenberger apparently expected us to fully agree with and testify in support of the Senate bill that delayed banning lead in gasoline until the mid-1990s. We could not support this provision and instead testified in favor of the House version of the bill, which banned lead on January 1, 1986. Senator Durenberger stopped the hearing to publicly reprimand me for not testifying in support of the Senate bill. Nevertheless, the next speaker, from Ashland Oil Company, threw down his prepared statement and agreed with us about the January 1, 1986 date. He said that his company had tooled up to meet the demand for unleaded fuel and was now paying a large monthly sum to store their unused lead-free fuel refining capabilities. The rapid phase-down became law on January 1, 1986. Unfortunately, the automobile had already spread the toxin. It played a role as a toxic-substance delivery system.

It is important to note that during recent decades public actions for lead exposure prevention have concentrated on screening children for elevated blood lead and treating children on a case-by-case basis for lead poisoning. This is secondary exposure prevention because a victim must be identified before treatment can proceed. Treatment invariably focused on limiting exposure to lead-based paint, usually by removing it. But the treatment often resulted in the release of lead dust, which exacerbated the lead poisoning.

There have, however, been phenomenal gains in the reduction of children's blood-lead levels for the whole population of U.S.

children, most of the gains taking place even before the federal lead-based paint intervention programs were in place. Although the gains have been remarkable, nevertheless, the poisoning problem continues to plague our society.

Our studies have shown that for most urban areas, decades of leaded paint and gasoline use have resulted in the accumulation of lead in the soil. The major conclusion from our New Orleans study is that a balanced prevention program must include activities to limit exposure to contaminated soil.

The Federal Residential Lead-Based Paint Hazard Reduction Act of 1992 includes soil and dust as sources of lead. Nevertheless, in practice, these vectors are not adequately acknowledged as the dangers they are. For example, Patrick Reagan of the Minnesota Environmental Research Association has documented that only 9 of the 26 member countries of the Organization for Economic Cooperation and Development regulate lead in soil, as opposed to 17 members that regulate lead in paint.

This is certainly true in the United States, where there exist more programs to limit exposure to lead paint than to leaded soil. In recently submitted rules, the EPA proposes 2,000 parts per million as an acceptable standard for soil lead. For children, this is not a protective standard. In Sweden, the standard soil-lead concentration for play areas is 80 parts per million. As guest scientist to Stockholm, I saw first hand the thoroughness of the implication of the standard. The attitude of the people of Stockholm is that a city safe for children is a city safe for everyone.

The U.S. Department of Housing and Urban Development (HUD) proposes less stringent regulation for lead in soil (500-1,000 parts per million) than for lead in paint (600 parts per million). These proposals come in spite of the fact that HUD among other federal agencies concluded that soil and dust are important sources of lead. HUD notes that "for infants and young children. . .* surface dust and soil are important pathways. . . ." The Agency for Toxic Substances and Disease Registry (ATSDR) specifically states that the two major sources of lead are paint and soil. And a recent CDC report states that "lead-based paint and lead-contaminated dusts and soil remain the primary sources. . . ." The EPA reports that "the three major sources of elevated blood lead are lead-based paint, urban soil and dust . . . and lead in drinking water."

There are more than 20 other government reports that recognize the major hazards posed by leaded soil and dust. When, and only when, the sources and pathways of lead contamination are explicitly recognized as entwined will measures to prevent childhood exposure improve the situation.

Recognizing that soil in play areas is a major source of environmental lead provides an obvious course for prevention: Parents can provide a clean play area for their children and they can teach their children to wash their hands after playing outside. Communities might consider joining together to resurface play areas.

In Minneapolis and St. Paul a pilot project to reduce children's exposure to lead was implemented during the summer, when children spend the most time outdoors. In the target communities, children had their blood-lead concentrations measured at the beginning of the project in May or early June and at the end of the project in late August and September. Adults in this community were taught how to reduce the children's contact with bare soil. At the end of the trial, 96 percent of the children in the target community either remained the same or experienced *decreases* of blood lead, and 4 percent experienced modest rises. In contrast, children within a control community had their blood-lead levels measured, but no further education was provided. At the end of the trial 53 percent of these children experienced an *increase* in blood-lead levels, while 47 percent either remained the same or experienced decreases. Eighteen percent of the children in the control group had blood-lead levels of 40 ug/dL or higher, which necessitated clinical intervention. This contrasted sharply with the target group, where none of the children required clinical intervention.

A second study was conducted in Minnesota during the winter to evaluate the changes in blood lead when the ground is covered with snow and ice. The target group received special house-cleaning assistance, which the control group did not receive. Nevertheless, at the end of the winter both groups of children displayed the same degree of reductions in blood lead, relative to levels at the beginning of the winter, regardless of the cleaning treatment in their houses. This result further reinforces the hypothesis that outdoor soils are a major reservoir of lead.

The community of Trail, British Columbia, the site of the largest lead-zinc smelter in North America, provides an example of another community that successfully dealt with the problem of toxic levels of lead in its soil. Initially community leaders feared that alerting the public to the potential hazards of the soil would drive people out of the community. In spite of those fears, a primary-prevention program was put in place. Parents and child-care providers were educated about the problem, and contaminated soils were covered with plants or wood chips. New play boxes containing clean sand were put in play areas as an alternative to bare soil. Service organizations such as the Kiwanis Club, the Knights of Columbus and the Rotary Club joined to carry out landscaping projects on properties where parents could not afford to do it themselves. Because of these efforts, the levels of lead in children's blood dropped significantly 2 to 4 ug/dL (in three out of five years) or remained unchanged (during the other two years). These measures—particularly the new landscaping—had the added benefit of improving the community's overall appearance. Ultimately, the city even attracted an influx of new residents. Now all residents are routinely provided information about the lead problem and are educated about preventive measures.

The quantity of lead in New Orleans inner-city soils is about the same as that of the lead-smelter community of Trail. There is reason to believe that the success achieved by Trail could be repeated in other urban areas with contaminated soil. To reduce lead exposure of New Orleans children, I envision a program similar to Trail's. The major stakeholders include public officials, community organizations, public health and health-care providers (including pharmacists who have daily and direct contact with customers), parents, teachers, church leaders, painters, homeowners and landlords. The overall goal would be to prevent children from exposure to lead dust and contaminated soil. The projects must include interior dust control combined with measures to prevent exposure to exterior soil hazards.

Lead paint can be prevented from entering the environment by preventing its unsafe removal from both interior and exterior surfaces of houses. Alternatively, surfaces coated with unsafe paints can be covered over with a physical barrier, such as wall board. Some New Orleans painters have already agreed to use wet scraping and modern sanding equipment that captures the paint dust, rather than releasing it to the environment.

To reduce exposure from soil-based contaminants, the safest measure would be to prevent children from playing in bare soil, especially soil in play areas next to residential dwellings in the inner city. Soils should also be covered and maintained with a tight and vigorous grass or alternative surface, such as rubberized matting, bark, gravel, decking or cement. New Orleans has an enormous supply of river sediment that originates from farmlands upstream, which can be used to cover over the currently contaminated soils. In addition, lead-safe play areas, such as sandboxes, can be erected. These measures can be paid for by a tax on gasoline. A program to relandscape children's play areas would also provide useful employment to city residents.

Ultimately, successful prevention programs can be achieved only through collaborative efforts that combine public education with community-based primary-prevention measures. These efforts must focus on the ideal of clean environments as a means to healthy communities. It took nearly 10 decades for lead to accumulate to its current levels in urban areas. With judicious planning, the problem can be resolved in much less time.

Bibliography

ATSDR. 1988. *The Nature and Extent of Lead Poisoning in Children in the United States: A Report to Congress.* Atlanta: U.S. Public Health Service.

Brody, D. J., J. L. Pirkle, R. A. Kramer, K. M. Flegal, T. D. Matte, E. W. Gunter and D. C. Paschal. 1994. Blood lead levels in the US population: Phase 1 of the third national health and nutritional examination survey, NHANES III 1988 to 1991. *Journal of the American Medical Association* 272:277–283.

Centers for Disease Control. 1991. *Preventing Lead Poisoning in Young Children.* Atlanta: Centers for Disease Control.

Charney, E., J. Sayre and M. Coulter. 1980. Increased lead absorption in inner city children: Where does the lead come from? *Pediatrics* 65:226–231.

Hammond, P. B., C. S. Clark, P. S. Gartside, O. Berger, A. Walker and L. W. Michael. 1980. Fecal lead excretion in young children as related to sources of lead in their environment. *International Archives of Occupational and Environmental Health* 46:191–202.

Hilts, S. R. 1996. A cooperative approach to risk management in an active lead/zinc smelter community. *Environmental Geochemistry and Health* 18: 17–24.

Hilts, S. R., S. E. Bock, T. L. Oke, C. L. Yats and R. A. Copes. 1998. Effect of interventions on children's blood lead levels. *Environmental Health Perspectives* 106:79–83.

Mielke, H. W., J. C. Anderson, K. J. Berry, P. W. Mielke and R. L. Chaney. 1983. Lead concentrations in inner city soils as a factor in the child lead problem. *American Journal of Public Health* 73:1366–1369.

Mielke, H. W., B. Blake, S. Burroughs and N. Hassinger. 1984. Urban lead levels in Minneapolis: The case of the Hmong children. *Environmental Research* 34:64–76.

Mielke, H. W., S. Burroughs, S. Wade, T. Yarrow and P. W. Mielke. 1984–1985. Urban lead in Minnesota: Soil transects of four cities. *Journal of Minnesota Academy of Science* 50:19–24.

Mielke, H. W., J. L. Adams, P. L. Reagan and P. W. Mielke, 1989. Soil-dust lead and childhood lead exposure as a function of city size and community traffic flow: The case for lead abatement in Minnesota. In "Lead in Soil," ed. B. E. Davies and B. G. Wixson. *Environmental Geochemistry and Health Supplement 9*, 253–271.

Mielke, H. W., J. E. Adams, B. Huff, J. Pepersack, P. L. Reagan, D. Stoppel and P. W. Mielke, Jr. 1992. Dust control as a means of reducing inner-city childhood Pb exposure. *Trace Substances and Environmental Health* 25:121–128.

Mielke, H. W. 1993. Lead dust contaminated USA cities: Comparison of Louisiana and Minnesota. *Applied Geochemistry* (Supplementary Issue 2):257–261.

Mielke, H. W. 1994. Lead in New Orleans soils: New images of an urban environment. *Environmental Geochemistry and Health* 16:123–128.

Mielke, H. W., D. Dugas, P. W. Mielke, K. S. Smith, S. L. Smith and C. R. Gonzales. 1997. Associations between soil lead and childhood blood lead in urban New Orleans and rural Lafourche Parishes of Louisiana USA. *Environmental Health Perspectives* 105:950–954.

Mielke, H. W., and P. L. Reagan. 1998. Soil is an important pathway of human lead exposure. *Environmental Health Perspectives* 106 (Suppl. 1):217–229.

Pirkle, J. L., D. J. Brody, E. W. Gunter, R. A. Kramer, D. C. Paschal, K. M. Flegal and T. D. Matt. 1994. The decline in blood lead levels in the United States: The national health and nutritional examination surveys NHANES. *Journal of the American Medical Association* 272:284–291.

Sayre, J. W., E. Charney, J. Vostal and I. B. Pless. 1974. House and hand dust as a potential source of childhood lead exposure. *American Journal of the Diseases of Children* 127:167–170.

Statement of the Ethyl Corporation. 1984. S 2609—A Bill to Amend the Clean Air Act with Regard to Mobile Source Emission Control. Hearings before the Committee on Environment and Public Works, U.S. Senate, 98th Contress 2nd Session, June 22.

Viverette, L., H. W. Mielke, M. Brisco, A. Dixon, J. Schaefer and K. Pierre. 1996. Environmental health in minority and other underserved populations: benign methods for identifying lead hazards at day care centres of new Orleans. *Environmental Geochemistry and Health* 18:41–45.

Unit 2

Unit Selections

Key Points to Consider

❖ What are the long-range implications of atmospheric pollution? Explain the greenhouse effect.

❖ How can the problem of regional transfer of pollutants be solved?

❖ The manufacture of goods needed by humans produces pollutants that degrade the environment. How can this dilemma be solved?

❖ Where in the world are there serious problems of desertification and drought? Why are these areas increasing in size?

❖ What will be the major forms of energy in the twenty-first century?

❖ How are you as an individual related to the land?

❖ Can humankind do anything to ensure the protection of the environment?

❖ What is your attitude toward the environment?

 **Links**   **www.dushkin.com/online/**

These sites are annotated on pages 6 and 7.

The home of humankind is Earth's surface and the thin layer of atmosphere enveloping it. Here the human populace has struggled over time to change the physical setting and to create the telltale signs of occupation. Humankind has greatly modified Earth's surface to suit its purposes. At the same time, we have been greatly influenced by the very environment that we have worked to change.

This basic relationship of humans and land is important in geography and, in unit 1, William Pattison identified it as one of the four traditions of geography. Geographers observe, study, and analyze the ways in which human occupants of Earth have interacted with the physical environment. This unit presents a number of articles that illustrate the theme of human-environment relationships. In some cases, the association of humans and the physical world has been mutually beneficial; in others, environmental degradation has been the result.

At the present time, the potential for major modifications of Earth's surface and the atmosphere is greater than at any other time in history. It is crucial that the consequences of these modifications for the environment be clearly understood before such efforts are undertaken.

The first selection in this unit deals with the devastating outcomes in Latin America of the 1997–98 El Niño. "The Great Climate Flip-Flop" suggests that there may be global cooling following the global warming trend and offers ways to prevent it. William Stevens's article provides evidence of human influence on climate change. Two articles follow which deal with new attitudes toward dams in the United States. Changing land use is examined by William Meyer, and urban sprawl is addressed in two articles. "Can This Swamp Be Saved?" deals with attempts to remedy water flow problems in Florida's Everglades. Finally, the question of NAFTA's impact on economics and the environment is covered.

This unit provides a small sample of the many ways in which humans interact with the environment. The outcomes of these interactions may be positive or negative. They may enhance the position of humankind and protect the environment, or they may do just the opposite. We human beings are the guardians of the physical world. We have it in our power to protect, to neglect, or to destroy.

Human-Environment Relationships

THE AMERICAS

The season of El Niño

At last, the El Niño of 1997–98 is returning to its cradle, after scarring Latin America with drought and fire, storm and flood. But it is not over yet, and the fall-out, economic, social and political, will not be cleared up for many months or even, in some places and activities, for years

It swept away roads and bridges, homes and farms, lives and livelihoods. It created a vast lake in a north Peruvian desert, and ruined the fisheries of Chile. It battered Mexico's Pacific coast resort of Acapulco, and lowered water levels, enforcing draught restrictions on ships in the Panama canal. Across stretches of northern Brazil, Guyana and Suriname, even in the island of Trinidad, it dried up vegetation and often with human help—burned up forests. It parched crops too, bringing Brazil's always dry north-east food shortages that had the president scurrying there this week to see relief efforts. Farther south, it swelled the mighty Parana river to eight metres above its normal height, flooding millions of hectares and driving hundreds of thousands of people from their homes in Paraguay, Uruguay and Argentina.

Overall, it killed at least 900 human beings and livestock by the hundreds of thousands. It hit economies for perhaps $20 billion, much of that still to be paid. And, in many countries, it exposed governments to harsh tests and harsher criticism.

And still no one knows just why it happened. Every year, usually around December—hence the name El Niño, the Christ-child—a warm Pacific current flows east to the coasts of Ecuador and Peru. But at random intervals of years, this current becomes a flood. The mechanisms are known. But the root cause is not. And though satellite observations and new weather buoys now enable a fierce El Niño to be foreseen months in advance—as the last big one, in 1982–83, was not—no one can be sure exactly in what form and where the extremes of weather that it brings will strike. Latin America and the Caribbean will not swiftly forget the El Niño of 1997–8.

Drought, flood and uncertainty

Of his two faces, El Niño this time broadly (see the map on the next page) showed drought to the north of the equator, flood—or at least torrential storms—to the south. But little was predictable.

Central America largely escaped the 1982–83 El Niño. This time most of it was hit by drought. So were Mexico's centre and south—but its Pacific coast was struck by hurricanes of wind and rain, rare on that side; the third of these in October smashed Acapulco. The Caribbean has been short of rain: Cuba recently had forest fires, and some east Caribbean islands have had to cut water supplies at night. Colombia has had half its normal rain, and the drought has stretched eastward across Venezuela into northern Brazil, Guyana, Suriname and down into Brazil's north-east.

In contrast, much of Ecuador and Peru have had downpours. It was Peru's Sechura desert, on its far-northern coast, that was converted into a lake, of several thousand square kilometres at its largest, though it is now degenerating into a stagnant swamp. Peru's government had done much to prepare against the rains, building dykes, cleaning waterways and strengthening bridges. El Niño from January to March made light of its efforts, badly damaging 600 kilometres (370 miles) of Peru's main roads and 30 bridges along them. And mocked its forecasts too. The authorities had expected the normally dry south to grow drier still. Not so: a swollen river three months ago flooded the southern city of Ica, near the coast, and it was in the southern mountains that 20 people died when a village was buried by a landslide.

Bolivia for six months has had highland droughts and lowland floods, an unlikely mixture due to unusually high temperatures: what would normally fall as snow on the peaks fell as rain, and ran straight off to drown areas below. Chile had a week of fierce floods last June, midwinter there. Argentina now has floods in the south. But all these pale beside the floods that for four weeks now have swamped northern Argentina and parts of Paraguay and Uruguay, as several big rivers, headed by the giant Parana, have burst their banks. The 250,000 people of Resistencia, capital of Argentina's Chaco province, spent May day wondering whether the Parana, already eight metres above normal, would summon up the extra metre to overwhelm the dyke hurriedly built around their city. It did not; but, in all, some 80,000 square kilometres (31,900 square miles) of Argentina were under water this week—and that excludes areas flooded earlier but now starting to dry out.

Costs, short-term and long

The economic costs imposed by El Niño on the region have been huge. Most of the figures, of course, are heroic estimates, and many are heroic forecasts, at that. Cattle drowned can be crudely counted (140,000, says one estimate for Argentina's Santa Fe province, stretched along 800km of the Parana), and the loss of crops ruined by flood or drought can be reckoned, but the long-term effects may take years to show up.

Not all will be bad. As now in Argentina and Uruguay, one rancher's drowned beast is another's higher beef price, and, more widely helpful, today's disaster can be tomorrow's rebuilding, maybe for the better. There have even been one or two instant winners. Chile's 1997 skiing season was prolonged by abundant snow into October. Tourists have flocked to see its Atacama desert in bloom. June's flash floods there made 80,000 people homeless and hit some farmers, especially vegetable growers. Later, excessive rains were too much for some crops. But both also filled reservoirs for the many Chilean farmers who artificially irrigate their land and had suffered three years of low rainfall. So they did for the hydro-power companies, which were within days of enforced electricity rationing. In many countries, road-builders and suppliers of concrete can look forward to a profitable year.

But these are the exceptions. The first to suffer were the fishermen on the Pacific coasts. Peru's boats as early as mid-1997 started finding the shrimps and lobsters usually caught in Ecuador's waters. The Peruvians' usual prey had swum south, looking for food in colder waters off Chile. Peru normally exports $1 billion a year of fishmeal from *anchoveta*, the Pacific pilchard. That business has collapsed, and jobs at sea and in processing plants with it. Chile's jack mackerel in turn went farther south, cutting its catch by 40% in the first quarter of this year. In the fishing port of San Antonio, nine of ten fishmeal plants have closed, and the tenth is handling 20 tonnes a day instead of its usual 100. Only now are Peru's coastal waters starting to cool again.

Farmers' losses have been huge. Drought in Central America has cost Guatamala perhaps 10% of its grain and hard-hit El Salvador 30% of its coffee. Parts of central Panama have been reduced to a dustbowl, and thousands of cattle have had to be slaughtered (not wholly a loss: they were already helping to make the dustbowl).

In Colombia drought has parched 5,000 square kilometres of pasture and 2,000 of maize, sorghum and soya. Overall, says the ministry of agriculture, El Niño has cost 7% of normal output, with the cotton harvest cut by a quarter and milk currently 30% down. Pests have flourished in the coffee plantations east of Bogota. In Brazil's north-east, SUDENE, the regional development agency, talks of drought losses above $4 billion. Cattle have begun to die and the many subsistence farmers have seen their crops wither. Even in this area, not famous for its

grain, 2½m tonnes of that are reckoned to have gone.

Argentina's economy ministry this week estimated the total costs of El Niño's floods at $3 billion, almost 1% of GDP. Farming will bear much of that: cattle apart, perhaps 40% of the expected 1½m tonnes of cotton has been lost, 30% of the rice, 50% of the tobacco. This week the wheat and soya crops in the *pampa humida*, Argentina's bread-

basket, west of Buenos Aires, were in danger. In the Andean region of Mendoza, home of the country's—rapidly improving and valuable—wineries, vineyards have been damaged by storms. Uruguay's cattle have largely escaped, but 10% of its main grain crop, rice, has gone.

Not all the losses are direct or immediate. Bolivia has seen its soya and cotton hit by floods. But even if their fields were untouched, growers of tropical fruit and vegetables—key crops in the government's efforts to wean farmers off coca-growing—have often seen their roads to market washed away. In the high *altiplano*, poor peasants depend for

much of their water on the slow melting of snow. With snowfall lacking, many have lost their crops and will find livestock dying in the coming winter.

Floods can also do long-lasting damage to soil. They wash out nutrients. They may raise salt levels. Irrigation, especially during drought, can also have that effect. And in Guyana and Suriname, coastal farmers irrigating with water from river estuaries have had to stop: because of the rivers' weak flow, it was sea-water.

Roads, power and output

The losses, of course, extend much wider than farming. In Panama, famous for its tropical downpours, the big river Chagres shrank almost to a brook. Lack of rain not only meant drought restrictions on ships (and loss of toll revenue) in the canal. It has also brought power cuts: like most of Central America (except Nicaragua), Panama gets about 70% of its power from hydro-electricity. Several countries, taught by a sharp drought in 1994, have invested in thermal plants, avoiding what they suffered then. But the price, in capital and fuel costs, has been heavy. Colombia too has seen power suppliers struggling to keep up, and river transport, not least of oil, disrupted on its big Magdalena river, which runs north-south deep into the country.

Floods too do more than destroy and kill. Ecuador has been savaged by this El Niño, recording more rain—7,000mm (275 inches)—in the past 14 months than in 17 of the El Niño of 1982–83. It will have to rebuild 19 large bridges and repair 2,500km of roads, at a cost put at $800m. Besides that, farmers have lost $1 billion in output. But on top come the unquantifiable costs of disruption to transport. Lorries stand in interminable queues where roads and bridges have been washed away. So do pedestrians, waiting to pay to cross makeshift plank-and-rope bridges. As in much of Latin America, especially its many mountain areas, a road blocked is movement blocked: without an enormous detour, if at all, traffic simply has no way round.

That has been convenient for Bolivia's weekend travellers to the tropical valleys east of La Paz, who have found "landslide" a neat excuse for not being at the office on Monday morning. No complaint came—except from her father—when a young Ecuadorean was forced by a sudden flood to pass the night at the house of her boyfriend. But the economy has suffered, as raw materials or components (or workers) did not arrive, and goods (or crops or ores) could not be shipped out. Service industries too have lost, as telephones have gone dead and tourist trips been cancelled. The storm that smashed Acapulco spared its glossy hotels, but—even arriving in low season—menaced its $1 billion a year and 150,000 jobs based on tourism.

The human cost

The heaviest cost, though not the most noticed, has been to human happiness, health and lives. Amid all the averages, many people, mostly poor ones, have lost everything: crops, jobs, homes and often hope. In Ecuador, some stand begging at the roadside, crying from shame as much as grief.

Peru has recorded nearly 350,000 people driven from their homes, Argentina 150,000. Some, though by no means all, have had their houses simply swept away. Many of Ecuador's 50,000 have taken refuge in schools left empty during the holidays. But many in all these countries have clung on to their homes, come what might, surrounded by polluted floodwaters; drinking and cooking with them too, with the predictable results—diarrhoea and other intestinal diseases, leptospirosis (caused by animal urine in the water), even cholera. Mosquito-borne malaria and dengue fever have multiplied.

Just how much is El Niño's fault is uncertain: dengue was already alarming Brazil. The death toll too is uncertain; a rough count says around 1,000. Peru blames 300 deaths on El Niño, Ecuador 250, with another 150 missing. Chile lost 20 in the June floods, Argentina about as many in the floods now going on, which were well signalled in advance. One landslip in a coastal resort killed 17 Ecuadoreans

last weekend. The Acapulco hurricane cost Mexico 200–400 lives in a day, and at least 19 people died in a single tragedy there this week, when a forest fire—one of thousands that have burned more than 2,000 square kilometres—suddenly overwhelmed those fighting it. But South America's worst reported disaster happily never occurred: a Bolivian goldmine was indeed buried in a mudslide, but earlier flooding had already driven the 70–80 supposed victims to leave.

Yet the true toll is surely far worse. Drought and malnutrition, added to poverty, must have killed far greater numbers than any disaster, but leaving no record.

The natural world

Wildlife too has paid a price. Nature will quickly replace the starving sea-lions and gulls found on Peru's beaches. She will take longer with the burnt savannahs on the Venezuelan-Brazilian border, and far longer with the forests that have burned in so many drought-hit areas, not least the Amazon rainforest destroyed in Roraima state, in northern Brazil, when fires—often lit by farmers clearing land—spread from the savannahs. Even a swamp "of international importance" has burned in Trinidad.

It was the Amazon blaze, engulfing some 3,250 square kilometres of forest, that aroused the outside world, largely indifferent to human suffering, to the fact of El Niño. There may be more to come: the southern edge of the forest too is dry; and this is the area subject to logging and, from June on, to the yearly burning by migrant would-be farmers desperate for land. Yet big fires in any hilly region—as in southern Mexico, in Colombia, even in Trinidad—can do enduring damage. Today's treeless hillside is tomorrow's one lashed by tropical rains, and tomorrow's landslips, instant run-off of rain and flash floods.

Leaders put to the test

The authorities' response to warnings of El Niño, and to its arrival, has varied widely—and been widely, but

not always fairly, criticised. Mexicans were quick to denounce the lack of precise forewarning of the Acapulco storm, and then the local theft of relief supplies. Yet action had been fast.

Brazil's rulers, local and federal, were accused of slow and inept response to the Roraima fires; and, more damningly, of ignoring months of drought warnings. President Fernando Henrique Cardoso rushed to the north-east this week, having (correctly) blamed churchmen and the MST landless movement for encouraging raids on food stores (and being told, as correctly, that it was not they who invented hunger). The leading newsweekly, Veja, castigated him—the Veja that in a December story recommended this as just the year for a north-eastern holiday, with El Niño promising "an exceptional season, with lots of sun, blue sky, warm water and soft breezes".

Colombia's government—it has other worries—did little to prepare for El Niño, and even let farmers set fires across huge areas of land, supposedly to enrich the soil, before it banned the practice. Governments in Ecuador, Peru, Bolivia and Chile declared states of emergency in parts of their territory; not so Colombia's, which has been much blamed for that. Paraguay's electioneering politicians have largely ignored its floods; Ecuador's are vying in plans for reconstruction. Argentine politicians—and the public, truly Latin American in this, whatever else—have shown solidarity with victims of the floods, and President Carlos Menem is promising $1 billion to clear up. But critics like Domingo Cavallo, once his economy minister, have damned the lack of advance precautions.

The true artist of El Niño has been Peru's President Alberto Fujimori. His government spent $300m in advance (not all in the right places, but at least the ones that looked right at the time); and since El Niño struck he has rushed about frenetically taking personal charge of relief efforts, even rescue attempts. Too frenetically, say some critics, who claim presidential efforts are muddling those of people on the spot. Maybe, maybe not; but his poll ratings, 30% in mid–1997, now stand at 45%.

The Great Climate Flip-flop

by WILLIAM H. CALVIN

ONE of the most shocking scientific realizations of all time has slowly been dawning on us: the earth's climate does great flip-flops every few thousand years, and with breathtaking speed. We could go back to ice-age temperatures within a decade—and judging from recent discoveries, an abrupt cooling could be triggered by our current global-warming trend. Europe's climate could become more like Siberia's. Because such a cooling would occur too quickly for us to make readjustments in agricultural productivity and supply, it would be a potentially civilization-shattering affair, likely to cause an unprecedented population crash. What paleoclimate and oceanography researchers know of the mechanisms underlying such a climate flip suggests that global warming could start one in several different ways.

For a quarter century global-warming theorists have predicted that climate creep is going to occur and that we need to prevent greenhouse gases from warming things up, thereby raising the sea level, destroying habitats, intensifying storms, and forcing agricultural rearrangements. Now we know—and from an entirely different group of scientists exploring separate lines of reasoning

> **"Climate change" is popularly understood to mean greenhouse warming, which, it is predicted, will cause flooding, severe windstorms, and killer heat waves. But warming could lead, paradoxically, to drastic cooling—a catastrophe that could threaten the survival of civilization**

and data—that the most catastrophic result of global warming could be an abrupt cooling.

We are in a warm period now. Scientists have known for some time that the previous warm period started 130,000 years ago and ended 117,000 years ago, with the return of cold temperatures that led to an ice age. But the ice ages aren't what they used to be. They were formerly thought to be very gradual, with both air temperature and ice sheets changing in a slow, 100,000-year cycle tied to changes in the earth's orbit around the sun. But our current warm-up, which started about 15,000 years ago, began abruptly, with the temperature rising sharply while most of the ice was still present. We now know that there's nothing "glacially slow" about temperature change: superimposed on the gradual, long-term cycle have been dozens of abrupt warmings and coolings that lasted only centuries.

The back and forth of the ice started 2.5 million years ago, which is also when the ape-sized hominid brain began to develop into a fully human one, four times as large and reorganized for language, music, and chains of inference. Ours is now a brain able to anticipate outcomes well enough to practice ethical behavior, able to head off disasters in the making by extrapolating trends. Our civilizations began to emerge right after the continental ice sheets melted about 10,000 years ago. Civilizations accumulate knowledge, so we now know a lot about what has been going on, what has

---

William H. Calvin is a theoretical neurophysiologist at the University of Washington at Seattle.

From *The Atlantic Monthly*, January 1998, pp. 47-50, 52-54. © 1998 by William H. Calvin. Reprinted by permission.

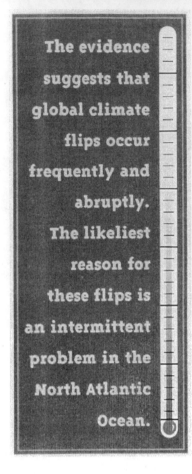

The evidence suggests that global climate flips occur frequently and abruptly. The likeliest reason for these flips is an intermittent problem in the North Atlantic Ocean.

made us what we are. We puzzle over oddities, such as the climate of Europe.

Keeping Europe Warm

EUROPE is an anomaly. The populous parts of the United States and Canada are mostly between the latitudes of 30° and 45°, whereas the populous parts of Europe are ten to fifteen degrees farther north. "Southerly" Rome lies near the same latitude, 42°N, as "northerly" Chicago—and the most northerly major city in Asia is Beijing, near 40°N. London and Paris are close to the 49°N line that, west of the Great Lakes, separates the United States from Canada. Berlin is up at about 52°, Copenhagen and Moscow at about 56°. Oslo is nearly at 60°N, as are Stockholm, Helsinki, and St. Petersburg; continue due east and you'll encounter Anchorage.

Europe's climate, obviously, is not like that of North America or Asia at the same latitudes. For Europe to be as agriculturally productive as it is (it supports more than twice the population of the United States and Canada), all those cold, dry winds that blow eastward across the North Atlantic from Canada must somehow be warmed up. The job is done by warm water flowing north from the tropics, as the eastbound Gulf Stream merges into the North Atlantic Current. This warm water then flows up the Norwegian coast, with a westward branch warming Greenland's tip, 60°N. It keeps northern Europe about nine to eighteen degrees warmer in the winter than comparable latitudes elsewhere—except when it fails. Then not only Europe but also, to everyone's surprise, the rest of the world gets chilled. Tropical swamps decrease their production of methane at the same time that Europe cools, and the Gobi Desert whips much more dust into the air. When this happens, something big, with worldwide connections, must be switching into a new mode of operation.

The North Atlantic Current is certainly something big, with the flow of about a hundred Amazon Rivers. And it sometimes changes its route dramatically, much as a bus route can be truncated into a shorter loop. Its effects are clearly global too, inasmuch as it is part of a long "salt conveyor" current that extends through the southern oceans into the Pacific.

I hope never to see a failure of the northernmost loop of the North Atlantic Current, because the result would be a population crash that would take much of civilization with it, all within a decade. Ways to postpone such a climatic shift are conceivable, however—old-fashioned dam-and-ditch construction in critical locations might even work. Although we can't do much about everyday weather, we may nonetheless be able to stabilize the climate enough to prevent an abrupt cooling.

Abrupt Temperature Jumps

THE discovery of abrupt climate changes has been spread out over the past fifteen years, and is well known to readers of major scientific journals such as *Science* and *Nature*. The abruptness data are convincing. Within the ice sheets of Greenland are annual layers that provide a record of the gases present in the atmosphere and indicate the changes in air temperature over the past 250,000 years—the period of the last two major ice ages. By 250,000 years ago *Homo erectus* had died out, after a run of almost two million years. By 125,000 years ago *Homo sapiens* had evolved from our ancestor species—so the whiplash climate changes of the last ice age affected people much like us.

In Greenland a given year's snowfall is compacted into ice during the ensuing years, trapping air bubbles, and so paleoclimate researchers have been able to glimpse ancient climates in some detail. Water falling as snow on Greenland carries an isotopic "fingerprint" of what the temperature was like en route. Counting those tree-ring-like layers in the ice cores shows that cooling came on as quickly as droughts. Indeed, were another climate flip to begin next year, we'd probably complain first about the drought, along with unusually cold winters in Europe. In the first few years the climate could cool as much as it did during the misnamed Little Ice Age (a gradual cooling that lasted from the early Renaissance until the end of the nineteenth century), with tenfold greater changes over the next decade or two.

The most recent big cooling started about 12,700 years ago, right in the midst of our last global warming. This cold period, known as the Younger Dryas, is named for the pollen of a tundra flower that turned up in a lake bed in Denmark when it shouldn't have. Things had been warming up, and half the ice sheets covering Europe and Canada had already melted. The return to ice-age temperatures lasted 1,300 years. Then, about 11,400 years ago, things suddenly warmed up again, and the earliest agricultural villages were established in the Middle East. An abrupt cooling got started 8,200 years ago, but it aborted within a century, and the temperature changes since then have been gradual in comparison. Indeed, we've had an unprecedented period of climate stability.

Coring old lake beds and examining the types of pollen trapped in sediment layers led to the discovery, early in the twentieth century, of the Younger Dryas. Pollen cores are still a primary means of seeing what regional climates were doing, even though they suffer from poorer resolution than ice cores (worms churn the sediment, obscuring records of all but the longest-lasting temperature changes). When the ice cores demonstrated the abrupt onset of the Younger Dryas, researchers wanted to know how widespread this event was. The U.S. Geological Survey took old lake-bed cores out of storage and re-examined them.

Ancient lakes near the Pacific coast of the United States, it turned out, show a shift to cold-weather plant species at roughly the time when the Younger Dryas was changing German pine forests into scrublands like those of modern Siberia. Subarctic ocean currents were reaching the southern California coastline, and Santa Barbara must have been as cold as Juneau is now. (But the regional record is poorly understood, and I know at least one reason why. These days when one goes to hear a talk on ancient climates of North America, one is likely to learn that the speaker was forced into early retirement from the U.S. Geological Survey by budget cuts. Rather than a vigorous program of studying regional climatic change, we see the shortsighted preaching of cheaper government at any cost.)

In 1984, when I first heard about the startling news from the ice cores, the implications were unclear—there seemed to be other ways of interpreting the data from Greenland. It was initially hoped that the abrupt warmings and coolings were just an oddity of Greenland's weather—but they have now been detected on a worldwide scale, and at about the same time. Then it was hoped that the abrupt flips were somehow caused by continental ice sheets, and thus would be unlikely to recur, because we now lack huge ice sheets over Canada and Northern Europe. Though some abrupt coolings are likely to have been associated with events in the Canadian ice sheet, the abrupt cooling in the previous warm period, 122,000 years ago, which has now been detected

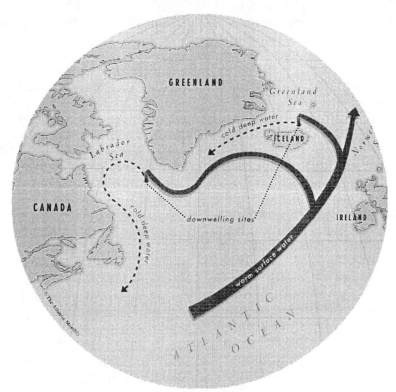

THE NORTHERN LOOP OF THE NORTH ATLANTIC CURRENT

even in the tropics, shows that flips are not restricted to icy periods; they can also interrupt warm periods like the present one.

There seems to be no way of escaping the conclusion that global climate flips occur frequently and abruptly. An abrupt cooling could happen now, and the world might not warm up again for a long time: it looks as if the last warm period, having lasted 13,000 years, came to an end with an abrupt, prolonged cooling. That's how our warm period might end too.

Sudden onset, sudden recovery—this is why I use the word "flip-flop" to describe these climate changes. They are utterly unlike the changes that one would expect from accumulating carbon dioxide or the setting adrift of ice shelves from Antarctica. Change arising from some sources, such as volcanic eruptions, can be abrupt—but the climate doesn't flip back just as quickly centuries later.

Temperature records suggest that there is some grand mechanism underlying all of this, and that it has two major states. Again, the difference between them amounts to nine to eighteen degrees—a range that may depend on how much ice there is to slow the responses. I call the colder one the "low state." In discussing the ice ages there is a tendency to think of warm as good—and therefore of warming as better. Alas, further warming might well kick us out of the "high state." It's the high state that's good, and we may need to help prevent any sudden transition to the cold low state.

Although the sun's energy output does flicker slightly, the likeliest reason for these abrupt flips is an intermittent problem in the North Atlantic Ocean, one that seems to trigger a major rearrangement of atmospheric circulation. North-south ocean currents help to redistribute equatorial heat into the temperate zones, supplementing the heat transfer by winds. When the warm currents penetrate farther than usual into the northern seas, they help to melt the sea ice that is reflecting a lot of sunlight back into space, and so the earth becomes warmer. Eventually that helps to melt ice sheets elsewhere.

The high state of climate seems to involve ocean currents that deliver an extraordinary amount of heat to the vicinity of Iceland and Norway. Like bus routes or conveyor belts, ocean currents must have a return loop. Unlike most ocean currents, the North Atlantic Current has a return loop that runs deep beneath the ocean surface. Huge amounts of seawater sink at known downwelling sites every winter, with the water heading south when it reaches the bottom. When that annual flushing fails for some years, the conveyor belt stops moving and so heat stops flowing so far north—and apparently we're popped back into the low state.

Flushing Cold Surface Water

SURFACE waters are flushed regularly, even in lakes. Twice a year they sink, carrying their load of atmospheric gases downward. That's because water density changes with temperature. Water is densest at about 39°F (a typical refrigerator setting—anything that you take out of the refrigerator, whether you place it on the kitchen counter or move it to the freezer, is going to expand a little). A lake surface cooling down in the autumn will eventually sink into the less-dense-because-warmer waters below, mixing things up. Seawater is more complicated, because salt content also helps to determine whether water floats or sinks. Water that evaporates leaves its salt behind; the resulting saltier water is heavier and thus sinks.

The fact that excess salt is flushed from surface waters has global implications, some of them recognized two centuries ago. Salt circulates, because evaporation up north causes it to sink and be carried south by deep currents. This was posited in 1797 by the Anglo-American physicist Sir Benjamin Thompson (later known, after he moved to Bavaria, as Count Rumford of the Holy Roman Empire), who also posited that, if merely to compensate, there would have to be a warmer northbound current as well. By 1961 the oceanographer Henry Stommel, of the Woods Hole Oceanographic Institution, in Massachusetts, was beginning to worry that these warming currents might stop flowing if too much fresh water was added to the surface of the northern seas. By 1987 the geochemist Wallace Broecker, of Columbia University, was piecing together the paleoclimatic flip-flops with the salt-circulation story and warning that small nudges to

> Huge amounts of seawater sink every winter in the vicinity of Iceland and Norway. When that flushing fails for some years, apparently we're popped into an abrupt cooling.

our climate might produce "unpleasant surprises in the greenhouse."

Oceans are not well mixed at any time. Like a half-beaten cake mix, with strands of egg still visible, the ocean has a lot of blobs and streams within it. When there has been a lot of evaporation, surface waters are saltier than usual. Sometimes they sink to considerable depths without mixing. The Mediterranean waters flowing out of the bottom of the Strait of Gibraltar into the Atlantic Ocean are about 10 percent saltier than the ocean's average, and so they sink into the depths of the Atlantic. A nice little Amazon-sized waterfall flows over the ridge that connects Spain with Morocco, 800 feet below the surface of the strait.

Another underwater ridge line stretches from Greenland to Iceland and on to the Faeroe Islands and Scotland. It, too, has a salty waterfall, which pours the hypersaline bottom waters of the Nordic Seas (the Greenland Sea and the Norwegian Sea) south into the lower levels of the North Atlantic Ocean. This salty waterfall is more like thirty Amazon Rivers combined. Why does it exist? The cold, dry winds blowing eastward off Canada evaporate the surface waters of the North Atlantic Current, and leave behind all their salt. In late winter the heavy surface waters sink en masse. These blobs, pushed down by annual repetitions of these late-winter events, flow south, down near the bottom of the Atlantic. The same thing happens in the Labrador Sea between Canada and the southern tip of Greenland.

Salt sinking on such a grand scale in the Nordic Seas causes warm water to flow much farther north than it might otherwise do. This produces a heat bonus of perhaps 30 percent beyond the heat provided by direct sunlight to these seas, accounting for the mild winters downwind, in northern Europe. It has been called the Nordic Seas heat pump.

Nothing like this happens in the Pacific Ocean, but the Pacific is nonetheless affected, because the sink in the Nordic Seas is part of a vast worldwide salt-conveyor belt. Such a conveyor is needed because the Atlantic is saltier than the Pacific (the Pacific has twice as much water with which to dilute the salt carried in from rivers). The Atlantic would be even saltier if it didn't mix with the Pacific, in long, loopy currents. These carry the North Atlantic's excess salt southward from the bottom of the Atlantic, around the tip of Africa, through the Indian Ocean, and up around the Pacific Ocean.

There used to be a tropical shortcut, an express route from Atlantic to Pacific, but continental drift connected North America to South America about three million years ago, damming up the easy route for disposing of excess salt. The dam, known as the Isthmus of Panama, may have been what caused the ice ages to begin a short time later, simply because of the forced detour. This major change in ocean circulation, along with a climate that had already been slowly cooling for millions of years, led not only to ice accumulation most of the time but also to climatic instability, with flips every few thousand years or so.

Failures of Flushing

FLYING above the clouds often presents an interesting picture when there are mountains below. Out of the sea of undulating white clouds mountain peaks stick up like islands.

Greenland looks like that, even on a cloudless day—but the great white mass between the occasional punctuations is an ice sheet. In places this frozen fresh water descends from the highlands in a wavy staircase.

Twenty thousand years ago a similar ice sheet lay atop the Baltic Sea and the land surrounding it. Another sat on Hudson's Bay, and reached as far west as the foothills of the Rocky Mountains—where it pushed, head to head, against ice coming down from the Rockies. These northern ice sheets were as high as Greenland's mountains, obstacles sufficient to force the jet stream to make a detour.

Now only Greenland's ice remains, but the abrupt cooling in the last warm period shows that a flip can occur in situations much like the present one. What could possibly halt the salt-conveyor belt that brings tropical heat so much farther north and limits the formation of ice sheets? Oceanographers are busy studying present-day failures of annual flushing, which give some perspective on the catastrophic failures of the past.

In the Labrador Sea, flushing failed during the 1970s, was strong again by 1990, and is now declining. In the Greenland Sea over the 1980s salt sinking declined by 80 percent. Obviously, local failures can occur without catastrophe—it's a question of how often and how widespread the failures are—but the present state of decline is not very reassuring. Large-scale flushing at both those sites is certainly a highly variable process, and perhaps a somewhat fragile one as well. And in the absence of a flushing mechanism to sink cooled surface waters and send them southward in the Atlantic, additional warm waters do not flow as far north to replenish the supply.

There are a few obvious precursors to flushing failure. One is diminished wind chill, when winds aren't as strong as usual, or as cold, or as dry—as is the case in the Labrador Sea during the North Atlantic Oscillation. This El Niño-like shift in the atmospheric-circulation pattern over the North Atlantic, from the Azores to Greenland, often lasts a decade. At the same time that the Labrador Sea gets a lessening of the strong winds that aid salt sinking, Europe gets particularly cold winters. It's happening right now: a North Atlantic Oscillation started in 1996.

Another precursor is more floating ice than usual, which reduces the amount of ocean surface exposed to the winds, in turn reducing evaporation. Retained heat eventually melts the ice, in a cycle that recurs about every five years.

Yet another precursor, as Henry Stommel suggested in 1961, would be the addition of fresh water to the ocean surface, diluting the salt-heavy surface waters before they became unstable enough to start sinking. More rain falling in the northern oceans—exactly what is predicted as a result of global warming—could stop salt flushing. So could ice carried south out of the Arctic Ocean.

There is also a great deal of unsalted water in Greenland's glaciers, just uphill from the major salt sinks. The last time an abrupt cooling occurred was in the midst of global warming. Many ice sheets had already half melted, dumping a lot of fresh water into the ocean.

A brief, large flood of fresh water might nudge us toward an abrupt cooling even if the dilution were insignificant when averaged over time. The fjords of Greenland offer some dramatic examples of the possibilities for freshwater floods. Fjords are long, narrow canyons, little arms of the sea reaching many miles inland; they were carved by great glaciers when the sea level was lower. Greenland's east coast has a profusion of fjords between 70°N and 80°N, including one that is the world's biggest. If blocked by ice dams, fjords make perfect reservoirs for meltwater.

Glaciers pushing out into the ocean usually break off in chunks. Whole sections of a glacier, lifted up by the tides, may snap off at the "hinge" and become icebergs. But sometimes a glacial surge will act like an avalanche that blocks a road, as happened when Alaska's Hubbard glacier surged into the Russell fjord in May of 1986. Its snout ran into the opposite side, blocking the fjord with an ice dam. Any meltwater coming in behind the dam stayed there. A lake formed, rising higher and higher—up to the height of an eight-story building.

Eventually such ice dams break, with spectacular results. Once the dam is breached, the rushing waters erode an ever wider and deeper path. Thus the entire lake can empty quickly. Five months after the ice dam

> The Nordic Seas sink is part of a worldwide conveyor belt. There used to be a shortcut, but it was dammed up by the Isthmus of Panama, which may have begun the ice ages.

at the Russell fjord formed, it broke, dumping a cubic mile of fresh water in only twenty-four hours.

The Great Salinity Anomaly, a pool of semi-salty water derived from about 500 times as much unsalted water as that released by Russell Lake, was tracked from 1968 to 1982 as it moved south from Greenland's east coast. In 1970 it arrived in the Labrador Sea, where it prevented the usual salt sinking. By 1971–1972 the semi-salty blob was off Newfoundland. It then crossed the Atlantic and passed near the Shetland Islands around 1976. From there it was carried northward by the warm Norwegian Current, whereupon some of it swung west again to arrive off Greenland's east coast—where it had started its inch-per-second journey. So freshwater blobs drift, sometimes causing major trouble, and Greenland floods thus have the potential to stop the enormous heat transfer that keeps the North Atlantic Current going strong.

The Greenhouse Connection

OF this much we're sure: global climate flip-flops have frequently happened in the past, and they're likely to happen again. It's also clear that sufficient global warming could trigger an abrupt cooling in at least two ways—by increasing high-latitude rainfall or by melting Greenland's ice, both of which could put enough fresh water into the ocean surface to suppress flushing.

Further investigation might lead to revisions in such mechanistic explanations, but the result of adding fresh water to the ocean surface is pretty standard physics. In almost four decades of subsequent research Henry Stommel's theory has only been enhanced, not seriously challenged.

Up to this point in the story none of the broad conclusions is particularly speculative. But to address how all these nonlinear mechanisms fit together—and what we might do to stabilize the climate—will require some speculation.

Even the tropics cool down by about nine degrees during an abrupt cooling, and it is hard to imagine what in the past could have disturbed the whole earth's climate on this scale. We must look at arriving sunlight and departing light and heat, not merely regional shifts on earth, to account for changes in the temperature balance. Increasing amounts of sea ice and clouds could reflect more sunlight back into space, but the geochemist Wallace Broecker suggests that a major greenhouse gas is disturbed by the failure of the salt conveyor, and that this affects the amount of heat retained.

In Broecker's view, failures of salt flushing cause a worldwide rearrangement of ocean currents, resulting in—and this is the speculative part—less evaporation from the tropics. That, in turn, makes the air drier. Because water vapor is the most powerful greenhouse gas, this decrease in average humidity would cool things globally. Broecker has written, "If you wanted to cool the planet by 5°C [9°F] and could magically alter the water-

vapor content of the atmosphere, a 30 percent decrease would do the job."

Just as an El Niño produces a hotter Equator in the Pacific Ocean and generates more atmospheric convection, so there might be a subnormal mode that decreases heat, convection, and evaporation. For example, I can imagine that ocean currents carrying more warm surface waters north or south from the equatorial regions might, in consequence, cool the Equator somewhat. That might result in less evaporation, creating lower-than-normal levels of greenhouse gases and thus a global cooling.

To see how ocean circulation might affect greenhouse gases, we must try to account quantitatively for important nonlinearities, ones in which little nudges provoke great responses. The modern world is full of objects and systems that exhibit "bistable" modes, with thresholds for flipping. Light switches abruptly change mode when nudged hard enough. Door latches suddenly give way. A gentle pull on a trigger may be ineffective, but there comes a pressure that will suddenly fire the gun. Thermostats tend to activate heating or cooling mechanisms abruptly—also an example of a system that pushes back.

We must be careful not to think of an abrupt cooling in response to global warming as just another self-regulatory device, a control system for cooling things down when it gets too hot. The scale of the response will be far beyond the bounds of regulation—more like when excess warming triggers fire extinguishers in the ceiling, ruining the contents of the room while cooling them down.

Preventing Climate Flips

THOUGH combating global warming is obviously on the agenda for preventing a cold flip, we could easily be blindsided by stability problems if we allow global warming per se to remain the main focus of our climate-change efforts. To stabilize our flip-flopping climate we'll need to identify all the important feedbacks that control climate and ocean currents—evaporation, the reflection of sunlight back into space, and so on—and then estimate their relative strengths and interactions in computer models.

Feedbacks are what determine thresholds, where one mode flips into another. Near a threshold one can sometimes observe abortive responses, rather like the act of stepping back onto a curb several times before finally running across a busy street. Abortive responses and rapid chattering between modes are common problems in nonlinear systems with not quite enough oomph—the reason that old fluorescent lights flicker. To keep a bistable system firmly in one state or the other, it should be kept away from the transition threshold.

We need to make sure that no business-as-usual climate variation, such as an El Niño or the North Atlantic Oscillation, can push our climate onto the slippery slope and into an abrupt cooling. Of particular importance are combinations of climate variations—this winter, for example, we are experiencing both an El Niño and a North Atlantic Oscillation—because such combinations can add up to much more than the sum of their parts.

We are near the end of a warm period in any event; ice ages return even without human influences on climate. The last warm period abruptly terminated 13,000 years after the abrupt warming that initiated it, and we've already gone 15,000 years from a similar starting point. But we may be able to do something to delay an abrupt cooling.

Do something? This tends to stagger the imagination, immediately conjuring up visions of terraforming on a science-fiction scale—and so we shake our heads and say, "Better to fight global warming by consuming less," and so forth.

Surprisingly, it may prove possible to prevent flip-flops in the climate—even by means of low-tech schemes. Keeping the present climate from falling back into the low state will in any case be a lot easier than trying to reverse such a change after it has occurred. Were fjord floods causing flushing to fail, because the downwelling sites were fairly close to the fjords, it is obvious that we could solve the problem. All we would need to do is open a channel through the ice dam with explosives before dangerous levels of water built up.

Timing could be everything, given the delayed effects from inch-per-second circulation patterns, but that, too, potentially has a low-tech solution: build dams across the major fjord systems and hold back the meltwater at critical times. Or divert eastern-Greenland meltwater to the less sensitive north and west coasts.

Fortunately, big parallel computers have proved useful for both global climate modeling and detailed modeling of ocean circulation. They even show the flips. Computer models might not yet be able to predict what will happen if we tamper with downwelling sites, but this problem doesn't seem insoluble. We need more well-trained people, bigger computers, more coring of the ocean floor and silted-up lakes, more ships to drag instrument packages through the depths, more instrumented buoys to study critical sites in detail, more satellites measuring regional variations in the sea surface, and perhaps some small-scale trial runs of interventions.

It would be especially nice to see another dozen major groups of scientists doing climate simulations, discovering the intervention mistakes as quickly as possible and learning from them. Medieval cathedral builders learned from their design mistakes over the centuries, and their undertakings were a far larger drain on the economic resources and people power of their day than anything yet discussed for stabilizing the climate in the twenty-first century. We may not have centuries to spare, but any economy in which two percent of the population produces all the food, as is the case in the United States today, has lots of resources and many options for reordering priorities.

Three Scenarios

FUTURISTS have learned to bracket the future with alternative scenarios, each of which captures important features that cluster together, each of which is compact enough to be seen as a narrative on a human scale. Three scenarios for the next climatic phase might be called population crash, cheap fix, and muddling through.

The population-crash scenario is surely the most appalling. Plummeting crop yields would cause some powerful countries to try to take over their neighbors or distant lands—if only because their armies, unpaid and lacking food, would go marauding, both at home and across the borders. The better-organized countries would attempt to use their armies, before they fell apart entirely, to take over countries with significant remaining resources, driving out or starving their inhabitants if not using modern weapons to accomplish the same end: eliminating competitors for the remaining food.

This would be a worldwide problem—and could lead to a Third World War—but Europe's vulnerability is particularly easy to analyze. The last abrupt cooling, the Younger Dryas, drastically altered Europe's climate as far east as Ukraine. Present-day Europe has more than 50 million people. It has excellent soils, and largely grows its own food. It could no longer do so if it lost the extra warming from the North Atlantic.

There is another part of the world with the same good soil, within the same latitudinal band, which we can use for a quick comparison. Canada lacks Europe's winter warmth and rainfall, because it has no equivalent of the North Atlantic Current to preheat its eastbound weather systems. Canada's agriculture supports about 28 million people. If Europe had weather like Canada's, it could feed only one out of twenty-three present-day Europeans.

Any abrupt switch in climate would also disrupt food supply routes. The only reason that two percent of our population can feed the other 98 percent is that we have a well-developed system of transportation and middlemen—but it is not very robust. The system allows for large urban populations in the

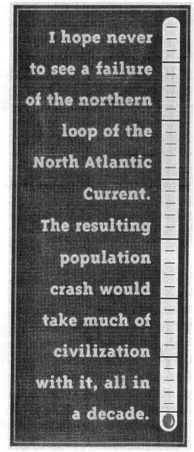

I hope never to see a failure of the northern loop of the North Atlantic Current. The resulting population crash would take much of civilization with it, all in a decade.

best of times, but not in the case of widespread disruptions.

Natural disasters such as hurricanes and earthquakes are less troubling than abrupt coolings for two reasons: they're short (the recovery period starts the next day) and they're local or regional (unaffected citizens can help the overwhelmed). There is, increasingly, international cooperation in response to catastrophe—but no country is going to be able to rely on a stored agricultural surplus for even a year, and any country will be reluctant to give away part of its surplus.

In an abrupt cooling the problem would get worse for decades, and much of the earth would be affected. A meteor strike that killed most of the population in a month would not be as serious as an abrupt cooling that eventually killed just as many. With the population crash spread out over a decade, there would be ample opportunity for civilization's institutions to be torn apart and for hatreds to build, as armies tried to grab remaining resources simply to feed the people in their own countries. The effects of an abrupt cold last for centuries. They might not be the end of *Homo sapiens*—written knowledge and elementary education might well endure—but the world after such a population crash would certainly be full of despotic governments that hated their neighbors because of recent atrocities. Recovery would be very slow.

A slightly exaggerated version of our present know-something-do-nothing state of affairs is know-nothing-do-nothing: a reduction in science as usual, further limiting our chances of discovering a way out. History is full of withdrawals from knowledge-seeking, whether for reasons of fundamentalism, fatalism, or "government lite" economics. This scenario does not require that the shortsighted be in charge, only that they have enough influence to put the relevant science agencies on starvation budgets and to send recommendations back for yet another commission report due five years hence.

A cheap-fix scenario, such as building or bombing a dam, presumes that we know enough to prevent trouble, or to nip a developing problem in the bud. But just as vaccines and antibiotics presume much knowledge about diseases, their climatic equivalents presume much knowledge about oceans, atmospheres, and past climates. Suppose we had reports that winter salt flushing was confined to certain areas, that abrupt shifts in the past were associated with localized flushing failures, *and* that one computer model after another suggested a solution that was likely to work even under a wide range of weather extremes. A quick fix, such as bombing an ice dam, might then be possible. Although I don't consider this scenario to be the most likely one, it is possible that solutions could turn out to be cheap and easy, and that another abrupt cooling isn't inevitable. Fatalism, in other words, might well be foolish.

A muddle-through scenario assumes that we would mobilize our scientific and technological resources well in advance of any abrupt cooling problem, but that the solution wouldn't be simple. Instead we would try one thing after another, creating a patchwork of solutions that might hold for another few decades, allowing the search for a better stabilizing mechanism to continue.

We might, for example, anchor bargeloads of evaporation-enhancing surfactants (used in the southwest corner of the Dead Sea to speed potash production) upwind from critical downwelling sites, letting winds spread them over the ocean surface all winter, just to ensure later flushing. We might create a rain shadow, seeding clouds so that they dropped their unsalted water well upwind of a given year's critical flushing sites—a strategy that might be particularly important in view of the increased rainfall expected from global warming. We might undertake to regulate the Mediterranean's salty outflow, which is also thought to disrupt the North Atlantic Current.

Perhaps computer simulations will tell us that the only robust solutions are those that re-create the ocean currents of three million years ago, before the Isthmus of Panama closed off the express route for excess-salt disposal. Thus we might dig a wide sea-level Panama Canal in stages, carefully managing the changeover.

Staying in the "Comfort Zone"

STABILIZING our flip-flopping climate is not a simple matter. We need heat in the right places, such as the Greenland Sea, and not in others right next door, such as Greenland itself. Man-made global warming is likely to achieve exactly the opposite—warming Greenland and cooling the Greenland Sea.

A remarkable amount of specious reasoning is often encountered when we contemplate reducing carbon-dioxide emissions. That increased quantities of greenhouse gases will lead to global warming is as solid a scientific prediction as can be found, but other things influence climate too, and some people try to escape confronting the consequences of our pumping more and more greenhouse gases into the atmosphere by supposing that something will come along miraculously to counteract them. Volcanos spew sulfates, as do our own smokestacks, and these reflect some sunlight back into space, particularly over the North Atlantic and Europe. But we can't assume that anything like this will counteract our longer-term flurry of carbon-dioxide emissions. Only the most naive gamblers bet against physics, and only the most irresponsible bet with their grandchildren's resources.

To the long list of predicted consequences of global warming—stronger storms, methane release, habitat changes, ice-sheet melting, rising seas, stronger El Niños, killer heat waves—we must now add an abrupt, catastrophic cooling. Whereas the familiar consequences of global warming will force expensive but gradual adjustments, the abrupt cooling promoted by man-made warming looks like a particularly efficient means of committing mass suicide.

We cannot avoid trouble by merely cutting down on our present warming trend, though that's an excellent place to start. Paleoclimatic records reveal that any notion we may once have had that the climate will remain the same unless pollution changes it is wishful thinking. Judging from the duration of the last warm period, we are probably near the end of the current one. Our goal must be to stabilize the climate in its favorable mode and ensure that enough equatorial heat continues to flow into the waters around Greenland and Norway. A stabilized climate must have a wide "comfort zone," and be able to survive the El Niños of the short term. We can design for that in computer models of climate, just as architects design earthquake-resistant skyscrapers. Implementing it might cost no more, in relative terms, than building a medieval cathedral. But we may not have centuries for acquiring wisdom, and it would be wise to compress our learning into the years immediately ahead. We have to discover what has made the climate of the past 8,000 years relatively stable, and then figure out how to prop it up.

> Those who will not reason
> Perish in the act:
> Those who will not act
> Perish for that reason.
>
> —W. H. Auden

Human Imprint on Climate Change Grows Clearer

By WILLIAM K. STEVENS

As evidence that the earth's atmosphere is warming continues to accumulate, scientists are making slow progress toward an answer to the big question raised by the evidence: how much of the warming is due to human activity and how much to natural causes?

The United Nations' Intergovernmental Panel on Climate Change, the group of scientists widely considered the most authoritative voice on the subject, has already concluded that there is a "discernible human influence" on the global climate. Now the panel is deep into another of its periodic full-scale scientific assessments of global climate change, to be finished in about 18 months. While the group's conclusions are unformed, some experts on the problem say the human imprint on climate is becoming clearer, and may even have been the dominant factor in the global warming of recent decades. Not everyone agrees—there is a range of judgments—and virtually all experts say that in any case, a reliable estimate of the human imprint's magnitude still remains some distance off.

A number of influences, both natural and man-made, cause the planet's temperature to vary. The natural ones include changes in solar radiation, and sulfate droplets called aerosols cast aloft by erupting volcanoes, which cool the atmosphere by reflecting sunlight. The human influence stems mostly from emissions of waste industrial gases like carbon dioxide, which trap heat in the atmosphere, and sulfate aerosols from industrial smokestacks.

The combined impact of industrial aerosols and greenhouse gases creates complex and distinctive temperature patterns. It was mostly an analysis of those patterns that led the intergovernmental panel in 1995 to abandon its previous position that global warming observed over the past century might as easily be natural as human-induced. Human factors appeared to be playing a part, the panel said then, but it offered no judgment on whether that part was big, small or in between.

One recent piece of evidence suggesting a strong human influence, which seems likely to carry some weight with the intergovernmental panel, appeared recently in the journal Nature. Scientists at the Hadley Center for Climate Prediction and Research, a British Government organization, analyzed the global climate record of the last century in an effort to isolate and quantify the major factors producing the century's rise of about 1 degree Fahrenheit in the earth's average surface temperature.

The research team led by Dr. Simon F. B. Tett found that in the earlier part of the century, the rise could be explained either by an increase in solar radiation or a combination of stronger solar radiation and heat-trapping greenhouse gases emitted by industrial economies. But they found that after the mid-1970's, when about half the century's warming took place, the warming resulted largely from the greenhouse gases.

Other researchers have lately come to a similar conclusion.

The Tett study represents "another jigsaw puzzle piece," said one expert, Dr. Tom M. L. Wigley of the National Center for Atmospheric Research in Boulder, Colo. "There is still a long way to go" in completing the puzzle, he said, "but we're beginning to see the smile on the face of the Mona Lisa, I think—or perhaps it should be a frown."

Dr. Wigley was a principal author of the section of the intergovernmental panel's 1995 report dealing with detection of the human imprint on climate, but not for the new report. He and his co-assessors of four years ago "would now make a stronger statement" if they were writing their report today, he said at a recent panel discussion sponsored in Washington by Resources for the Future, an independent research group.

But other participants in the discussion were more cautious. One, Dr. Michael E. Schlesinger, a climatologist at the University of Illinois at Urbana-Champaign, says that experts have insufficient knowledge of the magnitude of natural climatic variations, especially solar radiation, to gauge how large the human impact is by comparison.

Another, Dr. Ronald Prinn of the Massachusetts Institute of Technology, said that although there was "accumulating evidence that humans are having an influence on the climate system," it would not be possible to discern its magnitude until the degree of natural climate variability could be pinned down better.

Meanwhile, though, evidence of warming and its effects continues to mount. Earlier this year, scientists at the University of Massachusetts and the University of Arizona reconstructed the average annual surface temperature trend of the Northern Hemisphere for the last 1,000 years. While cautioning that the margin of error was large enough to render data from the early centuries untrustworthy, they found that the 20th century was the warmest of the millennium, by far.

This and other analyses have found that the warmest years of all occurred in the 1990's, with 1998 the warmest on record. El Niño, the great pool of warm water in the tropical Pacific Ocean that from time to time heats the atmosphere and disrupts weather patterns, was responsible for some of the 1998 heating. A preliminary analysis by Dr. Wigley, however, has shown that when El Niño's effects are filtered out of the global temperature record statistically, 1998 still ranks as the warmest year. (This year is also shaping up as unusually warm, but not as warm as 1998.)

Two studies reported in Nature this month suggest that the warming is being reflected in patterns of wildlife behavior and distribution. In one study, Dr. Camille Parmesan, a biologist at the National Center for Ecological Analysis in Santa Barbara, Calif., and 12 colleagues analyzed distribution patterns of 35 species of European butterflies. They found that for two-thirds of the species, their range of habitat had shifted northward by 22 to 150 miles, coincidentally with Europe's warming trend.

In the other study, Humphrey Q. B. Crick of the British Trust for Ornithology and Timothy H. Sparks of the Institute of Terrestrial Ecology in Cambridgeshire, England, analyzed the nesting habits of 20 species of birds in Britain. They found that, again coincident with a recent warming trend, the birds were laying their eggs earlier in the spring. This is the latest in a series of studies indicating that meteorological spring is coming earlier in the Northern Hemisphere. Some have also shown that fall is coming later.

A third study in Nature this month reports, on the basis of bubbles of atmospheric gas contained in ice cores extracted from the Antarctic ice sheet, that present-day atmospheric levels of heat-trapping carbon dioxide are higher than at any other time in the last 420,000 years. At 360 parts per million, they are 20 percent higher than in any previous warm period between ice ages, and double the typical concentrations during an ice age.

If greenhouse gas emissions are not reduced, the United Nations scientific panel has consistently said, atmospheric concentrations will continue to rise and warm the earth further. In its 1995 report, it projected a warming of 2 to 6 degrees Fahrenheit, with a best estimate of 3.5 degrees, by the year 2100. The warming would not end in 2100, the panel found, but would continue. By comparison, the earth has warmed by 5 to 9 degrees since the depths of the last ice age some 20,000 years ago.

The amount of warming projected by the panel, it said, would create widespread climatic and ecological changes, including a shift in climatic zones, an increase in heat waves, warmer northern winters, increased precipitation when it rains but worse droughts when it does not, and a rise in sea level that could inundate many small island nations and drive tens of millions of people away from the coasts when storm surges develop.

For a long time, the global warming debate focused on how much warming a given increase in green-house gases—say, a doubling of atmospheric concentrations—would produce. Though skeptics say it would be small, the dominant view for 20 years has been that a doubling would produce a warming of 3 to 8 degrees, other things being equal. This is a measure of the climate system's sensitivity to "forcing," as experts call it, by external heating and cooling influences, and many mainstream scientists say confidence that it is right has grown.

Now attention is shifting to the relative strength of the external forcings. This is the crux of the problem of figuring out the magnitude of humans' influence on the climate. There are basically three main forcings: greenhouse gases and solar radiation, which warm the atmosphere, and sulfate droplets, or aerosols, from both volcanoes and industrial sources, which cool it.

The amount of carbon dioxide and other greenhouse gases in the atmosphere has been firmly established and continuously monitored, though it is difficult if not impossible to predict how much there will be in the future. Once greenhouse gases diffuse throughout the entire lower part of the atmosphere they remain there for decades to centuries.

Sulfate aerosols from volcanoes sometimes spread to the stratosphere, where they, too, diffuse globally and cool the earth. But they dissipate in two or three years. That is what happened with Mount Pinatubo in the Philippines, which temporarily cooled the earth by about 1 degree Fahrenheit after it erupted in 1991.

Sulfate aerosols from industry generally rise only into the lower part of the atmosphere and fall out, as acid rain, within a few days. Moreover, their extent and impact is mostly regional rather than global. Until recently, they affected primarily the industrial countries of Europe and North America, but scientists now believe that they are diminishing in that part of the world because of controls on pollution. They are growing fast, however, in India, China and Southeast Asia.

Experts have difficulty in getting a handle on the constantly shifting amounts and patterns of industrial aerosols, and they have a similarly hard time establishing the varying strength of solar radiation.

Some mainstream scientists say that because the earth's average surface temperature has not varied by more than a degree or two since the last ice age, variations in the strength of the sun's radiation must be relatively small during the present interglacial period. Satellite measurements show some small variations, less than one-tenth of a percent, in concert with the 11-year sunspot cycle. But for the years before 1978, when satellite measurements began, scientists must estimate the variations from proxy indicators like sunspots. These estimates say that solar irradiance did not increase by more than 1 percent from 1908 to 1952.

In the recent Tett study, the British scientists fed into a computer model of the climate system varying estimates, based on different proxy studies, of past solar irradiance. They also fed in the observed and estimated changes in greenhouse gas concentrations and aerosols from both volcanoes and industrial sources. They ran many simulations of the effect of each of these forcings and then averaged them together to see if they explained the observed global temperature rise—and to determine what weight each should be given. Since the 1970's, the simulations revealed, global warming cannot be explained without a large impact from greenhouse gases.

Dr. Schlesinger, the University of Illinois climatologist, says he has "a couple of problems" with the study. First, he says, the data on solar variability before 1978 are unreliable. Second, the climate system's natural internal variability, apart from external forcings, is quite large relative to the observed warming and has not been well quantified. For example, he says, there is a growing body of evidence that the warming in the early part of the century resulted from a natural oscillation in the surface temperature of the North Atlantic Ocean.

The larger point, Dr. Schlesinger says, is that the climate system's internal variations may be so large that they render the Tett findings' statistical significance weaker. Nonetheless, Dr. Tett says he believes the findings strengthen the United Nations panel's 1995 conclusion about a human impact on climate.

The struggle to gauge that impact goes on. "The important question," Dr. Prinn says, "is numbers"—precise estimates of the relative weight of natural and human factors. "It could be there are big environmental issues involved here," he says, "or relatively modest ones."

Beyond the Valley of
THE DAMMED

*A strange alliance of fish lovers, tree huggers, and bureaucrats
say what went up must come down*

BY BRUCE BARCOTT

By God but we built some dams! We backed up the Kennebec in Maine and the Neuse in North Carolina and a hundred creeks and streams that once ran free. We stopped the Colorado with the Hoover, high as 35 houses, and because it pleased us we kept damming and diverting the river until it no longer reached the sea. We dammed our way out of the Great Depression with the Columbia's Grand Coulee; a dam so immense you had to borrow another fellow's mind because yours alone wasn't big enough to wrap around it. Then we cleaved the Missouri with a bigger one still, the Fort Peck Dam, a jaw dropper so outsized they put it on the cover of the first issue of *Life*. We turned the Tennessee, the Columbia, and the Snake from continental arteries into still bathtubs. We dammed the Clearwater, the Boise, the Santiam, the Deschutes, the Skagit, the Willamette, and the McKenzie. We dammed Crystal River and Muddy Creek, the Little River and the Rio-Grande. We dammed the Minnewawa and the Minnesota and we dammed the Kalamazoo. We dammed the Swift and we dammed the Dead.

One day we looked up and saw 75,000 dams impounding more than half a million miles of river. We looked down and saw rivers scrubbed free of salmon and sturgeon and shad. Cold rivers ran warm, warm rivers ran cold, and fertile muddy banks turned barren.

And that's when we stopped talking about dams as instruments of holy progress and started talking about blowing them out of the water.

Surrounded by a small crowd, Secretary of the Interior Bruce Babbitt stood atop McPherrin Dam, on Butte Creek, not far from Chico, California, in the hundred-degree heat of the Sacramento Valley. The constituencies represented—farmers, wildlife conservationists, state fish and game officials, irrigation managers—had been wrangling over every drop of water in this naturally arid basin for most of a century. On this day, however, amity reigned.

With CNN cameras rolling, Babbitt hoisted a sledgehammer above his head and—with "evident glee," as one reporter later noted—brought this tool of destruction down upon the dam. Golf claps all around.

The secretary's hammer strike in July 1998 marked the beginning of the end for that ugly concrete plug and three other Butte Creek irrigation dams. All were coming out to encourage the return of spring-run chinook salmon, blocked from their natural spawning grounds for more than 75 years. Babbitt then flew to Medford, Oregon, and took a swing at 30-year-old Jackson Street Dam on Bear Creek. Last year alone, Babbitt cracked the concrete at four dams on Wisconsin's Menominee River and two dams on Elwha River in Washington state; at Quaker Neck Dam on North Carolina's Neuse River; and at 160-year-old Edwards Dam on the Kennebec in Maine.

By any reckoning, this was a weird inversion of that natural order. Interior secretaries are supposed to christen dams, not smash them. Sixty years ago, President Franklin D. Roosevelt and his interior secretary, Harold Ickes, toured the West to dedicate four of the largest dams in the history of civilization. Since 1994, Babbitt, who knows his history, has been following in their footsteps, but this secretary is preaching the gospel of

What we know now that we didn't know then is that **a river isn't a water pipe.**

dam-going-away. "America overshot the mark in our dam-building frenzy," he told the Ecological Society of America. "The public is now learning that we have paid a steadily accumulating price for these projects. . . . We did not build them for religious purposes and they do not consecrate our values. Dams do, in fact, outlive their function. When they do, some should go."

Many dams continue, of course, to be invaluable pollution-free power plants. Hydroelectric dams provide 10 percent of the nation's electricity (and half of our renewable energy). In the Northwest, dams account for 75 percent of the region's power and bestow the lowest electrical rates in the nation. In the past the public was encouraged to believe that hydropower was almost free; but as Babbitt has been pointing out, the real costs can be enormous.

What we know now that we didn't know in 1938 is that a river isn't a water pipe. Dam a river and it will drop most of the sediment it carries into a still reservoir, trapping ecologically valuable debris such as branches, wood particles, and gravel. The sediment may be mixed with more and more pollutants—toxic chemicals leaching from abandoned mines, for example, or naturally occurring but dangerous heavy metals. Once the water passes through the dam it continues to scour, but it can't replace what it removes with material from upstream. A dammed river is sometimes called a "hungry" river, one that eats its bed and banks. Riverbeds and banks may turn into cobblestone streets, large stones cemented in by the ultrafine silt that passes through the dams. Biologists call this "armoring."

Naturally cold rivers may run warm after the sun heats water trapped in the reservoir; naturally warm rivers may run cold if their downstream flow is drawn from the bottom of deep reservoirs. Fish adapted to cold water won't survive in warm water, and vice versa.

As the toll on wild rivers became more glaringly evident in recent decades, opposition to dams started to go mainstream. By the 1990s, conservation groups, fishing organizations, and other river lovers began to call for actions that had once been supported only by environmental extremists and radical groups like Earth First! Driven by changing economics, environmental law, and most of all the specter of vanishing fish, government policy makers began echoing the conservationists. And then Bruce Babbitt, perhaps sensing the inevitable tide of history, began to support decommissioning as well.

So far, only small dams have been removed. Babbitt may chip away at all the little dams he wants, but when it comes to ripping major federal hydropower projects out of Western rivers, that's when the politics get national and nasty. Twenty-two years ago, when President Jimmy Carter suggested pulling the plug on several grand dam projects, Western senators and

representatives politically crucified him. Although dam opponents have much stronger scientific and economic arguments on their side in 1999, the coming dam battles are apt to be just as nasty.

Consider the Snake River, where a major confrontation looms over four federal hydropower dams near the Washington-Idaho state line. When I asked Babbitt about the Snake last fall, he almost seemed to be itching for his hammer. "The escalating debate over dams is going to focus in the coming months on the Snake River," he declared. "We're now face to face with this question: Do the people of this country place more value on Snake River salmon or on those four dams? The scientific studies are making it clear that you can't have both."

Brave talk—but only a couple of weeks later, after a bruising budget skirmish with congressional dam proponents who accused him of planning to tear down dams across the Northwest, Babbitt sounded like a man who had just learned a sobering lesson in the treacherous politics of dams. The chastened interior secretary assured the public that "I have never advocated, and do not advocate, the removal of dams on the main stem of the Columbia-Snake river system."

Showdown on the Snake

Lewiston, Idaho, sits at the confluence of the Snake and Clearwater Rivers. It's a quiet place of 33,000 solid citizens, laid out like a lot of towns these days: One main road leads into the dying downtown core, the other to a thriving strip of Wal-Marts, gas stations, and fast-food greaseries. When Lewis (hence the name) and Clark floated through here in 1805, they complained about the river rapids—"Several of them verry bad," the spelling-challenged Clark scrawled in his journal. Further downriver, where the Snake meets the Columbia, the explorers were amazed to see the local Indians catching and drying incredible numbers of coho salmon headed upriver to spawn.

The river still flows, though it's been dammed into a lake for nearly 150 miles. Between 1962 and 1975, four federal hydroelectric projects were built on the river by the Army Corps of Engineers: Ice Harbor Dam, Lower Monumental Dam, Little Goose Dam, and Lower Granite Dam. The dams added to the regional power supply, but more crucially, they turned the Snake from a whitewater roller-coaster into a navigable waterway. The surrounding wheat farmers could now ship their grain on barges to Portland, Oregon, at half the cost of overland transport, and other industries also grew to depend on this cheap highway to the sea.

Like all dams, however, they were hell on the river and its fish—the chinook, coho, sockeye, and steelhead. True, some salmon species still run up the river to spawn, but by the early 1990s the fish count had dwindled from 5 million to less than

20,000. The Snake River coho have completely disappeared, and the sockeye are nearing extinction.

In and around Lewiston, the two conflicting interests—livelihoods that depend on the dams on the one side, the fate of the fish on the other—mean that just about everyone is either a friend of the dams or a breacher. The Snake is the dam-breaching movement's first major test case, but it is also the place where dam defenders plan to make their stand. Most important, depending in part on the results of a study due later this year, the lower Snake could become the place where the government orders the first decommissioning of several big dams.

In the forefront of those who hope this happens is Charlie Ray, an oxymoron of a good ol'boy environmentalist whose booming Tennessee-bred baritone and sandy hair lend him the aspect of Nashville Network host. Ray makes his living as head of salmon and steelhead programs for Idaho Rivers United, a conservationist group that has been raising a fuss about free-flowing rivers since 1991. At heart he's not a tree hugger, but a steelhead junkie: "You hook a steelhead, man, you got 10,000 years of survival instinct on the end of that line."

Despite Ray's bluff good cheer, it's not easy being a breacher in Lewiston. Wheat farming still drives a big part of the local economy, and the pro-dam forces predict that breaching would lead to financial ruin. Lining up behind the dam defenders are Lewiston's twin pillars of industry: the Potlatch Corporation and the Port of Lewiston. Potlatch, one of the country's largest paper producers, operates its flagship pulp and paper mill in Lewiston, employing 2,300 people. Potlatch executives will tell you the company wants the dams mainly to protect the town's economy, but local environmentalists say the mill would find it more difficult to discharge warm effluent into a free-flowing, shallow river.

Potlatch provides Charlie Ray with a worthy foil in company spokesman Frank Carroll, who was hired after spending 17 years working the media for the U.S. Forest Service. Frankie and Charlie have been known to scrap. At an anti-breaching rally in Lewiston last September, Carroll stood off-camera watching Ray being interviewed by a local TV reporter. Fed up with hearing Ray's spin, Carroll started shouting "Bullshit, Charlie, that's bullshit!" while the video rolled. Ray's nothing more than a "paid operative," Carroll says. Ray's reaction: "Yeah, like Frankie's not."

"A lot of people are trying to trivialize the social and economic issues," Carroll says, "trying to tell us the lives people have here don't count, that we'll open up a big bait shop and put everyone to work hooking worms. We resent that. Right now, there's a blanket of prosperity that lies across this whole region, and that prosperity is due to the river in its current state—to its transportation."

Ever since the dams started going up along the Snake River, biologists and engineers have been trying to revive the rapidly declining salmon runs. Their schemes include fish ladders, hatcheries, and a bizarre program in which young smolts are captured and shipped downriver to the sea in barges. By the late 1980s, it was clear that nothing was working; the fish runs continued to plummet. In 1990, the Shoshone-Bannock Indians, who traditionally fished the Snake's sockeye run, successfully petitioned the National Marine Fisheries Service to list the fish as endangered. Every salmon species in the Snake River is now officially threatened or endangered, which means the agencies that control the river must deal with all kinds of costly regulations.

In 1995, under pressure from the federal courts, the National Marine Fisheries Service and the Army Corps of Engineers (which continues to operate the dams) agreed to launch a four-year study of the four lower Snake River dams. In tandem with the Fisheries Service, the Corps made a bombshell announcement. The study would consider three options: maintain the status quo, turbocharge the fish-barging operation, or initiate a "permanent natural river drawdown"—breaching. The study's final report is due in December, but whatever its conclusions, that initial statement marked a dramatic shift. Suddenly, an action that had always seemed unthinkable was an officially sanctioned possibility.

Two separate scientific studies concluded that breaching presented the best hope for saving the river. In 1997 the *Idaho Statesman,* the state's largest newspaper, published a three-part series arguing that breaching the four dams would net local taxpayers and the region's economy $183 million a year. The dams, the paper concluded, "are holding Idaho's economy hostage."

"That series was seismic," says Reed Burkholder, a Boise-based breaching advocate. Charlie Ray agrees. "We've won the scientific argument," he says. "And we've won the economic argument. We're spending more to drive the fish to extinction than it would cost to revive them."

In fact, the economic argument is far from won. The *Statesman*'s numbers are not unimpeachable. The key to their prediction, a projected $248 million annual boost in recreation and fishing, assumes that the salmon runs will return to pre-1960s levels. Fisheries experts say that could take up to 24 years, if it happens at all. The $34 million lost at the Port of Lewiston each year, however, would be certain and immediate.

The Northwest can do without the power of the four lower Snake River dams: They account for only about 4 percent of the region's electricity supply. The dams aren't built for flood control, and contrary to a widely held belief, they provide only a small amount of irrigation water to the region's farmers. What the issue comes down to, then, is the Port of Lewiston. You take the dams out, says port manager Dave Doeringsfeld, "and transportation costs go up 200 to 300 percent."

To breach or to blow?

The pro-dam lobbyists know they possess a powerful, not-so-secret weapon: Senator Slade Gorton, the Washington Republican who holds the commanding post of chairman of the Subcommittee on Interior Appropriations. Gorton has built his political base by advertising himself as the foe of liberal Seattle environmentalists, and with his hands on Interior's purse strings, he can back up the role with real clout. As determined as Bruce Babbitt is to bring down a big dam, Slade Gorton may be more determined to stop him.

During last October's federal budget negotiations, Gorton offered to allocate $22 million for removing two modest dams in the Elwha River on the Olympic Peninsula, a salmon-restoration project dear to the hearts of dam-breaching advocates. But Gorton agreed to fund the Elwha breaching if—and only if—the budget included language forbidding federal officials from unilaterally ordering the dismantling of any dam, including those in the Columbia River Basin. Babbitt and others balked at Gorton's proposal. As a result, the 1999 budget includes zero dollars for removal of the Elwah dams.

Gorton's Elwha maneuver may have been hardball politics for its own sake, but it was also a clear warning: If the Army Corps and the National Marine Fisheries Service recommend breaching on the Snake in their study later this year, there will be hell to pay.

Meanwhile, here's a hypothetical question: If you're going to breach, how do you actually do it? How do you take those behemoths out? It depends on the dam, of course, but the answer on the Snake is shockingly simple.

"You leave the dam there," Charlie Ray says. We're standing downstream from Lower Granite Dam, 35 million pounds of steel encased in concrete. Lower Granite isn't a classic ghastly curtain like Hoover Dam; it resembles nothing so much as an enormous half-sunk harmonica. Ray points to a berm of granite boulders butting up against the concrete structure's northern end. "Take out the earthen portion and let the river flow around the dam. This is not high-tech stuff. This is front-end loaders and dump trucks."

It turns out that Charlie is only a few adjectives short of the truth. All you do need are loaders and dump trucks—really, really big ones, says Steve Tatro of the Army Corps of Engineers. Tatro has the touchy job of devising the best way to breach his agency's own dams. First, he says, you'd draw down the reservoir, using the spillways and the lower turbine passages as drains. Then you'd bypass the concrete and steel entirely and excavate the dam's earthen portion. Depending on the dam, that could mean excavating as much as 8 million cubic yards of material.

Tatro's just-the-facts manner can't disguise the reality that there is something deeply cathartic about the act he's describing. Most environmental restoration happens at the speed of nature. Which is to say, damnably slow. Breaching a dam—or better yet, blowing a dam—offers a rare moment of immediate gratification.

The Glen Canyon story

From the Mesopotamian canals to Hoover Dam, it took the human mind about 10,000 years to figure out how to stop a river. It has taken only 60 years to accomplish the all-too-obvious environmental destruction.

Until the 1930s, most dam projects were matters of trial and (often) error, but beginning with Hoover Dam in 1931, dam builders began erecting titanic riverstoppers that approached an absolute degree of reliability and safety. In *Cadillac Desert*, a 1986 book on Western water issues, author Marc

Reisner notes that from 1928 to 1956, "the most fateful transformation that has ever been visited on any landscape, anywhere, was wrought." Thanks to the U.S. Bureau of Reclamation, the Tennessee Valley Authority, and the Army Corps, dams lit a million houses, turned deserts into wheat fields, and later powered the factories that built the planes and ships that beat Hitler and the Japanese. Dams became monuments to democracy and enlightenment during times of bad luck and hunger and war.

Thirty years later, author Edward Abbey became the first dissenting voice to be widely heard. In *Desert Solitaire* and *The Monkey Wrench Gang*, Abbey envisioned a counterforce of wilderness freaks wiring bombs to the Colorado River's Glen Canyon Dam, which he saw as the ultimate symbol of humanity's destruction of the American West. Kaboom! Wildness returns to the Colorado.

Among environmentalists, the Glen Canyon Dam has become an almost mythic symbol of riparian destruction. All the symptoms of dam kill are there. The natural heavy metals that the Colorado River used to disperse into the Gulf of California now collect behind the dam in Lake Powell. And the lake is filling up: Sediment has reduced the volume of the lake from its original 27 million acre-feet to 23 million. One million acre-feet of water are lost to evaporation every year—enough, environmentalists note, to revive the dying upper reaches of the Gulf of California. The natural river ran warm and muddy, and flushed its channel with floods; the dammed version runs cool, clear, and even. Trout thrive in the Colorado. This is like giraffes thriving on tundra.

Another reason for the dam's symbolic power can be traced to its history. For decades ago, David Brower, then executive director of the Sierra Club, agreed to a compromise that haunts him to this day: Conservationists would not oppose Glen Canyon and 11 other projects if plans for the proposed Echo Park and Split Mountain dams, in Utah and Colorado, were abandoned. In 1963, the place Wallace Stegner once called "the most serenely beautiful of all the canyons of the Colorado" began disappearing beneath Lake Powell. Brower led the successful fight to block other dams in the Grand Canyon area, but he remained bitter about the compromise. "Glen Canyon died in 1963," he later wrote, "and I was partly responsible for its needless death."

In 1981 Earth First! inaugurated its prankster career by unfurling an enormous black plastic "crack" down the face of Glen Canyon Dam. In 1996 the Sierra Club rekindled the issue by calling for the draining of Lake Powell. With the support of Earth Island Institute (which Brower now chairs) and other environmental groups, the proposal got a hearing before a subcommittee of the House Committee on Resources in September 1997. Congress has taken no further action, but a growing number of responsible voices now echo the monkey-wrenchers' arguments. Even longtime Bureau of Reclamation supporter Barry Goldwater admitted, before his death last year, that he considered Glen Canyon Dam a mistake.

Defenders of the dam ask what we would really gain from a breach. The dam-based ecosystem has attracted peregrine falcons, bald eagles, carp, and catfish. Lake Powell brings in $400 million

In 1963, the most beautiful of all the canyons of the Colorado **began disappearing beneath Lake Powell.**

a year from tourists enjoying houseboats, powerboats, and personal watercraft—a local economy that couldn't be replaced by the thinner wallets of rafters and hikers.

"It would be completely foolhardy and ridiculous to deactivate that dam," says Floyd Dominy during a phone conversation from his home in Boyce, Virginia. Dominy, now 89 years old and retired since 1969, was the legendary Bureau of Reclamation commissioner who oversaw construction of the dam in the early 1960s. "You want to lose all that pollution-free energy? You want to destroy a world-renowned tourist attraction—Lake Powell—that draws more than 3 million people a year?"

It goes against the American grain: the notion that knocking something down and returning it to nature might be progress just as surely as replacing wildness with asphalt and steel. But 30 years of environmental law, punctuated by the crash of the salmon industry, has shifted power from the dam builders to the conservationists.

The most fateful change may be a little-noticed 1986 revision in a federal law. Since the 1930s, the Federal Energy Regulatory Commission has issued 30- to 50-year operating licenses to the nation's 2,600 or so privately owned hydroelectric dams. According to the revised law, however, FERC must consider not only power generation, but also fish and wildlife, energy conservation, and recreational uses before issuing license renewals. In November 1997, for the first time in its history, FERC refused a license against the will of a dam owner, ordering the Edwards Manufacturing Company to rip the 160-year-old Edwards Dam out of Maine's Kennebec River. More than 220 FERC hydropower licenses will expire over the next 10 years.

If there is one moment that captures the turning momentum in the dam wars, it might be the dinner Richard Ingebretsen shared with the builder of Glen Canyon Dam, Floyd Dominy himself. During the last go-go dam years, from 1959 to 1969, this dam-building bureaucrat was more powerful than any Western senator or governor. Ingebretsen is a Salt Lake City physician, a Mormon Republican, and a self-described radical environmentalist. Four years ago, he founded the Glen Canyon Institute to lobby for the restoration of Glen Canyon. Ingebretsen first met Dominy when the former commissioner came to Salt Lake City in 1995 to debate David Brower over the issue of breaching Glen Canyon Dam. To his surprise, Ingebretsen found that he liked the man. "I really respect him for his views," he says.

Their dinner took place in Washington, D.C., in early 1997. At one point Dominy asked Ingebretsen how serious the move-ment to drain Lake Powell really was. Very serious, Ingebretsen replied. "Of course I'm opposed to putting the dam in mothballs," Dominy said. "But I heard what Brower wants to do." (Brower had suggested that Glen Canyon could be breached by coring out some old water bypass tunnels that had been filled in years ago.). "Look," Dominy continued, "those tunnels are jammed with 300 feet of reinforced concrete. You'll never drill that out."

With that, Dominy pulled out a napkin and started sketching a breach. "You want to drain Lake Powell?" he asked. "What you need to do is drill new bypass tunnels. Go through the soft sandstone around and beneath the dam and line the tunnels with waterproof plates. It would be an expensive, difficult engineering feat. Nothing like this has ever been done before, but I've done a lot of thinking about it, and it will work. You can drain it."

The astonished Ingebretsen asked Dominy to sign and date the napkin. "Nobody will believe this," he said. Dominy signed.

Of course, it will take more than a souvenir napkin to return the nation's great rivers to their full wildness and health. Too much of our economic infrastructure depends on those 75,000 dams for anyone to believe that large numbers of river blockers, no matter how obsolete, will succumb to the blow of Bruce Babbitt's hammer anytime soon. For one thing, Babbitt himself is hardly in a position to be the savior of the rivers. Swept up in the troubles of a lame-duck administration and his own nagging legal problems (last spring Attorney General Janet Reno appointed an independent counsel to look into his role in an alleged Indian casino-campaign finance imbroglio), this interior secretary is not likely to fulfill his dream of bringing down a really big dam. But a like-minded successor just might. It will take a president committed and powerful enough to sway both Congress and the public, but it could come to pass.

Maybe Glen Canyon Dam and the four Snake River dams won't come out in my lifetime, but others will. And as more rivers return to life, we'll take a new census of emancipated streams: We freed the Neuse, the Kennebec, the Allier, the Rogue, the Elwah, and even the Tuolumne. We freed the White Salmon and the Souradabscook, the Ocklawaha and the Genesee. They will be untidy and unpredictable, they will flood and recede, they will do what they were meant to do: run wild to the sea.

Bruce Barcott is the author of The Measure of a Mountain: Beauty and Terror on Mount Rainier (*Sasquatch, 1997*).

ENVIRONMENT

A River Runs Through It

75,000 dams were built. Now a few are coming down.

BY ANDREW MURR

AS CHURCH BELLS PEALED AND thousands cheered, the backhoe scooped out a dollop of dirt and gravel that had been packed against Maine's Edwards Dam. And suddenly the Kennebec River did something it hasn't done since Andrew Jackson sat in the White House: first in a trickle and then in a torrent, it flowed freely to the Atlantic, through a 60-foot hole that workers had earlier punched underneath the dirt-and-gravel bandage. The 917-foot dam is the first that federal regulators have ordered razed against the owner's wishes—because the environmental damage it did by preventing salmon, striped bass, shad and six other species from reaching spawning grounds outweighed the benefits of the .1 percent of Maine's power it provided. Interior Secretary Bruce Babbitt was on hand for last week's milestone. "If someone's got a dam that's going down," he tells friends, "I'll be there."

Babbitt can expect to rack up the frequent-flier miles. About 75,000 big dams block American rivers, testaments to the conviction that any river flowing to sea unimpeded is a waste of water and power. But that attitude is under attack. Many of the aging dams kill millions of valuable salmon migrating to sea. As a result, the Federal Energy Regulatory Commission (FERC) is refusing to relicense dams where the environmental

costs outweigh the value of the hydropower, or demanding that a dam be retrofitted with fish ladders. That's often so expensive that the owner opts to tear down the dam instead. Portland (Ore.) General Electric will raze dams on the Sandy and Little Sandy rivers, for instance, rather than make repairs. Demolition will free 22 miles of salmon and steelhead spawning grounds.

This year dams will fall from California to Connecticut (map). It shouldn't take long to see results. The Quaker Neck Dam on the Neuse River in North Carolina came down in 1997–98; bass and striped shad are already running again. On Butte Creek in northern California, removing three dams beginning in 1997 allowed the salmon run to jump from zero to 20,000. Those numbers have environmentalists and fishermen eyeing dams in the Olympic Peninsula, where dozens of populations of salmon are endangered or extinct. The Elwha and Glines Canyon dams, for example, cut the annual runs of salmon and

steelhead on the Elwha River from 380,000 early this century to zero. The biggest targets are four hydroelectrics on the Snake River. But many locals oppose razing these giants, which supply 5 percent of the region's electricity, because doing so could raise rates the same amount. Call it salmon vs. watts.

With SHARON BEGLEY

Tearing Down the Water Walls

This year, dams will come down from Maine to California. In August, tourists will even be able to watch the demolition of the Cascade Dam in Yosemite Valley.

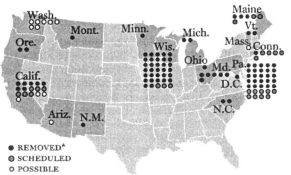

● REMOVED*
◉ SCHEDULED
○ POSSIBLE

*SINCE 1990. SOURCE FOR REMOVED DAMS: AMERICAN RIVERS

Sprawling, Sprawling . . .

Move to a suburb—and the world moves out with you. A case study in hypergrowth.

BY DANIEL PEDERSEN, VERN E. SMITH AND JERRY ADLER

THE SALES BROCHURES DON'T LIE, BECAUSE they show you a picture, a sweeping aerial view of forest stretching toward the distant skyline of Atlanta from the vicinity of VININGS ESTATES, FROM HIGH $200s-500s. The table of travel times puts downtown 20 minutes away (a footnote points out that "times may vary in rush-hour traffic"), and the photos of "historic Vinings" in the sales office are not meant to suggest that homebuyers will actually be shopping in a quaint general store. The houses are as crammed with luxuries as a Pharaoh's tomb; the lots are wooded; the road a peaceful two-lane blacktop. It all conspires to make you fall in love with the place—which requires only that you close your mind to what it might look like in five years, when you're battling your way toward it past the Wal-Marts and Waffle Houses that its very existence will call into being.

Here, just outside Atlanta's I-285 beltway, is the front line of the great lifestyle struggle of the next century. It is taking the paradoxical form of a war not on poverty, but on affluence—or the way affluence is typically realized in America, in suburban enclaves that eat away at the countryside and promote the triple evils of sprawl: air pollution, traffic congestion and visual blight. Al Gore has made an anti-sprawl "livability agenda" the centerpiece of his presidential campaign, warning darkly of commuters who arrive home "too late to read a child a bedtime story." Most of the $8.8 billion the administration was seeking for the next fiscal year—for projects such as mass transit, green-space acquisition and road improvements—has already been approved by Congress. New Jersey Gov. Christine Whitman, a moderate Republican, has ambitious plans to spend almost $1 billion to preserve undeveloped land in the most densely populated state in the union. Even builders here in Atlanta, those prodigious cultivators of affluence, are scrambling to meet an unexpected demand for "simplicity."

But translating this shift of sentiment into lines on the map and trees on the ground is a daunting task. Anyone who has flown over it can see that there is no shortage of empty land in the United States, 95 percent of which is undeveloped. Activists like to cite figures for the overall loss of rural countryside that vary from about 400,000 to 1.4 million acres a year. Those on the other side often repeat a statistic from Steven Hayward of the conservative Pacific Research Institute, which puts the rate of land development for the United States, excluding Alaska, at an infinitesimal 0.0006 percent a year. Last week Hayward admitted he had made "a stupid math mistake" and that the real figure is .07 percent, more than 100 times greater. That's still not a lot, although it means that over the lifetime of a child born today, the developed area of the nation will more than double.

But the problem is that this development is concentrated in a relatively few areas close to big cities—such as Atlanta, whose 10-county metropolitan area population has

> "I'm no tree-hugger . . . my pitch to the business community is, this is about money for schools and cities, to grow this state"
>
> **Gov. Roy Barnes** *His new agency can stop development anywhere in a 13-county area*

grown by roughly a quarter just since 1990, to about 3.1 million, as of a year ago. Reversing or even slowing this trend will require a whole new way of thinking by local planners, federal regulatory agencies, developers and homebuyers. Richard Moe, president of the National Trust for Historic Preservation, thinks that's going to happen: "People are willing to pay [for mass transit and open space]; they are willing to accept regulations," he says, "because this issue more than any other affects the quality of life in communities today." On the other side of the issue, Samuel Staley, an economist with the Reason Public Policy Institute in Los Angeles, points out that "suburbanization by other people is what's unpopular; people love living in the suburbs, they just don't want anyone else out there with them." He adds: "You can't develop a public policy around stopping people from moving to the communities and homes they want to live in, at least not in the United States. Not yet."

Atlanta will be a test of those competing views, and its citizens, business leaders and government officials provide a good window on how this struggle might be waged over the next few years. One who will be watching closely is Julie Haley, a 39-year-old mother of two who moved to a subdivision in Alpharetta back when the neighborhood was mostly horse farms and trees. That was in 1994. Today, she says, "people who visited us five years ago say, 'I couldn't find your house.' The roads have all gotten wider, they've knocked down all the trees, there's a million shopping centers." On business trips, her husband, Michael, leaves the house at 5:30 for a 9 a.m. flight to beat the traffic to Atlanta's Hartsfield airport. Her children, Kaitlin, 9, and Conor, 5, both suffer from asthma, which she attributes to sharing the air with cars whose drivers make the longest average round-trip commute in the country, 36.5 miles. "The kids cry when they see the bulldozers," she says. "They say, 'When I grow up, I'm gonna be president and I'm not going to let them cut down any more trees'."

Of course, Haley, a nonpracticing lawyer, knows it's more important to have the local

Oh No, It's Spreading!

More than 3 million people are spread throughout Metro Atlanta. Studies show that 1 percent in population growth results in 10 to 20 percent growth in land consumption. In Atlanta, momentum is so great that the North Georgia mountains, by one estimate, could become part of Metro Atlanta by 2002.

VANISHING FORESTS

▓ **Area with 50% or more tree cover**

■ **Area with less than 20% tree cover**

As Atlanta gains ground in miles, it loses ground of the green variety. Between 1974 and 1996, areas with 50 percent or more tree cover decreased by more than 44 percent, while areas with less than 20 percent tree cover increased by more than 60 percent.

ATLANTA URBANIZED AREA

Come on Down

Atlanta's welcome-y'all hospitality has contributed to a more than 25 percent increase in population since 1990. The city just keeps on growing.

Atlanta has grown "faster than any human settlement in history"—from 65 miles north to south to 110 miles since 1990. Some effects of the growth:

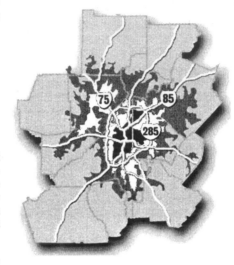

■ **1950**
□ **1970**
▓ **1990**
▓ **County boundaries**

■ **Longest Commute** Atlantans drive 36.5 daily miles round trip to work, more than Dallas's 29.5 and Los Angeles's 20.5

■ **Land Use** Every day, more than 50 acres of green space in the metro area is plowed under

■ **Pollution** Atlanta is in violation of clean-air standards

EFFECTS OF SPRAWL

As people leave the city, they develop the rural areas. Consequences of the sprawl:

■ **Environment** Fragments landscapes, disrupts wildlife habitats, alters rivers and streams, increases pollution

■ **Cities** Concentrates poverty in urban centers, hurts downtown commerce by pulling shoppers from locally owned stores and restaurants to large regional malls

■ **Traffic** Longer commutes that take time away from families and work

■ **Taxes** Increases taxes to pay for police and fire-department services, schools and infrastructure such as new roads and sewer construction

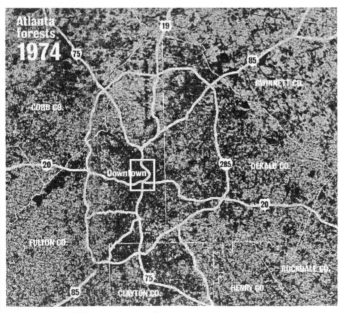

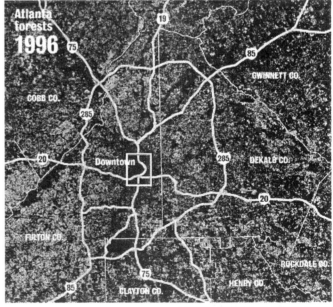

OVERLAY OF CURRENT HIGHWAY SYSTEM PLACED FOR GEOGRAPHIC ORIENTATION

LANDSAT IMAGERY—AMERICAN FORESTS

SPRAWL AROUND THE COUNTRY

Atlanta isn't the only U.S. city falling to sprawl. Other examples of large and medium-size sprawlers:

■ **Washington, DC.** Between 1982 and 1994, 69 percent increase in time commuters spent stuck in traffic

■ **Cincinnati, Ohio** Land area increase from 335 square miles in 1970 to 573 square miles in 1996

■ **Kansas City, Mo.** More freeway lane miles per capita than any other city

■ **Denver, Colo.** Farmland falling to sprawl at a rate of 90,000 acres per year

■ **Seattle, Wash.** More cars than people; twice the number of automobile trips each day per household than in 1990

Sprawl-Threatened Cities

● Population
1 million or more
● Population
500,000–1 million

■ **Austin, Texas** From 1982 to 1992, 35 percent increase in open space lost to development

■ **Las Vegas, Nev.** Fastest-growing population in the United States

■ **West Palm Beach, Fla.** Between 1990 and 1996, 25 percent decrease in urban population density

■ **Akron, Ohio** Between 1990 and 1996, 37 percent decrease in population density and land area increase of 65 percent

RESEARCH AND TEXT BY ELLEN FLORIAN. SOURCES: SIERRA CLUB, ATLANTA REGIONAL COMMISSION, AMERICAN FORESTS

---

zoning board on your side than the president. But that body, she has found, responds mostly to the well-connected developers. She has won a few battles, only to lose the wars. Fulton County officials recently declared a moratorium on new development in an area where the sewers were overflowing. But a day later the moratorium was lifted for developers who already had a "land disturbance permit." It turned out there were hundreds of them, she says; "We're in the middle of a moratorium, and there's twice as much building as before." She joined a boycott of a supermarket whose builders, she says, chopped down several oak trees of a protected species. The store went out of business, but that didn't bring the trees back.

Haley, though, may have a powerful ally in the statehouse. Gov. Roy Barnes, a moderate Democrat, was elected last fall after promising to do something about sprawl, and he promptly did, creating a new agency with unprecedented power over zoning, roads and transit in a 13-county area. Surprising even the governor, the law authorizing the Georgia Regional Transportation Authority rolled to a quick passage—despite some grumbling that the agency's initials should stand for Give Roy Total Authority. (Barnes gets to appoint all 15 members of the board.) But the legislature had little choice, since the 13 counties were out of compliance with clean-air standards and about to lose federal highway funds. Now the new agency faces the task of extending Atlanta's feeble mass-transit system to the suburban counties that have always resisted it, and deciding whether to proceed with the "northern arc" freeway connecting I-75 and I-85 25 miles beyond the existing beltway. The creation of GRTA was praised by Gore as a model environmental initiative, although Barnes is quick to disavow the idea that he acted out of concern for anything as intangible as the atmosphere. "I'm no tree-hugger," he insists. "I'm a busi-

> **"People who visited us five years ago can't find our house. The roads have gotten wider and they've knocked down all the trees."**
>
> **Julie Haley** *When her children wheeze from asthma, she blames air pollution*

nessman who thinks you can't let your prosperity slip through your fingers. The pitch I made to the business community is that this is about money for schools and cities, to grow this state."

Almost everyone in the region has a stake in how GRTA does its job, especially the leaders of the city of Atlanta, which has been mostly a bystander in the great land rush of the '90s. Until recently, Realtors helping families relocate would show off Atlanta's parks, museums and stadium, says Mayor Bill Campbell, but then head to the suburbs to look at houses. "Now, when they take them out there, they're getting stuck in traffic," he says gleefully. "We've had more construction building permits issued in the

last three years than any other time in our history." Campbell tends to regard the suburbs as a giant Ponzi scheme, in which taxes are kept low for the first arrivals by deferring the cost of roads and schools. "Now they're choking on their growth and they don't have the infrastructure to provide for it," he says—while the developers have moved on to the next interstate exit.

And there are powerful business interests lining up with Campbell, such as the developer John A. Williams. His company built thousands of suburban garden apartments up through 1987, but since then has worked almost exclusively inside the I-285 beltway, often on "brownfield" sites recycled from other uses. He has moved toward "live/work/walk" projects that mix housing with shops and offices, and housing for different economic classes. "I was sort of poor when I grew up," says Williams, an Atlanta native and Georgia Tech grad, "but there were people poorer than me, and I didn't think anything of it"—an attitude that, if it ever spreads, could subvert a major unspoken premise of suburban developments priced FROM THE HIGH $200s–500s.

Yet the businessman who may have the most impact on Atlanta's development over the next few years isn't a developer at all; he's F. Duane Ackerman, CEO of BellSouth, the second largest local employer after Delta Airlines. At a cost of some $750 million, Ackerman has ordained the relocation of 13,000 employees from 75 suburban offices to three new complexes within the beltway, each walking distance from a transit stop. The company will build four giant parking lots at the suburban ends of the transit lines, equipped with small business centers so that employees can keep working on the train and send off their e-mails and faxes before driving home. Ackerman, a hard-driving businessman whose personal formula for avoiding traffic is to get up at 5 a.m. and work until 6:30 at night, insists he is only trying to get more productivity out of a work force that was in danger of disappearing into daylong traffic jams. "I'm not a member of the Sierra Club," he assures interviewers. "But we are a company that's aware of its environment."

And there are others, of course, who are watching Barnes's great plans with more skepticism, like builder John Wieland, the brusque, tough-minded developer of Vinings Estates and dozens of other suburban projects over the last 29 years. He says what many probably think: that it's easy to be against "sprawl" in the abstract, until you're asked to weigh the logical alternative, which can only be "density." "I ask people, would they like to see sprawl controlled, and they answer, 'That would be good'," he says. "Then I ask, how would they like town homes next door? That wouldn't be so good." As for GRTA, he's in less awe of its dictatorial power than you might think. "As soon as somebody tells a well-connected developer he can't build where he wants to build," he predicts, "the governor's going to get a phone call." And

> ## "As soon as a well-connected developer is told he can't build where he wants, the governor is going to get a phone call"
>
> **John Wieland** *He thinks home buyers in the end would choose 'sprawl' over 'density'*

there are, of course, a few holdouts against the whole concept of state intervention in local planning. "The people in my neighborhood aren't complaining about dirty air," says Don Balfour, a Republican state senator who opposed Barnes's plan. "They're saying, 'Man, it takes me an hour and a half to go 30 miles'." The solution for that problem, he contends, is not more density—but more highways.

Which is just the attitude that drives Julie Haley nuts. "People keep saying we don't want to be another Gwinnett County," she says, referring to Balfour's home district, "because they've seen how ugly and desolate a landscape can look, devoid of trees and full of strip malls." Sometimes, she thinks of moving even farther out—say, to Athens, 55 miles from Atlanta—but she doubts she can outrun sprawl. "They're starting on what we've already been through," she muses. "I'm thinking, 'I've battled around here for so long, why don't I stick it out' "—and fight.

With STEPHEN TOTILO

Past and Present Land Use and Land Cover in the USA

WILLIAM B. MEYER

Dr. William B. Meyer is a geographer currently employed on the research faculty of the George Perkins Marsh Institute at Clark University in Worcester, Massachusetts. His principal interests lie in the areas of global environmental change with particular emphasis on land use and land cover change, in land use conflict, and in American environmental history.

Land of many uses," runs a motto used to describe the National Forests, and it describes the United States as a whole just as well. "Land of many covers" would be an equally apt, but distinct, description. *Land use* is the way in which, and the purposes for which, human beings employ the land and its resources: for example, farming, mining, or lumbering. *Land cover* describes the physical state of the land surface: as in cropland, mountains, or forests. The term land cover originally referred to the kind and state of vegetation (such as forest or grass cover), but it has broadened in subsequent usage to include human structures such as buildings or pavement and other aspects of the natural environment, such as soil type, biodiversity, and surface and groundwater. A vast array of physical characteristics—climate, physiography, soil, biota—and the varieties of past and present human utilization combine to make every parcel of land on the nation's surface unique in the cover it possesses and the op-portunities for use that it offers. For most practical purposes, land units must be aggregated into quite broad categories, but the frequent use of such simplified classes should not be allowed to dull one's sense of the variation that is contained in any one of them.

Land cover is affected by natural events, including climate variation, flooding, vegetation succession, and fire, all of which can sometimes be affected in character and magnitude by human activities. Both globally and in the United States, though, land cover today is altered principally by direct human use: by agriculture and livestock raising, forest harvesting and management, and construction. There are also incidental impacts from other human activities such as forests damaged by acid rain from fossil fuel combustion and crops near cities damaged by tropospheric ozone resulting from automobile exhaust.

Changes in land cover by land use do not necessarily imply a degradation of the land. Indeed, it might be presumed that any change produced by human use is an improvement, until demonstrated otherwise, because someone has gone to the trouble of making it. And indeed, this has been the dominant attitude around the world through time. There are, of course, many reasons why it might be otherwise. Damage may be done with the best of intentions when the harm inflicted is too subtle to be perceived by the land user. It may also be done when losses produced by a

From *Consequences*, Spring 1995, pp. 25-33. © 1995 by Saginaw Valley State University. Reprinted by permission.

change in land use spill over the boundaries of the parcel involved, while the gains accrue largely to the land user. Economists refer to harmful effects of this sort as *negative externalities*, to mean secondary or unexpected consequences that may reduce the net value of production of an activity and displace some of its costs upon other parties. Land use changes can be undertaken because they return a net profit to the land user, while the impacts of negative externalities such as air and water pollution, biodiversity loss, and increased flooding are borne by others. Conversely, activities that result in secondary benefits (or *positive externalities*) may not be undertaken by landowners if direct benefits to them would not reward the costs.

Over the years, concerns regarding land degradation have taken several overlapping (and occasionally conflicting) forms. *Conservationism* emphasized the need for careful and efficient management to guarantee a sustained supply of productive land resources for future generations. *Preservationism* has sought to protect scenery and ecosystems in a state as little human-altered as possible. Modern *environmentalism* subsumes many of these goals and adds new concerns that cover the varied secondary effects of land use both on land cover and on other related aspects of the global environment. By and large, American attitudes in the past century have shifted from a tendency to interpret human use as improving the condition of the land towards a tendency to see human impact as primarily destructive. The term "land reclamation" long denoted the conversion of land from its natural cover; today it is more often used to describe the restoration and repair of land damaged by human use. It would be easy, though, to exaggerate the shift in attitudes. In truth, calculating the balance of costs and benefits from many land use and land cover changes is enormously difficult. The full extent and consequences of proposed changes are often less than certain, as is their possible irreversibility and thus their lasting significance for future generations.

WHERE ARE WE?

The United States, exclusive of Alaska and Hawaii, assumed its present size and shape around the middle of the 19th century. Hawaii is relatively small, ecologically distinctive, and profoundly affected by a long and distinctive history of human use; Alaska is huge and little affected to date by direct land use. In this review assessment we therefore survey land use and land cover change, focusing on the past century and a half, only in the conterminous or lower 48 states. Those states

> *"The adjustments that are made in land use and land cover in coming years will in some way alter the life of nearly every living thing on Earth."*

cover an area of almost 1900 million acres, or about 3 million square miles.

How land is *used*, and thus how *land cover* is altered, depends on who owns or controls the land and on the pressures and incentives shaping the behavior of the owner. Some 400 million acres in the conterminous 48 states—about 21% of the total—are federally owned. The two largest chunks are the 170 million acres of western rangeland controlled by the Bureau of Land Management and the approximately equal area of the National Forest System. Federal land represents 45% of the area of the twelve western states, but is not a large share of any other regional total. There are also significant land holdings by state governments throughout the country.

Most of the land in the United States is privately owned, but under federal, state, and local restrictions on its use that have increased over time. The difference between public and private land is important in explaining and forecasting land use and land coverage change, but the division is not absolute, and each sector is influenced by the other. Private land use is heavily influenced by public policies, not only by regulation of certain uses but through incentives that encourage others. Public lands are used for many private activities; grazing on federal rangelands and timber extraction from the national forests by private operators are the most important and have become the most controversial. The large government role in land use on both government and private land means that policy, as well as economic forces, must be considered in explaining and projecting changes in the land. Economic forces are of course significant determinants of policy—perhaps the most significant—but policy remains to some degree an independent variable.

There is no standard, universally accepted set of categories for classifying land by either use or cover, and the most commonly used, moreover, are hybrids of land cover and land use. Those employed here, which are by and large those of the U.S. National Resources Inventory conducted every five years by relevant federal agencies, are cropland, forest, grassland (pasture and rangeland), wetlands, and developed land.

- *Cropland* is land in farms that is devoted to crop production; it is not to be confused with total farmland, a broad land use or land ownership category that can incorporate many forms of land cover.
- *Forest land* is characterized by a predominance of tree cover and is further divided by the U.S. Census into timberland and non-timberland. By definition, the former must be capable of producing 20 cubic feet of industrial wood per acre per year and remain legally open to timber production.
- *Grassland* as a category of land cover embraces two contrasting Census categories of use: pasture (enclosed and what is called improved grassland, often closely tied to cropland and used for intensive livestock raising), and range (often unenclosed or unimproved grazing land with sparser grass cover and utilized for more extensive production).
- *Wetlands* are not a separate Census or National Resources Inventory category and are included within other categories: swamp, for example, is wetland forest. They are defined by federal agencies as lands covered all or part of the year with water, but not so deeply or permanently as to be classified as water surface *per se*.
- The U.S. government classifies as *developed* land urban and built-up parcels that exceed certain size thresholds. "Developed" or "urban" land is clearly a use rather than a cover category. Cities and suburbs as they are politically defined have rarely more than half of their area, and often much less, taken up by distinctively "urban" land cover such as buildings and pavement. Trees and grass cover

Land Use and Cover in the Conterminous U.S.

Land Class	Area in Million Acres	Fraction of Total Area
Privately Owned (shown in diagram below)		
Cropland	422	22.4%
Rangeland	401	21.3
Forest	391	20.8
Pasture	129	6.9
Developed	77	4.1
Other Catagories[1]	60	3.2
Federally Owned[2]	404	21.4
TOTAL[3]	1884	100%

[1] Other minor covers and surface water
[2] Federal land is approximately half forest and half rangeland
[3] Included in various catagories is about 100 million acres of wetland, covering about 5% of the national area

Table 1 Source: U.S. 1987 National Resources Inventory, published in 1989. U.S. Government Printing Office.

substantial areas of the metropolitan United States; indeed, tree cover is greater in some settlements than in the rural areas surrounding them.

By the 1987 U.S. National Resources Inventory, non-federal lands were divided by major land use and land cover classes as follows: cropland, about 420 million acres (22% of the entire area of the 48 states); rangeland about 400 million (21%); forest, 390 million (21%); pasture, 130 million (7%); and developed land, 80 million (4%). Minor covers and uses, including surface water, make up another 60 million acres (Table 1). The 401 million acres of federal land are about half forest and half range. Wetlands, which fall within these other Census classes, represent approximately 100 million acres or about five percent of the national area; 95 percent of them are freshwater and five percent are coastal.

These figures, for even a single period, represent not a static but a dynamic total, with constant exchanges among uses. Changes in the area and the location of cropland, for example, are the result of the *addition* of new cropland from conversion of grassland, forest, and wetland and its *subtraction* either by abandonment of cropping and reversion to one of these less intensive use/cover forms or by conversion to developed land. The main causes of forest *loss* are clearing for agriculture, logging, and clearing for development; the main cause of forest *gain* is abandonment of cropland followed by either passive or active reforestation. Grassland is converted by the creation of pasture from forest, the interchange of pasture and cropland, and the conversion of rangeland to cropland, often through irrigation.

Change in wetland is predominantly loss through drainage for agriculture and construction. It also includes natural gain and loss, and the growing possibilities for wetland creation and restoration are implicit in the Environmental Protection Agency's "no *net* loss" policy (emphasis added). Change in developed land runs in only one direction: it expands and is not, to any significant extent, converted to any other category.

Comparison of the American figures with those for some other countries sets them in useful perspective. The United States has a greater relative share of forest and a smaller relative share of cropland than does Europe as a whole and the United Kingdom in particular.

Though Japan is comparable in population density and level of development to Western Europe, fully two-thirds of its area is classified as forest and woodland, as opposed to ten percent in the United Kingdom; it preserves its largely mountainous forest area by maintaining a vast surplus of timber imports over exports, largely from the Americas and Southeast Asia.

Regional patterns within the U.S. (using the four standard government regions of Northeast, Midwest, South, and West) display further variety. The Northeast, though the most densely populated region, is the most heavily wooded, with three-fifths of its area in forest cover. It is also the only region of the four in which "developed" land, by the Census defi-

nition, amounts to more than a minuscule share of the total; it covers about eight percent of the Northeast and more than a quarter of the state of New Jersey. Cropland, not surprisingly, is by far the dominant use/cover in the Midwest, accounting for just under half of its expanse. The South as a whole presents the most balanced mix of land types: about 40 percent forest, 20 percent each of cropland and rangeland, and a little more than ten percent pasture. Western land is predominantly rangeland, with forest following and cropland a distant third. Wetlands are concentrated along the Atlantic seaboard, in the Southeast, and in the upper Midwest. Within each region, of course, there is further variety at and below the state level.

WHERE HAVE WE BEEN?

The public domain, which in 1850 included almost two-thirds of the area of the present conterminous states, has gone through two overlapping phases of management goals. During the first, dominant in 1850 and long thereafter, the principal goal of management was to transfer public land into private hands, both to raise revenue and to encourage settlement and land improvements. The government often attached conditions (which were sometimes complied with) to fulfill other national goals, such as swamp drainage, timber planting, and railroad construction in support of economic development.

The second phase, that of federal retention and management of land, began with the creation of the world's first national park, Yellowstone, shortly after the Civil War. It did not begin to be a significant force, however, until the 1890s, when 40 million acres in the West were designated as federal forest reserves, the beginning of a system that subsequently expanded into other regions of the country as well. Several statutory vestiges of the first, disposal era remain (as in mining laws, for example), but the federal domain is unlikely to shrink noticeably in coming decades, in spite of repeated challenges to the government retention of public land and its regulation of private land. In recent years, such challenges have included the "Sagebrush Rebellion" in the rangelands of the West in the 1970s and 1980s calling for the withdrawal of federal control, and legal efforts to have many land use regulations classified as "takings," or as exercises of the power of eminent domain. This classification, where it is

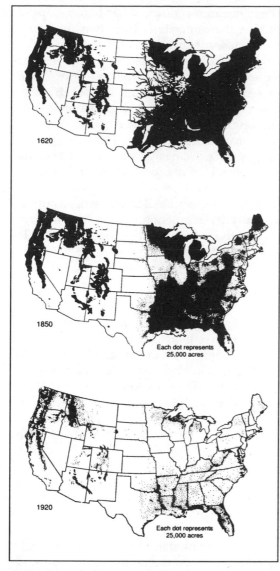

Figure 1 Area of virgin forest: top to bottom 1620, 1850, and 1920 as published by William B. Greeley, "The Relation of Geography to Timber Supply," Economic Geography, vol. 1, pp. 1–11 (1925). The depiction of U.S. forests in the later maps may be misleading in that they show only old-growth forest and not total tree cover.

granted, requires the government to compensate owners for the value of development rights lost as a result of the regulation.

Cropland

Total cropland rose steadily at the expense of other land covers throughout most of American history. It reached a peak during the 1940s and has subsequently fluctuated in the neighborhood of 400 million acres, though the precise figure depends on the definition of cropland used. Long-term regional patterns have displayed more variety. Cropland abandonment in some areas of New England began to be significant in some areas by the middle of

the nineteenth century. Although total farmland peaked in the region as late as 1880 (at 50%) and did not decline sharply until the turn of the century, a steady decline in the subcategory of cropland and an increase in other farmland covers such as woodland and unimproved pasture was already strongly apparent. The Middle Atlantic followed a similar trajectory, as, more recently, has the South. Competition from other, more fertile sections of the country in agricultural production and within the East from other demands on land and labor have been factors; a long-term rise in agricultural productivity caused by technological advances has also exerted a steady downward pressure on total crop acreage even though population, income, and demand have all risen.

Irrigated cropland on a significant scale in the United States extends back only to the 1890s and the early activities in the West of the Bureau of Reclamation. Growing rapidly through about 1920, the amount of irrigated land remained relatively constant between the wars, but rose again rapidly after 1945 with institutional and technological developments such as the use of center-pivot irrigation drawing on the Ogallala Aquifer on the High Plains. It reached 25 million acres by 1950 and doubled to include about an eighth of all cropland by about 1980. Since then the amount of irrigated land has experienced a modest decline, in part through the decline of aquifers such as the Ogallala and through competition from cities for water in dry areas.

Forests

At the time of European settlement, forest covered about half of the present 48 states. The greater part lay in the eastern part of the country, and most of it had already been significantly altered by Native American land use practices that left a mosaic of different covers, including substantial areas of open land.

Forest area began a continuous decline with the onset of European settlement that would not be halted until the early twentieth century. Clearance for farmland and harvesting for fuel, timber, and other wood products represented the principal sources of pressure. From an estimated 900 million acres in 1850, the wooded area of the entire U.S. reached a low point of 600 million acres around 1920 (Fig. 1).

It then rose slowly through the postwar decades, largely through abandonment of cropland and regrowth on cutover areas, but around 1960 began again a modest decline, the result of settlement expansion and of higher rates of timber extraction through mechanization. The agricultural censuses recorded a drop of 17 million acres in U.S. forest cover between 1970 and 1987 (though data uncertainties and the small size of the changes relative to the total forest area make a precise dating of the reversals difficult). At the same time, if the U.S. forests have been shrinking in area they have been growing in density and volume. The trend in forest biomass has been consistently upward; timber stock measured in the agricultural censuses from 1952 to 1987 grew by about 30%.

National totals of forested area again represent the aggregation of varied regional experiences. Farm abandonment in much of the East has translated directly into forest recovery, beginning in the mid- to late-nineteenth century (Fig. 2). Historically, lumbering followed a regular pattern of harvesting one region's resources and moving on to the next; the once extensive old-growth forest of the Great Lakes, the South, and the Pacific Northwest represented successive and overlapping frontiers. After about 1930, frontier-type exploitation gave way to a greater emphasis on permanence and management of stands by timber companies. Wood itself has declined in importance as a natural resource, but forests have been increasingly valued and protected for a range of other services, including wildlife habitat, recreation, and streamflow regulation.

Grassland

The most significant changes in grassland have involved impacts of grazing on the western range. Though data for many periods are scanty or suspect, it is clear that rangelands have often been seriously overgrazed, with deleterious consequences including soil erosion and compaction, increased streamflow variability, and floral and faunal biodiversity loss as well as reduced value for production. The net value of grazing use on the western range is nationally small, though significant locally, and pressures for tighter management have increasingly been guided by ecological and preservationist as well as production concerns.

Wetland

According to the most recent estimates, 53% of American wetlands were lost between the 1780s and the 1980s, principally to drainage for agriculture. Most of the conversion presumably took place during the twentieth century; between the 1950s and the 1970s alone, about 11 million acres were lost. Unassisted private action was long thought to drain too

little; since mid-century, it has become apparent that the opposite is true, that unfettered private action tends to drain too much, i.e., at the expense of now-valued wetland. The positive externalities once expected from drainage—improved public health and beautification of an unappealing natural landscape—carry less weight today than the negative ones that it produces. These include the decline of wildlife, greater extremes of streamflow, and loss of a natural landscape that is now seen as more attractive than a human-modified one. The rate of wetland loss has now been cut significantly by regulation and by the removal of incentives for drainage once offered by many government programs.

Developed land

As the American population has grown and become more urbanized, the land devoted to settlement has increased in at least the same degree. Like the rest of the developed world, the United States now has an overwhelmingly non-farm population residing in cities, suburbs, and towns and villages. Surrounding urban areas is a classical frontier of rapid and sometimes chaotic land use and land cover change. Urban impacts go beyond the mere subtraction of land from other land uses and land covers for settlement and infrastructure; they also involve the mining of building materials, the disposal of wastes, the creation of parks and water supply reservoirs, and the introduction of pollutants in air, water, and soil. Long-term data on urban use and cover trends are unfortunately not available. But the trend in American cities has undeniably been one of residential dispersal and lessened settlement densities as transportation technologies have improved; settlement has thus required higher amounts of land per person over time.

WHERE ARE WE GOING?

The most credible projections of changes in land use and land cover in the United States over the next fifty years have come from recent assessments produced under the federal laws that now mandate regular national inventories of resource stocks and prospects. The most recent inquiry into land resources, completed by the Department of Agriculture in 1989 (and cited at the end of this article), sought to project their likely extent and condition a half-century into the future, to the year 2040. The results indicated that only slow

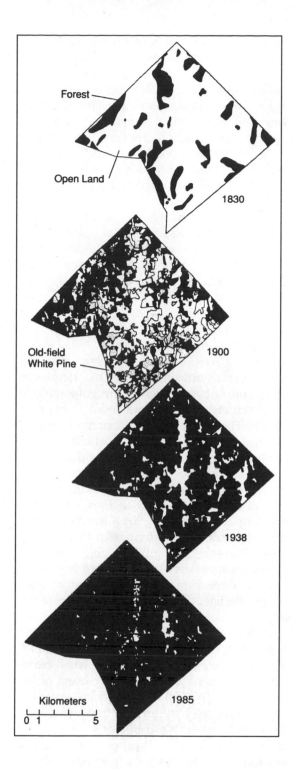

Figure 2 Modern spread of forest (shown in black) in the township of Petersham, Massachusetts, 1830 through 1985. White area is that considered suitable for agriculture; shaded portions in 1900 map indicate agricultural land abandoned between 1870 and 1900 that had developed forest of white pine in this period. From "Land-use History and Forest Transformations," by David R. Foster, in *Humans as Components of Ecosystems*, edited by M. J. McDonnell and S.T.A. Pickett, Springer-Verlag, New York, pp. 91–110, 1993.

changes were expected nationally in the major categories of land use and land cover: a loss in forest area of some 5% (a slower rate of loss than was experienced in the same period before); a similarly modest decline in cropland; and an increase in rangeland of about 5% through 2040. Projections are not certainties, however: they may either incorrectly identify the consequences of the factors they consider or fail to consider important factors that could alter the picture. Because of the significant impacts of policy, its role—notoriously difficult to forecast and assess— demands increased attention, in both its deliberate and its inadvertent effects.

Trends in the United States stand in some contrast to those in other parts of the developed world. While America's forest area continues to decline somewhat, that of many comparable countries has increased in modest degree, while the developing world has seen significant clearance in the postwar era. There has been substantial stability, with slow but fluctuating decline, in cropland area in the United States. In contrast, cropland and pasture have declined modestly in the past several decades in Western Europe and are likely to decline sharply there in the future as longstanding national and European Community agricultural policies subsidizing production are revised; as a result, the European countryside faces the prospect of radical change in land use and cover and considerable dislocation of rural life.

WHY DOES IT MATTER?

Land use and land cover changes, besides affecting the current and future supply of land resources, are important sources of many other forms of environmental change. They are also linked to them through synergistic connections that can amplify their overall effect.

Loss of plant and animal biodiversity is principally traceable to land transformation, primarily through the fragmentation of natural habitat. Worldwide trends in land use and land cover change are an important source of the so-called greenhouse gases, whose accumulation in the atmosphere may bring about global climate change. As much as 35% of the increase in atmospheric CO_2 in the last 100 years can be attributed to land use change, principally through deforestation. The major known sources of increased methane—rice paddies, landfills, biomass burning, and cattle—are all related to land use. Much of the increase in nitrous oxide is now thought due to a collection of sources that also depend upon the use of the land, including biomass burning, livestock raising, fertilizer application and contaminated aquifers.

Land use practices at the local and regional levels can dramatically affect soil condition as

> *In most of the world, both fossil fuel combustion and land transformation result in a net release of carbon dioxide to the atmosphere.*

well as water quality and water supply. And finally, vulnerability or sensitivity to existing climate hazards and possible climate change is very much affected by changes in land use and cover. Several of these connections are illustrated below by examples.

Carbon emissions

In most of the world, both fossil fuel combustion and land transformation result in a net release of carbon dioxide to the atmosphere. In the United States, by contrast, present land use and land cover changes are thought to absorb rather than release CO_2 through such processes as the rapid growth of relatively youthful forests. In balance, however, these land-use-related changes reduce U.S. contributions from fossil fuel combustion by only about 10%. The use of carbon-absorbing tree plantations to help diminish global climate forcing has been widely discussed, although many studies have cast doubt on the feasibility of the scheme. Not only is it a temporary fix (the trees sequester carbon only until the wood is consumed, decays, or ceases to accumulate) and requires vast areas to make much of a difference, but strategies for using the land and its products to offset some of the costs of the project might have large and damaging economic impacts on other land use sectors of the economy.

Effects on arable land

The loss of cropland to development aroused considerable concern during the 1970s and early 1980s in connection with the 1981 National Agricultural Lands Study, which estimated high and sharply rising rates of conversion. Lower figures published in the 1982 National Resource Inventory, and a number of associated studies, have led most experts to regard the conversion of cropland to other land use categories as representing something short of a genuine crisis, likely moreover to continue at slower rather than accelerating rates into the future. The land taken from food and fiber production and converted to developed land has been readily made up for by conversion of land from grassland and forest. The new lands are not necessarily of the same quality as those lost, however, and some measures for the protection of prime farmland are widely considered justified on grounds of economics as well as sociology and amenities preservation.

Vulnerability to climate change

Finally, patterns and trends in land use and land cover significantly affect the degree to which countries and regions are vulnerable to climate change—or to some degree, can profit from it. The sectors of the economy to which land use and land cover are most critical—agriculture, livestock, and forest products—are, along with fisheries, among those most sensitive to climate variation and change. How vulnerable countries and regions are to climate impacts is thus in part a function of the importance of these activities in their economies, although differences in ability to cope and adapt must also be taken into account.

These three climate-sensitive activities have steadily declined in importance in recent times in the U.S. economy. In the decade following the Civil War, agriculture still accounted for more than a third of the U.S. gross domestic product, or GDP. In 1929, the agriculture-forest-fisheries sector represented just under ten percent of national income. By 1950, it had fallen to seven percent of GDP, and it currently represents only about two percent. Wood in 1850 accounted for 90 percent of America's total energy consumption; today it represents but a few percent. These trends suggest a lessened macroeconomic vulnerability in the U.S. to climate change, though they may also represent a lessened ability to profit from it to the extent that change proves beneficial. They say nothing, however, about primary or secondary impacts of climate change on other sectors, about ecological, health, and amenity losses, or about vulnerability in absolute rather than relative terms, and particularly the potentially serious national and global consequences of a decline in U.S. food production.

The same trend of lessening vulnerability to climate changes is apparent even in regions projected to be the most exposed to the more harmful of them, such as reduced rainfall. A recent study examined agro-economic impacts on the Missouri-Iowa-Nebraska-Kansas area of the Great Plains, were the "Dust Bowl" drought and heat of the 1930s to recur today

Shifting patterns of land use in the U.S. and throughout the world are a proximate cause of many of today's environmental concerns.

or under projected conditions of the year 2030. It found that although agricultural production would be substantially reduced, the consequences would not be severe for the regional economy overall: partly because of technological and institutional adaptation and partly because of the declining importance of the affected sectors, as noted above. The 1930s drought itself had less severe and dramatic effects on the population and economy of the Plains than did earlier droughts in the 1890s and 1910s because of land use, technological, and institutional changes that had taken place in the intervening period.

Shifting patterns in human settlement are another form of land use and land cover change that can alter a region's vulnerability to changing climate. As is the case in most other countries of the world, a disproportionate number of Americans live within a few miles of the sea. In the postwar period, the coastal states and counties have consistently grown faster than the country as a whole in population and in property development. The consequence is an increased exposure to hazards of hurricanes and other coastal storms, which are expected by some to increase in number and severity with global warming, and to the probable sea-level rise that would also accompany an increase in global surface temperature. It is unclear to what extent the increased exposure to such hazards might be balanced by improvements in the ability to cope, through better forecasts, better construction, and insurance and relief programs. Hurricane fatalities have tended to decline, but property losses per hurricane have steadily increased in the U.S., and the consensus of experts is that they will continue to do so for the foreseeable future.

CONCLUSIONS

How much need we be concerned about changes in land use and land cover in their own right? How much in the context of other anticipated environmental changes?

As noted above, shifting patterns of land use in the U.S. and throughout the world are a proximate cause of many of today's environmental concerns. How land is used is also among the human activities most likely to feel the effects of possible climate change. Thus if we are to understand and respond to the challenges of global environmental change we need to understand the dynamics of land transformation. Yet those dynamics are notoriously difficult to predict, shaped as they are by patterns of individual decisions and collective human behavior, by history and geography, and by tangled economic and political considerations. We should have a more exact science of how these forces operate and how to balance them for the greatest good, and a more detailed and coherent picture of how land in the U.S. and the rest of the world is used.

The adjustments that are made in land use and land cover in coming years, driven by worldwide changes in population, income, and technology, will in some way alter the life of nearly every living thing on Earth. We need to understand them and to do all that we can to ensure that policy decisions that affect the use of land are made in the light of a much clearer picture of their ultimate effects.

FOR FURTHER READING

Americans and Their Forests: A Historical Geography, by Michael Williams. Cambridge University Press, 599 pp, 1989.

An Analysis of the Land Situation in the United States: 1989–2040. USDA Forest Service General Technical Report RM-181. U.S. Government Printing Office, Washington, D.C., 1989.

Changes in Land Use and Land Cover: A Global Perspective. W. B. Meyer and B. L. Turner II, editors. Cambridge University Press, 537 pp, 1994.

"Forests in the Long Sweep of American History," by Marion Clawson. *Science,* vol. 204, pp 1168–1174, 1979.

INFRASTRUCTURE

Operation Desert Sprawl

The biggest issue in booming Las Vegas isn't growth. It's finding somebody to pay the staggering costs of growth.

BY WILLIAM FULTON AND PAUL SHIGLEY

On a late spring afternoon, the counters at the Las Vegas Development Services Center are only slightly less crowded than those at the nearby McDonald's. Here, in a nondescript office building some eight blocks from City Hall, a small army of planners occupies counters and cubicles, standing ready to process the daily avalanche of building projects. Several times a minute, people with blueprints tucked under their arms hurry in or out the door.

Upstairs, Tim Chow, the city's planning and development director, shakes his head and smiles. He says he has "the toughest planning job in the country," and he may be right. Two hundred new residents arrive in Las Vegas every day; a house is built every 15 minutes. Last year alone, the city issued 7,700 residential building permits, plus permits for $200 million worth of commercial construction—enough to build a good-sized Midwestern county seat from scratch.

Before he came to Nevada in April, Chow held a similar position in a smaller county in California. There, he points out, the planning process grinds slowly. State law—and local politics—require extensive environmental studies and public hearings before planners approve subdivisions and retail centers.

But this isn't California. It's Las Vegas, the nation's fastest-growing community, and the policy is to build first and ask questions later. "We just don't have time to do the kinds of rigorous analysis done in other places with regard to compatibility, impacts, infrastructure, coordination," Chow says. "Sometimes you make mistakes, and the impacts of those mistakes are felt many years later."

Tim Chow's planning counter isn't the only one in this area that looks like a fast-food restaurant. Fifteen miles to the southeast, his counterpart in Henderson, Mary Kay Peck, is presiding over the rapid creation of the second-largest city in Nevada. Henderson officials claim to have "the high-

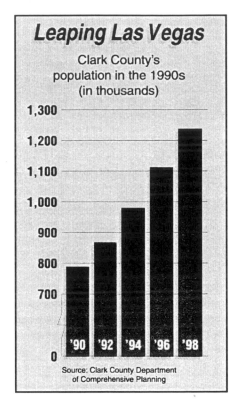

Leaping Las Vegas

Clark County's population in the 1990s (in thousands)

Source: Clark County Department of Comprehensive Planning

est development standards in the Las Vegas Valley," but that hasn't stopped the city from growing seven-fold in the past two decades, to 170,000 people. The city recently annexed 2,500 acres to accommodate a new Del Webb project, and is lining up a federal land exchange that will allow the addition of 8,000 more acres near McCarran International Airport.

Henderson is in a running argument with well-established Reno over which city is second most populous in the state after Las Vegas. But that argument won't last much longer. "We have room," Peck says, "to grow

and grow and grow." Henderson is expected to add another 100,000 people in the next decade—pushing it far past Reno. By 2010, Henderson will be as big as Las Vegas was in 1990.

The new residential subdivisions built all over the Las Vegas Valley in the past few years have created an enormous unsatisfied demand for parks, transportation, and water delivery systems. "Traffic is probably 100 times worse than it was 10 years ago," says Bobby Shelton, spokesman for the Clark County Public Works Department.

Local governments are trying to cope with the onslaught. A rail system connecting downtown, the casino-lined Strip and the airport is on the drawing board. Voters recently approved two tax increases—one for transportation projects, one for water projects—that together will produce some $100 million a year. But even with these projects moving forward Vegas-style—ready, fire, aim—the problem is getting worse, not better. Even the local building industry acknowledges that growth has gotten far ahead of the infrastructure that is needed to support it.

"Someone back in the '70s and '80s should have said, Hey, we are going to need parks and schools and roads,' " says Joanne Jensen, of the Southern Nevada Home Builders, who moved to Las Vegas from Chicago in 1960. "We're playing catch-up." After two months in town, Planning Director Chow uses the same words to describe the situation.

Take the infrastructure needs faced by any fast-growing American community, multiply by a factor of about 20, and you get a rough idea of what is going on in Las Vegas. It is similar to other places in that it has come to realize that residential development does not pay for itself. The difference is in the magnitude of its problem. No other city of comparable size is taking on people at anything remotely close to the Las Vegas rate. No other city is being challenged to build so much so quickly.

The infrastructure will be built. That is not really the issue. Las Vegas and Henderson will continue to grow. The question is who will pay the bill.

As in other American cities, property owners have paid for community infrastructure during most of Las Vegas' history. But they have grown weary of the expense, and the one obvious way to grant them relief is to make developers cover more of the cost. Local homebuilders estimate that "impact fees"—fees paid by developers to cover the cost of community infrastructure—already account for a quarter of the cost of a new house in Las Vegas.

It was no surprise this year that a budding politician decided to make a name for himself by proposing that Las Vegas solve its growth problems by soaking its developers. The surprise was who that politician turned out to be, and how potent his message proved.

Nine months ago, Oscar Goodman was known around town as the classic Vegas mob lawyer—a veteran criminal defense attorney, given to wearing dark, double-breasted suits and a short-cropped silver beard. Today, on the strength of his soak-the-developer campaign rhetoric, Oscar Goodman is mayor of Las Vegas.

Ever since the mid 1980s, civic leaders here have worked hard to bury the Bugsy Siegel image—downplaying the gangster past, building suburban-style homes and office parks at a furious pace, and generally trying to re-position Vegas as an affordable, high-energy Sun Belt city attractive to everyone from retirees to young families. A few years ago, the *New York Times Magazine* heralded this new era in Las Vegas by reporting the city's transformation "from vice to nice."

It's true that the gambling industry remains the local economy's bedrock foundation. Las Vegas boasts 110,000 hotel rooms—one for every 11 residents of the region—and attracts 30 million visitors a year. But a fresh-scrubbed image was necessary to catapult Vegas past its sleazy resort-town reputation and support a new wave of mainstream urban growth. And so even as it tried to remain affordable, Las Vegas began to go upscale as well.

Perhaps the most highly publicized success story is Summerlin, the Howard Hughes Corp.'s 35-square-mile development on the valley's west side. On a tract of land that was nothing but empty desert when Hughes bought it, 35,000 people now live in the earth-tone houses and apartments lining cul-de-sacs. The company predicts the population could reach 180,000 by buildout. And lower-end Summerlin knock-offs line the roads leading from the big development toward downtown Las Vegas and the Strip.

Summerlin's sales literature boasts of a school and recreation system that starts with T-ball fields and continues on up to college scholarships for the local youth. "Summerlin," one of the brochures says, "offers a surprising number of public schools and private academies, offering a full range of close-to-home choices for preschool/kindergarten through high school education." It is clever promotion like this that has enabled Las Vegas to become the nation's fastest-growing Sun Belt city without sacrificing its economic base of tourism and gambling. But the cost of all of that growth has been high nevertheless. There was no way to finesse the need to build public infrastructure at an exponential rate. And for most of the past decade, the local governments swallowed hard and arranged for that infrastructure in the old-fashioned way: by raising taxes.

Three years ago, for example, traffic congestion had become so bad that it was decided to speed up completion of the 53-mile Las Vegas Beltway—by 17 years—so it could open in 2003. In Henderson, giant belly scrapers building the road rumble back and forth just beyond the walled-in backyards of brand-new houses.

Amazingly, the $1.5 billion beltway is being finished without federal funds. A 1 percent motor vehicle privilege tax and a "new home fee" of about $500 per house are generating $50 million a year, which goes toward bonds issued to raise capital for the construction.

Seven years ago, Las Vegas had no bus system at all. Today, the Citizen Area Transit system carries 128,000 passengers a day—the same volume as the busiest part of the beltway. Passengers pay 50 percent of the system's costs at the farebox—a higher percentage than in almost any other American city. The Regional Transportation System hopes to begin construction of a 5.2-mile rail line around downtown in 2001—mostly with federal funds and sales-tax revenues—and tie it to a privately funded monorail along the Strip.

Meanwhile, the Southern Nevada Water Authority, which serves as a wholesaler to cities and water districts in the Las Vegas area, is building a $2 billion water delivery project. The project involves a "second straw" from Lake Mead, which Hoover Dam creates about 30 miles southeast of Las Vegas, and 87 miles of large water mains in the Las Vegas Valley. The water agency's goal is to provide enough water for an additional 2 million people.

There is one catch so far as water supply is concerned: Las Vegas still lacks the legal right to draw additional water from Lake Mead. Changing this arrangement will require a massive political deal to overturn the 75-year-old agreement among seven Western states along the Colorado River. But that didn't stop Clark County voters from approving a quarter-cent sales-tax increase last

year—expected to generate close to $50 million a year—to pay for the new straw and the other improvements.

So it can't be said that Las Vegas area residents have been unwilling to open their wallets and spend money for growth—they have spent heavily for it. But they realize all too well how much of the bill remains to be paid. This is the issue that is driving Las Vegas politics, and producing its unexpected results.

Even before this year's mayoral campaign, critics of the growth machine began speaking more loudly in the Las Vegas Valley, and their demand that developers pay impact fees had begun to resonate with the voters.

The ringleaders of this new movement have been Jan Laverty Jones, who was Goodman's predecessor as mayor, and Dina Titus, a political science professor at the University of Nevada–Las Vegas, who is Democratic leader in the Nevada Senate. Because Las Vegas has a city manager form of government, the mayor's job was never viewed as important. But Jones, a former Vegas businesswoman who used to appear as Little Bo Peep in television car ads, adopted a high profile—especially on growth issues—at exactly the moment when growth was creating massive frustration.

Jones pushed to require developers to bear more of the cost of parks, roads and other community improvements. Meanwhile, Titus took an aggressive approach in the legislature. In 1997, she floated a "Ring Around the Valley" urban growth boundary concept. That bill failed, but last year she managed to push through a bill requiring local governments to conduct an impact analysis for large commercial and residential projects. Although the bill doesn't require mitigation of impacts, Titus hopes it will encourage local officials in the Las Vegas area to impose mitigation requirements more often and more rationally. "Right now, it's very capricious," she says. "Somebody has to put in street lights, another developer doesn't have to put in anything. It all depends on the political situation at hand."

Into this combustible situation stepped two mayoral candidates with exactly the wrong credentials to deal with it. Arnie Adamsen, a three-term city council member, was a title company executive; Mark Fine was one of the creators of Summerlin. Their ties to the real estate industry could not have been more conspicuous. Both Fine and Adamsen acknowledged that growth was a problem, but refused to recommend that developers pay more of its cost. That opened the door for Oscar Goodman, an unlikely political leader even in a town as unusual as Las Vegas.

At age 59, Goodman had lived in Las Vegas far longer than most of the voters—more than 30 years. A native of Philadelphia, he once clerked for Arlen Specter, then a prosecutor, now a U.S. senator. In the early '60s, however, Goodman headed for Las Vegas, looking for a place where he could make a reputation on his own.

Before long, he gained a reputation as a colorful—and effective—defender of accused mobsters. He once persuaded a judge to drop mob financier Meyer Lansky as a defendant in a case, apparently because of Lansky's failing health. In his most celebrated victory, he kept Tony "The Ant" Spilotro out of jail in the face of multiple murder and racketeering charges. Spilotro became the model for the character played by Joe Pesci in the movie "Casino"; Goodman played himself in the same movie.

Of course, Goodman always insisted that he had a wide-ranging criminal law practice, and only 5 percent of his clients were alleged mobsters. Furthermore, Goodman and his wife were regarded around town as good citizens who often contributed to, or even spearheaded, civic causes. Even so, when Goodman first announced his candidacy for mayor, it was surprising that he even wanted the job.

Given the structure of Las Vegas' government, most mayors have been part-timers who served as glorified presiding officers for the city council. On the other hand, the most recent occupant of the seat had shown that it could be more than a ceremonial position. "Jones made it a more important job," says Eugene Moehring, author of *Resort City in the Sunbelt,* a history of Las Vegas. "She showed that you could use it as a platform. I think Oscar saw that and decided to go for it."

Goodman announced that, if elected mayor, he would give up his law practice. Even so, he wasn't taken seriously at first by the Las Vegas political establishment. Then he started talking about sprawl, real estate developers and impact fees. "Growth has to pay for growth," Goodman said in his radio and television commercials, using his fast-talking West Philadelphia accent to get his message across. "Either it has to come from taxpayers or developers. . . . It's time to make developers pay impact fees when they build new homes."

Goodman argued that even a modest impact fee would bring in more than $15 million a year to help pay for sprawl costs and revitalize Las Vegas' downtown. He recommended a fee of roughly $2,000 for each new home, and said he would use it to revitalize neglected downtown neighborhoods. Such a scheme would require a change in state law—and a big fight with the real estate lobby—but Goodman promised to take on both the legislature and the builders. Polls showed that 80 percent of Las Vegans supported higher development fees.

> # It wasn't ties to organized crime that concerned voters in the Las Vegas mayor's race; it was ties to real estate.

Goodman was widely perceived in Las Vegas as a "Jesse Ventura candidate"—a celebrity from another field with a populist message and enough personal wealth to assure that he would not be bought by special interests. His opponents essentially played into Goodman's hands. Adamsen, his opponent in the June runoff, insisted that the fees would merely be passed on to homebuyers. "Development fees are very popular with the public because they want someone else to pay for it," he said. Adamsen focused on Goodman's criminal connections, saying they would harm the city's new, family-oriented image. If anybody in town knew about extracting money from a community through nefarious means, he charged, it was Goodman, not the developers.

In the end, though, Goodman's contacts with organized crime were a non-issue. In the Las Vegas of 1999, impact fees are a bigger issue than crime, organized or otherwise. What mattered to most voters was that the candidate didn't have any close ties to the real estate business. Goodman won 64 percent of the vote in his runoff with Adamsen. The Jesse Ventura approach worked so well that even many leading members of the Las Vegas political establishment embraced Goodman during the runoff.

On the last Monday in June, Oscar Goodman took the oath of office before an overflow crowd in the Las Vegas City Council chambers, then stepped outside for his first press conference as mayor. At 10:30 in the morning, it was 95 degrees, with a hot, dry desert wind blowing across the City Hall courtyard. Wearing a trademark double-breasted suit, Goodman was unfazed by what was, for Las Vegas, a typical June morning.

And he also seemed unfazed by his quick introduction to the real world of municipal politics in America's most transient city. In the three short weeks between the election and the inauguration, one of Goodman's campaign aides was accused of attempting to charge $150,000 to arrange an appointment to the City Council. The new mayor was faced with the question of what to do about renewing the contract of the city's waste hauler, who had contributed heavily to his campaign. And, of course, during the Goodman interregnum, 4,000 new residents had arrived in Las Vegas.

Goodman insists he will take his plan for a $2,000-a-house impact fee to the legislature. But the next regular session does not begin until January of 2001, and even with 18 months to build support, his task will not be easy. Current state law prevents any such scheme from being implemented—as it does many other development fees—and the real estate lobby has vowed to oppose it.

Meanwhile, says Goodman with his typical glib charm, "I've sat down with the developers, and I've told them we're going to run an efficient City Hall operation that will meet their day-to-day needs. Those guys are going to *want* impact fees to pay us back for all the good things we're going to do for them in the next two years."

Even Goodman sympathizers agree that Goodman—like that other populist, Jesse Ventura—faces a difficult challenge in learning how to govern on the job. "Growth is a tough issue, and he's going to have to hit the ground running on it," says former Mayor Jan Jones. "It's going to be a big learning curve." In particular, Goodman will have to cultivate regional agencies and suburban governments in Las Vegas—with which Jones often clashed—in order to curb growth on the metropolitan fringe and encourage renewal of older areas in Las Vegas itself.

Increasingly, however, there is agreement among both politicians and developers that the future of Las Vegas—like the future of most American communities—involves developers paying for the impact of growth somehow: There simply is no other way to finance the infrastructure costs. Whatever Goodman does, the Southern Nevada Water Authority is planning to levy a hookup fee of several thousand dollars a unit to help pay for the second straw from Lake Mead. That will only increase the average home price in the metro area, which is already up to $142,000.

Oscar Goodman may or may not succeed as mayor, but it's clear that the political sentiment that he tapped into is here to stay. In wide-open Las Vegas—as in so many of the more conventional cities across the country—the crucial votes no longer lie with the upwardly mobile families desperate for a place to live. They lie with the established middle-class residents who are tired of paying the bill.

Can This Swamp Be Saved?

Bold Everglades-protection strategies may revive the river of grass

By LAURA HELMUTH

In 1905, Florida elected Governor Napoleon Bonaparte Broward, who campaigned on a promise to drain the Everglades. He didn't, but over the next century, others almost did.

Dams, canals, and levees have carved up most of the Everglades, which once covered almost 9 million acres. Everglades National Park protects only about a sixth of the historic Everglades area. Much of the rest has been planted with sugarcane, housing developments, and amusement parks.

Today, the Everglades is at the beginning of the largest ecological restoration effort in history. Many public and private projects are already under way, and in July, the Department of the Interior and the U.S. Army Corps of Engineers will submit to Congress a proposal for a 20-year, $8 billion massive replumbing of South Florida.

The plan, optimistic and desperate at the same time, includes some of the most ambitious public works projects ever. This venture is designed to repair the damage from another one of the world's largest public works projects, the corps' midcentury effort to reengineer the Everglades.

"We're at a crossroads right now," says Michael L. Davis, deputy assistant secretary of the Army for civil works.

"We have the opportunity to reverse 50 to 75 years of degradation of the Everglades ecosystem."

The Everglades once meandered over most of South Florida. The flat state is rimmed with slight rises on the east and west coasts, creating a wide, shallow valley. To the north, and slightly uphill, Lake Okeechobee released water that mingled with rain to form a wide, slow-moving "river of grass," as early conservationist Marjory Stoneman Douglas named the Everglades.

Imagine a grassy sheet of water 60 miles wide and 6 inches deep. A given drop of rain could take a year to glide south from Lake Okeechobee to the Florida Keys.

In the age of ecological awareness, air conditioning, and DEET mosquito repellent, the Everglades inspire awe. Early settlers, however, saw the Everglades as one big, soggy, malaria-infested impediment to prosperity. In the mid-1880s, the state offered Everglades land, cheap, to anyone who would drain it.

Natural disasters, too, impelled Floridians to tame the Everglades, says Lance Gunderson of Emory University in Atlanta. Flooding from a 1928 hurricane drowned thousands around Lake Okeechobee, some 40 miles from the coast. An extraordinary wet season in 1947 dumped 12 to 13 feet of rain on the Everglades watershed, flooding 90 percent of southeastern Florida. The next year, Congress sent the U.S. Army Corps of Engineers to war.

The corps accepted two missions: to control flooding and to supply freshwater, primarily from Lake Okeechobee, to the state's growing population. Twenty years of building canals, dams, levees, and locks fragmented the Everglades' natural flow.

The engineers mounded dams to break the Everglades up into small, easily drained plots. They built a network of canals to sweep water east to the Atlantic Ocean and west to the Gulf of Mexico. The corps diked seeping Lake Okeechobee into a reservoir to supply freshwater to Florida's farms and booming cities during the dry season and to dump excess water into the canals during the wet season.

"The [Army corps] project was very successful," says Davis. "It protected against floods and provided water for the population and for agricultural needs. It also had tremendous unintended consequence—draining the Everglades."

An invasive species lurks at the gates

Subtropical southern Florida, with its moist, warm, teeming ecosystems, is a natural target for plagues of nonnative species, says Nick Aumen of the South Florida Water Management District in West Palm Beach. The Asian swamp eel, which may have been released from hobbyists' aquariums, is poised to become the Everglades' fiercest invader.

The eel breathes air, grows up to 3 feet long, slithers across land, and voraciously eats fish, frogs, and crayfish. It now swarms in Miami canals, just one flood-control gate away from entering the Everglades. That gate is scheduled to open in June, when the rainy season starts.

No one knows how to stop the eel, but Florida officials may try to block it with a barricade of electricity coursing in front of the gates. Wildlife managers in the Pacific Northwest have prevented nonnative lamprey eels from migrating into the Columbia River basin by using electric barriers. —*L.H.*

U.S. Geological Survey

The hearty, 3-foot-long Asian swamp eel could devour many of the Everglades' native species.

Fifty years after the original South Florida water management plan, the corps and the South Florida Water Management District (SFWMD) have completed a new study that will form the core of continuing Everglades restoration efforts.

The 3,700-page draft of a report to be officially presented to Congress this summer proposes ways to mimic natural water-flow patterns in what remains of Florida's wetlands. It recommends essentially replumbing the southern end of the state.

"We get 55 inches of rain in an average year," says ecologist Nick Aumen, research director of SFWMD in West Palm Beach. During monsoon season, from May to October, 85 percent of the year's rain falls. The hundreds of water-control structures in and around the Everglades have disrupted the levels, timing, and flow of freshwater, Aumen says.

Everglades wildlife depends on this seasonal cycle, says ornithologist Stuart L. Pimm of the University of Tennessee at Knoxville. During nesting season, wading birds need relatively low water levels that concentrate fish in shallow pools. The birds situate their nests near these reliable food sources. Too much drainage, however, has created frequent bone-dry periods from which fish populations can't bounce back. Both the fish and the birds suffer.

Artificially high water levels can also harm wildlife. The endangered Cape Sable seaside sparrow, for example, builds its nest a few inches above the water. When the water in a parcel of Everglades rises too rapidly, the nestlings drown.

Populations of wading birds in the Everglades are down at least 90 percent in the past 50 years, largely because of changes in water flow. Even in the 15 years that Pimm has been doing aerial surveys of the park, he has seen the number of wading birds decline.

Problems stemming from the artificial water control extend beyond the Everglades. During the wet season, Lake Okeechobee overflows into canals that shunt its water into brackish coastal estuaries. The influx of freshwater kills oyster beds and sea grasses there.

In order to restore the Everglades and still provide enough freshwater for South Florida's growing population, the corps must reclaim the water flowing from Lake Okeechobee into the ocean each year. Under the state's Everglades Forever Act, passed in 1994, Florida has been buying up land ringing the lake. The state is creating new marshes to reserve and filter water for wetlands year-round.

One of the most creative, untested water-storage techniques proposes plumbing on the grandest scale. One thousand feet below Florida's surface lies the briny Floridan aquifer. The corps suggests drilling hundreds of passages into this aquifer and pumping freshwater into it during the wet season.

The less dense freshwater should float in recoverable pockets above the native brine. When the dry season begins, engineers can reverse the pumps and bring the stored water back to the surface to replenish wetlands, farms, and city water supplies.

In effect, South Florida would create a deep, massive, underground parking garage for its freshwater. A complicated system, indeed, but with the high price of South Florida real estate and a population expected to double in the next 50 years, this injection-well solution may be cheaper than ground-level reservoir alternatives, says Aumen.

Other, less futuristic projects are already restoring the state's original waterways. For the sake of wildlife habitat, the corps, which once straightened the Kissimmee River, is helping it snake back to its natural course. Engineers have begun filling in the trench that their predecessors had dug to speed the water flow. Locks on the two ends of the river, where it meets the Kissimmee Reservoir and Lake Okeechobee, will let the intervening stretch slowly get its kinks back, says Tom Adams, senior policy advisor for the National Audubon Society in Washington, D.C.

The old stereotype about the U.S. Army Corps of Engineers, Adams says, is that they "dig, ditch, drain, and dam, and to hell with everything else." Under their current environmental protection and restoration mission, says Adams, "clearly, that doesn't apply."

The Everglades is a fussy system. It needs very clean water. Excess phosphorus from agricultural runoff, primarily from sugarcane fields, throws the natural plant systems out of balance. The Everglades carpet a limestone bedrock, which historically has bound up most phosphorus and kept it out of the ecosystem. Native plants and animals that evolved in this low-phosphorus condition have been poisoned by runoff or crowded out by phosphorus-friendly species.

Cattails, for example, flourish in phosphorus-rich waters, and the land downstream of agricultural areas has turned into a "monoculture of cattails," says Aumen.

Faced with state and federal lawsuits, the sugarcane industry has drastically reduced the amount of phosphorus flowing into the Everglades during the past few years, says Aumen. Sugarcane growers now use less fertilizer and have altered their crop-rotation schedules and irrigation ditches.

The most dramatic change in phosphorus runoff, however, came when growers changed their procedures for draining cane fields. Rather than beginning to pump water off their fields as soon as it starts to rain, they now wait until 1 inch of rain has fallen, says Aumen.

The reduced pumping hasn't hurt sugarcane productivity, and it allows more of the phosphorus to settle into the soils rather than flow across the state.

The state restoration fund has invested $800 million over 6 years to buy land and develop marshes that filter agricultural runoff before it reaches the Everglades. One 3,500-acre prototype of a filtering marsh began operating 4 years ago. SFWMD plans to open three more artificial marshes this summer and is designing two more.

"This [complex of marshes] is the largest constructed wetlands in the world," says Aumen. The prototype marsh filters out about 70 to 80 percent of the phosphorus in the water—not enough, says Aumen, but a good start.

The water-management district is evaluating technologies to clean out even more phosphorus before the water flows into the Everglades. Options range from highly engineered plans, such as building chemical treatment plants among the wetlands, to relatively natural strategies, such as submerging mats of algae and vegetation in constructed marshes to trap phosphorus.

The corps' ambitious plan recognizes three basic objectives: to improve the quality of water, to increase the amount of water available for natural systems, and to "reconnect the parts" of the Everglades.

In South Florida's patchwork of pools, levees, dams, and canals, water moves through pipes and gates, rather than flowing freely over the land. One side of Tamiami Trail, the road that cuts through the Everglades from Naples to Miami, can be knee-deep in backed-up swamp water while the other side is dry enough to burn.

The barriers also confine animal populations. "Alligators don't cross Tamiami Trail," says hydrologist Tom Van Lent of the Everglades National Park in Homestead, Fla. The northern gators don't mingle in the southern gators' gene pool, a situation that eventually could lead to genetically isolated populations. Researchers may already be seeing genetic drift in mosquito fish on different sides of levees, says Van Lent.

Pimm worries that the proposed restoration of the Everglades doesn't go far enough. It will create "a series of very small, managed pools instead of letting this wonderful river of grass flow along natural flow patterns."

The proposed plan calls for ripping up some barriers and filling in some canals, but it's not yet clear whether there will be a net reduction in the number of structures controlling water flow through the Everglades. Tamiami Trail may be raised to allow water to flow underneath, but no one can say yet when it will happen or how much of its expanse will be lifted.

Reconnecting the compartmentalized Everglades is "like putting Humpty Dumpty back together again," says Adams.

Development hasn't shattered the west coast of Florida into as many pieces as the east coast, says David E. Guggenheim, president of the Conservancy of Southwest Florida in Naples. "Right now, we're trying to avoid [this] unwanted destiny—canals, berms, and expensive restoration efforts. Over here, we still have natural flow-ways."

In one of the conservancy's most ambitious efforts, it's tracking down the owners of Golden Gate Estates. These seemingly worthless parcels of land were sold in the 1950s and 1960s, largely sight unseen, to folks who hadn't yet gotten the joke "I've got a piece of swampland in Florida I'd like to sell ya."

The conservancy has begun to buy back and protect these 53,000 acres of prime habitat for black bears and endangered Florida panthers.

Politically, the Everglades has won over many friends during the past century. In Florida's last election cycle, all the candidates for state and federal offices supported Everglades restoration. Several counties, including most recently Palm Beach County, have passed bond initiatives to buy land for water storage and buffering the Everglades from urban areas.

The corps' replumbing plan already has widespread support in Congress, says Rep. Dan Miller (R–Fla.). "This should not be a partisan issue," he says.

The full $8 billion plan may be a hard sell, however. "There's going to be some sticker shock on this, no doubt," says Adams, "but nobody is saying we shouldn't be doing this."

Even Broward County, named for Florida's swamp-phobic governor and famous for the boundless development in suburban Miami, has begun to limit its sprawl, says Adams.

Bumper stickers protesting overdevelopment say, "Stop the Browardization"—a slogan that also applies to the effort to reverse a century's worth of Everglades decline.

A greener, or browner, Mexico?

CIUDAD JUAREZ

NAFTA purports to be the world's first environmentally friendly trade treaty, but its critics claim it has made Mexico dirtier. There is evidence on both sides

GIVEN the industrial invasion that the North American Free-Trade Agreement has brought to the cities that line Mexico's border with the United States, one might expect the skies of Ciudad Juarez to be brown with pollution, and its watercourses solid with toxic sludge. But no. The centre of Ciudad Juarez looks like a poorer version of El Paso, Texas, its cross-border neighbour: flat, dull and full of shopping malls. Since NAFTA took effect on January 1st 1994, Mexico has passed environmental laws similar to those of the United States and Canada, its NAFTA partners; has set up a fully-fledged environment ministry; and has started to benefit from several two- and three-country schemes designed to fulfil "side accords" on the environment—the first such provisions in a trade agreement.

But the debate over NAFTA's impact on Mexico's environment remains polarised. Certainly NAFTA has had the hoped-for result of encouraging industries to move to Mexico. Since 1994, the number of *maquiladoras* (factories using imported raw materials to make goods for re-export, usually to the United States) has doubled, and half

of the new factories are in the 100km-wide (60 mile) northern border region.

But that has put pressure on water supplies. The influx of migrant workers is slowly drying out cities like Juarez, which shares its only water source, an underground aquifer, with El Paso (which has other water supplies). Growth is aggravating old deficiencies: 18% of Mexican border towns have no drinking water, 30% no sewage treatment, and 43% inadequate rubbish disposal, according to Franco Barreno, head of the Border Environment Co-operation Commission (BECC), a bilateral agency set up under NAFTA.

Bubbling Mexico's *maquiladora* plants:

Sources: Mexican trade and industry ministry; United States Environmental Protection Agency

*to the United States

Putting that right, which is the BECC's job, will cost $2 billion–3 billion.

The problem of toxic waste is more serious still, and is being ignored. Mexico has just one landfill site for hazardous muck, which can take only 12% of the estimated waste from non-*maquila* industry; the rest is dumped illegally. Mexican law says *maquiladoras* must send their toxic waste back to the country the raw materials came from but, with enforcement weak and repatriation expensive, many are probably ignoring the law, says Victoriano Garza, of the Autonomous University of Ciudad Juarez.

Mexico has long been a dumping-ground for unwanted rubbish from the United States. But in fact less than 1% of the country's toxic gunge is generated in the border region. After 1994, despite the growth in *maquiladoras*, Mexico's exports of toxic waste to the United States dropped sharply—suggesting that tougher rules were making companies adopt greener manufacturing policies. But then, in 1997, waste exports mysteriously shot up again (see chart).

Even greens are split about NAFTA. "In 15 years as an environmentalist, I haven't noticed any change in policy," laments Homero Aridjis, a writer. Quite the contrary, says Tlahoga Ruge, of the North American Centre for Environmental Information and Communication: "NAFTA has brought the environment into the mainstream in Mexico, as something to be taken seriously."

Such is the complexity of the issue, only in June did the Commission for Environmental Co-operation (CEC), a Montreal-based body created in 1994 to implement the environmental side-agreement, publish its methodology for assessing NAFTA's impact. An accompanying case study, on maize farming in Mexico, illustrates the difficulties. Lower protective tariffs are forcing Mexico's subsistence maize farmers to modernise, change their crops or look for other work; each choice has potential impacts, such as soil deterioration, depletion of the maize gene pool, or migration to cities. Social and environmental impacts go hand-in-hand.

Critics say that, as well as being slow, the CEC has too few powers. It can investigate citizens' complaints about breaches of each NAFTA country's national environmental laws, but governments are free to ignore its recommendations. Moreover, the side-agreement's gentle urgings are outgunned by the main accord's firm admonition, in its Chapter 11, to protect foreign investors from uncertainty. A recent report by the International Institute for Sustainable Development, a group based in Winnipeg, Canada, warns that this provision is increasingly being used by business to the detriment of environmental protection. And unlike the CEC, the arbitration bodies that settle Chapter 11 disputes make binding decisions.

Companies have challenged environmental laws in all three NAFTA countries, but Mexico is particularly vulnerable. Its laws are new, its institutions untested and its environmental culture undeveloped. To these handicaps was added the collapse of Mexico's peso in 1995, from which it is still trying to recover. Beefing up environmental institutions has not been its top priority. In Juarez, for example, there are only 15 federal environmental inspectors (and they cover the whole of Chihuahua state). Mexico also lags in data-collection: next week the CEC is due to publish a report on industrial emissions and air pollution, but in the United States and Canada only, because Mexican factories do not have to report their emissions.

These things need to change. NAFTA is supposed to lead to completely free trade after 15 years; its full impact has yet to be felt. In 2001, the *maquiladoras* will lose some of their tax breaks. Increasingly, American firms may opt to set up ordinary factories in Mexico, which would not have to send their waste back home. Mexico badly needs somewhere to put this stuff, but so far attempts to find sites for new toxic-waste landfills have been scuppered by not-in-my-backyard opposition (and, ironically, by the strict stipulations of the new environmental laws).

But not all is murk. Besides better laws and institutions, Mexico has seen a rapid rise in non-governmental organisations, themselves partly an import from the north, which work on the education and awareness-raising that they say the government is neglecting. Ciudad Juarez may be running out of water to drink, but Mexico City's aquifer is so low that the whole place is slowly sinking. And, unlike the capital, Juarez benefits from all kinds of two-country efforts to fix its problem, thanks mainly to NAFTA.

Unit 3

Key Points to Consider

❖ To what regions do you belong?

❖ Why are maps and atlases so important in discussing and studying regions?

❖ What major regions in the world are experiencing change? Which ones seem not to change at all? What are some reasons for the differences?

❖ What regions in the world are experiencing tensions? What are the reasons behind these tensions? How can the tensions be eased?

❖ Why are regions in Africa suffering so greatly?

❖ Discuss whether or not the nation-state system is an anachronism.

❖ Why is regional study important?

 **Links**   **www.dushkin.com/online/**

These sites are annotated on pages 6 and 7.

The region is one of the most important concepts in geography. The term has special significance for the geographer, and it has been used as a kind of area classification system in the discipline.

Two of the regional types most used in geography are "uniform" and "nodal." A uniform region is one in which a distinct set of features is present. The distinctiveness of the combination of features marks the region as being different from others. These features include climate type, soil type, prominent languages, resource deposits, and virtually any other identifiable phenomenon having a spatial dimension.

The nodal region reflects the zone of influence of a city or other nodal place. Imagine a rural town in which a farm-implement service center is located. Now imagine lines drawn on a map linking this service center with every farm within the area that uses it. Finally, imagine a single line enclosing the entire area in which the individual farms are located. The enclosed area is defined as a nodal region. The nodal region implies interaction. Regions of this type are defined on the basis of banking linkages, newspaper circulation, and telephone traffic, among other things.

This unit presents examples of a number of regional themes. These selections can provide only a hint of the scope and diversity of the region in geography. There is no limit to the number of regions; there are as many as the researcher sets out to define.

Paul Starrs's thought-provoking essay on the importance of place and the concept of region leads this unit. Then, "The Rise of the Region State" suggests that the nation-state is an unnatural and even dysfunctional unit for organizing human activity.

"Europe at Century's End" and "Russia's Fragile Union" both deal with changing geopolitical situations. James Hooper discusses the ongoing dilemma in Kosovo and highlights the importance of the United States in that country. "Sea Change in the Arctic" analyzes the impact of global warming in the region. The continuing rise of Greenville, South Carolina, is documented in the next article. John Cobourn summarizes flood types in the arid western United States. Then, "Does It Matter Where You Are?" considers aspects of geographical location principles in the context of the new global economic systems.

The Region

The Importance of Places, or, a Sense of Where You Are[1]

In this mobile society where every shopping mall has the same stores, can we really say there are differences among the country's regions?

Paul F. Starrs

Paul F. Starrs is professor of geography at the University of Nevada.

"A time and a place for everything" goes a well-weathered adage. The meaning of time in this spare phrasing has never warranted a second thought. Unvarying since the sundial, time is a comfortable given governed by planetary rotation and counted through the cycle of seconds, hours, days and months. If the exact interval captured by a throwaway term like "just a moment" fluctuates depending on whether you're Hopi, Huron, Hasidic, a homeboy, 3 years old or 102, a clock nonetheless ticks through 24 inviolable and orderly hours in a day.

But consider again that initial sentence: While time is known, it is without doubt "a place" that is less defined. Can the region—a place broadly construed—really have any universal meaning, or is it just a vaguely named sacrament in the church of location? What are regions, and how should we receive them in an ultra-modern era of mass communication?

A thoughtful mention of any "region" immediately takes on a burden of complicated assumptions. Europe is a vexatious and fitting example. That it exists. is incontestable. And yet, what exactly is Europe? A separate continent surely Europe is not; a uniform economic organization also assuredly no. True, Europe counts as its own a motley collection of highly assorted, if generally Caucasian,

peoples (Basque, Catalan, Friesian, Welsh, Walloon), with a presumptive (if hardly universal) Christian religious heritage, and claims can be made to a very grossly familiar linguistic heritage. Beyond that, Europe is an eclectic assemblage whose political sovereignty is in essence a status deigned by United Nations recognition—Norway, Ireland, Portugal, the Netherlands, San Marino, Bosnia, Macedonia, Greece. These disparate parts amount to a decidedly unseemly whole. So if Europe is genuinely a single region, then what is the binder for its constituent parts? Obvious answer: "Europe" is, fact, a convention, a useful ploy hearkening to a generally common history that includes the World Wars, the Holy Roman Empire, Jenghis Khan's depredations, the Neanderthal and the European Economic Community (except for the countries that have not been allowed to join or elect not to). With such a messy match, Europe as a region is more organizing thought than any demonstrable fact.

While the "new regionalist" geographers, the breed of scholar whose domain generally is said to include places and regions, have of late been self-eviscerating like a sea cucumber over just what constitutes "a region," most essayists and geographical scholars agree that the region as a unit of analysis and description is basic and not to be rubbished. Remaining as a category fluid, elusive and mutable, regions are entrenched in common thought and vernacular speech.[2]

Reprinted with permissions from *Spectrum: The Journal of State Government*, Summer 1994, pp. 5–17. © 1994 by The Council of State Government.

As with fashions for couture and academe, regions erratically fall in and out of favor. They are today's hot ticket, especially as renewed in the guise of "regionalism," which last peaked in the United States during the 1940s heyday of the Tennessee Valley Authority, when economic needs were thought best explored in a great restructuring of the country into socioeconomic regions that had their own distinctive personalities (Odum & Jocher 1973; Campbell 1968; Dorman 1993; Archer 1993).[3]

Global forces both dismember and contribute to regional identity. In the 1990s, a major point of contention in academic geography is how to parse, map and understand the connections of regions to booming inter-regional phenomena: The globalization of finance (capital), information (cyberspace), communications (cellular, FAX, satellite), transportation (frequent flier flights and high-speed rail), language (the hegemony of American English), or culture otherwise construed (MTV, grunge, CNN, Rupert Murdoch's newspapers, music).[4] An accurate rendering would show the globe slashed by vast arrows of movement that reshape and deform, describing essential and ongoing patterns of geographical change. While post-modernists like Doreen Massey are able to write intelligently of "a new burst of time/space compression," a skeptic would reply that this "burst" is merely an acceleration of the same foreshortening of the world's borders, a death of certainty and control, which started in 1492.

For all the glitz of global systems and parlance about the world economy, regions remain vital. Teasing out the singularities and generalities of places are among the foremost skills that anyone, from politician to market researcher to planner, can trot out to make sense of the planet. If the Earth is not simply made up of nugget-like places that embody perfectly consistent traits, the region amounts to the most resourceful, utilitarian and creative of lies.

WHEN YOU'RE THERE, YOU'LL KNOW

Regions are, in general, more useful than real. What often are taken to be time-honored physical wholes—like the Southwest, the short-grass prairie, the Mid-west and the Middle East—turn out to have sloppy edges (Meinig 1971, Said 1979, Shortridge 1985). If a mental map, an image of shared traits and cultural stereotypes, is firm, when it comes actually to mapping boundaries and attributes the cartography of any region loses exactness. An old-time re-

gional geographer could counter by saying that ephemera like planning districts, culture areas and homelands are hardly regions at all. By conservative reckoning, the only region is something with a self-evident physical presence. While for some of the tire-kicking school there remains a real pleasure toiling to map the Basin and Range physiographic province, imprecision is hardly a fatal flaw. Mathematicians working with fractals, physicists with sub-atomic particles and philosophers charting the ideological canyons of the mind are quick to admit that fuzziness can be more useful than the obsession of an accountant or a scientist with calculator exactitude. As with the edges of forests, deserts, tundra or other ecotones, the margins of a region frequently have the highest diversity and interest, and all borderlines say much of what lies inside.

As organization structures rather than geographical certainties, regions embody events, emotions, physical similarities, human activity, or history and economy. Find one unity or several, and the geographic elaboration of a region begins. Many of the "new regionalists" argue regions are social constructions, revealing economic or class practices. Regions, which are interesting primarily as a physical setting for social interactions, suffice, in essence, as a game board (Goldfield 1984, Thrift 1990, Pudup 1988). Others are content if a regional description stockpiles plenty of room for change, which regions do (Gilbert 1988). All regions are effective manipulations of nature, casting and organizing human practice and gathering up bundles of traits to make sense of our presence on earth. If regions did not exist, they would have to be invented. And because we need to know where and what the manifold human creations are, we have regions. There is poignancy to this metaphysics, as with any discussion of scholarship's fastidious fashions.[5] The reality is straightforward: Regions exist because we want, need and relish them.

The inherent problem with regions is as simple as it is true. Geographers, like the sociologists, conservation biologists, anthropologists and historians who have joined in the traffic of places, worry about criteria used to identify a region (Hough 1990). Are the keys solely economic? Religious? Ethnic? Territorial? Do regions originate in history or are they entirely contemporary? If an area where overlapping attributes meet forms a region, then the edges are a pronounced problem. Are places defined by watersheds and biological life—bioregions—a valid means of defining a region? Do the world's nations, discrete groupings of peoples that exist within and among the larger political entities that convention calls "countries" (think of Navajo, Provencal, Scots or Zulu territory), constitute regions? (Nietschmann 1993).

Predictably, questions raised are more persistent than easy. How should we contend with larger social and structural forces defining a region? Each place is linked to others. From the Internet to the global economy, from the tracing of transnational money flows to the Hollywood movies that bully the French cinematic self-identify, even the smallest Appalachian hollow, Mormon village and South Central Los Angeles gang are connected to additional parts of the world. The most erudite sort of formulations discussing the relationship between the local and the global admit that regions are indispensable (Lipietz 1993).

There is space aplenty for floruishing skepticism. Whenever a region is identified it is easy to carp about the criteria used to single it out.[6] Regions are ultimately a state of mind, a convention. They exist in untold numbers, interwoven and overlapping. And while scholars squirm about the squishy boundaries, if you're there, you generally know. Having a sense of where you are is not just street smart, it is a survival trait through human history and geography.[7]

Debate over regional structure turns around a problem that social theorists call reification, or, in clinical terms, setting a region in stone (Jameson 1981). Places are no more uniform than the people within them. Talking about the West or the Bible Belt or the Colorado Plateau or the Chicken Fried Steak Line is no more defensible than prattling on about what "American Catholics believe," or "Cuban Americans argue," or "environmentalists claim." Catholics, Cubans and environmentalists are many and diverse. By the same token, people who fall into a designated region do not all hold to whatever is being attributed to them. But regional characterizations allow each and every person to make sense of the world, and the categories developed often are remarkably acute in their capturing what is important about people who are, after all, geographically located on a discrete part of the earth.

Regions literally hold together. But each region is also part of larger processes and interactions. A secondary lesson is, then, simplicity itself: Always examine the meaning and assumptions behind any region. Almost certainly the region is real. How have those come to be, and what do they mean? Much elegant prose is devoted to the description of regions. In fact, some of the best writing, going from Joel Garreau to Jan Morris to John McPhee to Gretel Ehrlich, delves deeply into regions and their sumptuous character. But analyze always, for a region can be developed as a scientific fact as easily as an emotional and poetic necessity.

THE USES OF REGIONAL IDENTITY

Modern politics is replete with examples of regions created in an attempt to control problems too large for the traditional political structures of towns, cities and countries. Thrown together of opportunism and necessity, these utilitarian creations have become essential political and economic facts. There are regions, however, that are literary and spiritual so much as real. Places do have life, and regions have identity—as, for example, has the American South (Cash 1941; Odum 1947; Wilson & Ferris 1989). The range of forms and purposes behind singling out regions illustrates problems of the age. Regions meet an impressive assortment of ends. The past uses, boundaries and creators of regions prove as illuminating as where and how regions today are constituted. A region always accommodates its particular time and is build within the limitations of available space. For all that, there are six regional categories that are characteristically contemporary, and a few words on each will suggest why regions are likely to stick around. What distinguishes them is that each has to be mapped to have any meaning: Regions are pure and simple geographic creations, and however their charter may be construed, for each, the map is the territory, and the accuracy of the map directly reflects on the capacity and vigilance of the creator.

THE ECOSYSTEM

After a century of trying to manage natural resources according to boundaries that have far more to do with political accident, land division history and convenience than with biological necessities, government resource managers and scientists are attempting to piece together large coherent natural bodies. Unsurprisingly, this generally occurs in areas that are in some degree of crisis. Designated "ecosystems" have obvious value. While two—the Greater Yellowstone Ecosystem and the Everglades Ecosystem—are built around national parks, there are other prominent attempts, including the Forest Service-developed "Sierra Nevada Ecosystem Project" (SNEP, more commonly), which is struggling to bound and understand a swarming human presence in the Sierra Nevada foothills of California and Nevada. Ecosystem studies try to transcend politics, seeking biological and planning alternatives to the threats of unbridled growth in environments that are subject to an unusual degree of hazard, and which cannot be planned within the older city or country boundaries.

These vaunted "ecosystems" are by and large the creations of scientists-cum-managers, and they therefore fulfill and reinforce their own public credibility. Ironically, what the scientific land managers keep rediscovering is an old Alfred Korzybski line, "the map is not the territory, and the name is not the thing named" (Bateson 1979). Sadly, naming an ecosystem and designating experts to study it goes only a teeny distance toward solving crises often political, social and economic in nature. In being so vexedly human, problems are rarely readily accessible to the gimlet eye of control by scientific edict. Other recognizable ecosystems have not fared well—Amazonia, the Sahara, the Aral or the Caspian sea—in part because they extend into multinational space. Considering that earnest efforts like the "Biosphere Preserves" program of UNESCO are well-intentioned, it is obvious that conservation efforts within countries are far easier to develop and enforce than protections offered under international treaties or covenants.

THE REGIONAL AUTHORITY

Among the most successful regions are those embraced by regional authorities, which are created to handle questions ranging from transit problems, sewage, and hazard abatement to conservation districts, water systems, and comprehensive area planning. These umbrella agencies, as they are often called, have on occasion met with surprisingly good results. An experiment extending back decades, successful regional authorities accommodate themselves to political reality, but also can weather storms that might dislodge more local agencies.

Like corporations after passage of the 14th Amendment to the U.S. Constitution, regional authorities have come to be as real and authoritative as elected officials. They became, in fact, nearly human; physically real through legislative midwifery. In general, regional authorities are created when a job is too large (read costly) for a single political entity to handle, or when what is sought is a planning window directed far enough in the future that there is no apparent harm in many different political entities banding together for discussions. Politics and planning creates the region.

There are wonderful examples of regional authorities. The Tennessee Valley Authority and the Bonneville Power Authority are two historic cases, but others are less clouded with time. The Metropolitan Water District of greater Los Angeles is prime territory; born of the successes of the Los Angeles Department of Water and Power; which preceded the Metropolitan Water District. Together, these regional entities have done the impossible, bringing reliable and unchallengeable water to 12 million in the middle of a desert. Public utilities often are blended into regional authorities, thanks in part to their being unreachable under antitrust statutes. Their boundaries are political creations: Census tract lines, county divides, the foreseeable edges of planning areas or unincorporated city limits. Other examples of these sorts of authorities include regional park districts, regional planning authorities and the occasional super-agency, like the California Coastal Commission and its allied Coastal Conservancy. They have charters buried in time, but have assumed as charge ruling on the environmental and development future of a thousand miles of California coastline. The authority of these regional entities ebbs and flows with the urgency of constituency concerns, but can be vast.

THE VERNACULAR REGION

For all the unnaturalness of some planning and political regions, there also are valiant and ongoing attempts to recognize and name places that are plainly geographically independent and coherent. The clearest attempt comes with analyzing common speech and asking the residents of different areas where they see themselves living. This leads to the bounding of "vernacular regions," areas where much of the population has little doubt about where they live and who they are. When cast in large terms, this can establish carefully documented boundaries around common places like "The Midwest," of "The Southwest," or it can lead to the recognition of sub-regions like the Panhandle, the "Oil Patch," or the German Hill Country of Texas (Jordan 1979, Shortridge 1987).

A vernacular approach is the simplest regional take—people in and around a large area are asked to toss forward names, and the answers mapped. The resulting lines, or isonyms, designate areas where there is common acceptance of a regional name. The subtleties of use and meaning in this method can be telling—while residents of El Paso, Presidio, Pecos or Wink are given to consider themselves Southwesterners, the same can hardly be said of Texans who hail from Texarkana, Tyler, Pineland or Port Arthur, who hearken to the South. In matters of outlook on race or ethnicity or religion, in where people go for major shopping, or how area residents self-identify with a broad range of political, economic and social issues, these things matter.

BIOREGIONS AND WATERSHEDS

Offshoots of the physiographic regions of the 19th century, bioregions and watersheds have found a great and growing constituency among a number of contemporary essayists and poets like Gary Snyder; who press for a return to closer forms of community than anything favored by large cities (Parsons 1985). The bioregionalists hold that the good fight is best begun at home.[8] At its best, the reasoning has echoes of Pestalozzi and the reformist educators of the 1800s, who argued that learning was best conducted using the terrain, plants and animals—a local habitat—as ledger and classroom (Pestalozzi & Green 1912). Speaking (and especially writing) with passion and great literacy for attending to geographically immediate needs and understanding the physical whole of the watershed or bioregion ahead of vainly attempting to control the world's whole surface, bioregionalists pledge that community comes first. Although James Malin issued similar plaints in decades past, and Ray Dasmann placed the bioregionalist and "ecosystem" traditions in able contrast nearly 20 years ago, the rhetoric is hard to counter (Dasmann 1976; Opie 1983, Bogue 1981.)

In essence, bioregionalists, many of whom also preach for a "watershed consciousness," note that too much of humanity's destructive exploitation of the Earth is driven by quick movements between dissimilar areas, which feeds growth by acquiring and exploiting colonized places, emphasizing profit over knowledge. Know a place, and you are less likely to abuse it, goes the ethos. To understand environmental history is to grapple with local needs. Citing examples from traditional peoples who have managed lands around them for centuries with but modest deterioration, bioregional advocates voice a limited respect for the ambitious architects who preach for global environmental harmony. Their resolute business, however, is to get on with saving their own nest before instructing others in how to clean up the Earth. The intelligence and influence of bioregionalists is far in excess of their numbers.[9]

CULTURAL AREAS, ETHNIC REGIONS & HOMELANDS

Interest in coherent bodies of cultural attributes goes back thousands of years; thematic mapping thrives on locating collections of traits, and early thematic maps often plotted fairly singular behaviors. Geographers have reveled in tight areas where distinctive traits are uniquely preserved, and have mapped these for generations (Gastil 1975; Rooney et al 1982, Garreau 1981, Hart 1972). The Mormon Culture Region, Amish or Cajun country, the Bible Belt, the Sun Belt, are typical tips of a vast iceberg. The traits can be religious, political, economic, racial or linguistic. That this form of region is durable goes almost without saying. In effect, residents within the region are singled out for preserving traits that can be singular or archaic. The degree of uniformity varies hugely, place to place, region to region. By and large, the areas mapped in such regions are so distinctive as to be almost beyond argument (Arreola 1993; Conzen 1993).

Mapping regions is far more than an ideal scholarly exercise. Where the overlaps between regions are pronounced, or where there is quick shrinking, conflict or the loss of distinguishing traits can occur. Where the boundaries alter little with time, the area is unlikely soon to disappear to dilution or disturbance. And yet there are oddities. Some culture regions persist despite migration and a loss of language and economy; the Basque, if anything, have grown stronger in the last two decades, largely thanks to political liberalization. What nudges change is never entirely easy to say, but these regions can and will persist. To a point, they are self-selected and self-sustaining.

NODES IN THE GLOBAL EXCHANGE

Finally, regions are important parts of the global system. Whether the connections are those of politics, empire, economics, culture, or religion, there is no doubt that any region, in this era of mass communication, is both beholden and at times hapless before global forces, the powers that Fernand Braudel, Immanuel Wallerstein, William Appleman Williams and Donald W. Meinig have traced with rare wisdom. This bears comment, but also requires the remark. "So what?" In many a sense, this is nothing new; the links are just more obvious and better drawn than 100 years ago, and it requires more will to insist on self-reliance, to put down the cable remote, than before the World Wars. But the articulations between places are old hat, in either a historic or a geographic sense.

The existence of larger connections in no way eliminates the need or sensibility of regions. If anything, the distractions of a beckoning world act to strengthen regions. The emphasis on learning traditional languages (Welsh, Catalan, Hopi) and relearning ways of life once on the verge of extinction likely have never been so strong. Differences have become precious, and to be distinct and have a separate identity is more than faddish. It is to know and have a sense of yourself, even if the adopted identity is more fancied than honestly come by. This is the oddest part of the region as one small segment of the known world—as the boundaries really should be falling, and homogenization growing rampant, instead the ultimate luxury of 1990s society is to "discover your place."

CONCLUSION

. . . time is absorbed into place, and place into mind. The land becomes history, and history becomes thought as people cross space in awareness.
—Henry Glassie
Passing the Time in Ballymenone

The precision of time, whether dispensed by the clockwork grace of gnomic Swiss watchmakers or driven by quartz crystal Swatches, is not the stuff of which regions and places are made. Time is a certainty, and for all that, a subject of hate and dread. "Saving time" is no small act of desperation. But "saving a place" is geography, history and environmental preservation writ large. We live in time, but for places; they are our communities, where horses are trotted or buses colorfully "tagged" by graffiti territorialists. Here historic preservation committees solemnize over the sanctity of the past, arguing about whether "that stone house over by Fleming's" should be torn down; planning commissioners debate regional futures and decide whether a community garden ought to be fostered in the projects. Regions are among the most intelligent acts that we can work with as humans. As Buckaroo Banzai reminds us: "Remember: No matter where you go, there you are."[10] Of many stripes, places matter. And that, most of all, is why regions are relevant.

NOTES

1. Prepared with support from the S. V. Ciriacy-Wantrup Postdoctoral Fellowship in Natural Resource Economics at the University of California at Berkeley, 1993–94.

2. As good a measure as any flows from Current Contents, a vast data base for contemporary journal articles, book reviews and commentaries. Although "bioregion" appears in only one title, "culture region" is found in 44, "watershed" in 321 and "ecosystem" appears in 784 citations. Alas, a voguish term like "post-modern" is referred to in a paltry 166 items. On the other hand, "place" is cited 1,981 times, the word "region" is in 14,890 references, and when "regional" and "regionalism" are added, the count rises to 29,811 articles. Computers offer a false precision, but the bottom line is evident: Places matter.

3. The discipline of sociology had never before, and has never since, ridden so high. That the regionalist project grew moribund with economic recovery after World War II is history. However, the importance of regions (James Madison called them "sections" in his famous Federalist Number 10) is axiomatic through American history, from early days of the Republic to the Civil War. Joel Garreau's *Nine Nations of North America* suggests some of the current interest, but there is more. In California, for example, proposals regularly float through the Legislature and increasingly onto ballot measure referenda, asking for the state to be split into more "rational" divisions. So far, no joy; check in next year.

4. The globalization of information and mass communications is quixotic—as *The Economist* has noted, "And a network is more than links between places, it is itself a place" (Editors, *The Economist*, 1993). Often the changes imposed by such technology are not so direct or self-evident as the technologically-addicted might argue. William Gibson has put it nicely: "She was a courier in the city. . . . Was it significant that Skinner shared his dwelling with one who earned her living at the archaic intersection of information and technology? The offices the girl rode between were electronically conterminous—in effect, a single desktop, the map of distances obliterated by the seamless and instantaneous nature of communication. Yet this very seamlessness, which had rendered physical mail an expensive novelty, might as easily be viewed as porosity, and as such created the need for the service the girl provided." (*Virtual Light*, Bantam Books, New York, 1993): p. 93.

5. Academic debate over the intersections of place, time and space at their worst can toss even a hardened stomach. Perhaps fortunately, Patricia Nelson Limerick has dealt an elegant swat, if sadly unlikely to carry the force of a death blow, to the gamboling semi-literacy of academic fashion in her *New York Times Book Review* essay, "Dancing with Professors: The Trouble with Academic Prose," 31 October 1993, pp. 3, 23–24.

6. Take, for example, the sumptuous category of "the Southland" that has been the name for Southern California for at least five decades. How can Southern California be bounded by any single term—it now reaches into the Mojave Desert, nearly to the Nevada and Arizona boundaries, and is virtually across the Techachapi Ranges, into the San Joaquin Valley, and runs south to San Diego and Tijuana. Yet the term "Southland" is a sufficiently flexible regional label to encompass it all. See Starrs 1988, Davis 1990, and *The Economist* 1994.

7. I will apologize here to John McPhee for borrowing the great phrase that he used to describe the proxemics of Bill Bradley, who was a Princeton basketball player when McPhee first used the phrase. I hope the original author will take no exception to the usurpation of titles.

8. "It is not the ecologists, engineers, economists or earth scientists who will save spaceship earth, but the poets, priests, artists and philosophers," is how Lawrence Hamilton has put it in the Introduction to a volume he edited, *Ethics, Religion, and Biodiversity: Relations between Conservation and Cultural Values.* (Knapwell, Cambridge: The White Horse Press, 1993).

9. The bioregionalists are assuredly NOT to be confused with the "biospherians" whose "space capsule" existence in Biosphere 2 near Oracle, Arizona, shows far more en-

thusiasm for the ecosystem model than a modest biosphere consciousness.

10. The line is Peter Weller's in "The Adventures of Buckaroo Banzai Across the 8th Dimension," Twentieth Century Fox, 1984.

SOURCES

Archer, Kevin, 1993. "Regions As Social Organism: The Lamarckian Characteristics of Vidal de la Blache's Regional Geography," *Annals of the Association of American Geographers,* September; 83(3): pp. 498–513.

Arreola, Daniel D., 1993. "The Texas-Mexican Homeland," *Journal of Cultural Geography,* Spring-Summer; 13(2): pp. 61–74.

Bogue, Allan G., 1981. "The Heirs of James C. Malin: A Grassland Historiography," *Great Plains Quarterly,* Spring, 1(2): pp. 105–131.

Campbell, Robert D., 1968. "Personality as an Element of Regional Geography," *Annals of the Association of American Geographers,* December; 58(4): pp. 748–759.

Cash, W. J. [Wilbur Joseph], 1941. *The Mind of the South;* (New York: Alfred A. Knopf); 429 pages.

Conzen, Michael. 1993. "Culture Regions, Homelands, and Ethnic Archipelagos in the United States: Methodological Considerations," *Journal of Cultural Geography,* Spring-Summer; 13(2): pp. 13–30.

Dasmann, Raymond, 1976. "Future Primitive: Ecosystem People versus Biosphere People," *The CoEvolution Quarterly,* Fall, 11: pp. 26–31.

Davis, Mike. 1990. *City of Quartz: Excavating the Future in Los Angeles* (New York & London, Verso).

Dorman, Robert L., 1993. *Revolt of the Provinces: The Regionalist Movement in America, 1920–1945;* (Chapel Hill: University of North Carolina Press); 366 pages.

The Economist, 1994. "The Point of Los Angeles," [Editorial], *The Economist* [London], 22 Jan., p. 14.

The Economist, 1993. "Make Way for Multimedia," [Lead Editorial] *The Economist* [London], 16 October: pp. 15–16.

Entrikin, J. Nicholas, 1991. *The Betweenness of Place: Towards a Geography of Modernity;* [Critical Human Geography]; (London: Macmillan).

Garreau, Joel, 1981. *The Nine Nations of North America.* (New York: Avon Books).

Gastil, Raymond, 1975. *Cultural Regions of the United States* (Seattle, University of Washington), 366 pages.

Gilbert, Anne. 1988. "The New Regional Geography in English and French-speaking Countries," *Progress in Human Geography* 12:2, June, pp. 208–228.

Goldfield, David R., 1984. "The New Regionalism [Review Essay]," *Journal of Urban History,* February, 10(2): pp. 171–186.

Hart, John Fraser, 1991. "The Perimetropolitan Bow Wave," *Geographical Review,* January, 81(1): pp. 35–51.

Hart, John Fraser [editor], 1972. *Regions of the United States;* (New York: Harper & Row).

Hough, Michael, 1990. *Out of Place: Restoring Identity to the Regional Landscape;* (New Haven, Connecticut: Yale University Press); 230 pages.

Jameson, Frederic, 1981. *The Political Unconscious: Narrative as a Socially Symbolic Work* (London, Methuen).

Jordan, Terry G., 1978. "Perceptual Regions in Texas," *Geographical Review,* July, 68(3): pp. 293–307.

Lewis, Peirce, 1979. "Defining a Sense of Place," *The Southern Quarterly: A Journal of the Arts in the South,* Spring-Summer; 27(3 & 4): pp. 24–46.

Lipietz, Alain. 1993. "The Local and the Global: Regional Individuality or Interregionalism?" *Transactions of the Institute of British Geographers, New Series;* Vol. 18, pp. 8–18.

Massey, Doreen, 1992. "A Place Called Home?" *New Formations* Number 17, pp. 3–15.

Meinig, D. W. [Donald William], 1971. *Southwest: Three Peoples in Geographical Change, 1600–1970;* (New York: Oxford University).

Nietschmann, Bernard, 1993. "Authentic, State, and Virtual Geography in Film," *Wide Angle: A Quarterly Journal of Film History, Theory, Criticism, & Practice,* October, 15(4): pp. 5–12.

Odum, Howard W., 1947. *The Way of the South; Toward the Regional Balance of America;* (New York: Macmillan); 350 pages.

Odum, Howard W. and Katharine C. Jocher; [editors], 1973. *In Search of The Regional Balance of America;* [The University of North Carolina Sesquicentennial Publications]; (Westport, Conn.: Greenwood Press; orig. copyright 1945); 162 pages.

Opie, John, 1983. "Environmental History: Pitfalls and Opportunities," *Environmental Review,* 7(1): pp. 8–16.

Parsons, James J., 1985. "On "Bioregionalism" and "Watershed Consciousness" *The Professional Geographer,* February, 37(1): pp. 1–6.

Pestalozzi, Johann Heinrich and John Alfred Green, 1912. *Pestalozzi's Educational Writings.* (New York, London: Longmans, Green & Co.; E. Arnold.) 328 pages.

Pudup, Mary Beth, 1988. "Arguments Within Regional Geography," *Progress in Human Geography,* September, 12(3): pp. 369–390.

Rooney, John F., Jr., Wilbur Zelinsky, and Dean R. Louder, [General editors], 1982. *This Remarkable Continent: An Atlas of United States and Canadian Society and Cultures;* Cartographic editor John D. Viteck; (College Station, Texas: Texas A & M University Press for The Society for the North American Cultural Survey); 321 pages.

Said, Edward W., 1979. *Orientalism;* (New York: Vintage Books, Random House); 368 pages.

Shortridge, James R., 1985. "The Vernacular Middle West," *Annals of the Association of American Geographers,* March, 75(1): pp. 48–57.

Shortridge, James R., 1987. "Changing Usage of Four American Regional Labels," *Annals of the Association of American Geographers,* September 1987, 77(3): pp. 325–336.

Starrs, Paul F., 1988. "The Navel of California and Other Oranges: Images of California and the Orange Crate," *The California Geographer,* Vol. 28, pp. 1–42.

Thrift, Nigel, 1990. "For a New Regional Geography 1," in *Progress in Human Geography,* June, 14(2): pp. 272–279.

Wilson, Charles Reagan, and William Ferris, [co-editors], 1989. *The Encyclopedia of Southern Culture;* (Chapel Hill: University of North Carolina Press for the Center for the Study of Southern Culture at the University of Mississippi); 1634 pages.

THE RISE OF THE REGION STATE

Kenichi Ohmae

Kenichi Ohmae is Chairman of the offices of McKinsey & Company in Japan.

The Nation State Is Dysfunctional

THE NATION STATE has become an unnatural, even dysfunctional, unit for organizing human activity and managing economic endeavor in a borderless world. It represents no genuine, shared community of economic interests; it defines no meaningful flows of economic activity. In fact, it overlooks the true linkages and synergies that exist among often disparate populations by combining important measures of human activity at the wrong level of analysis.

For example, to think of Italy as a single economic entity ignores the reality of an industrial north and a rural south, each vastly different in its ability to contribute and in its need to receive. Treating Italy as a single economic unit forces one—as a private sector manager or a public sector official—to operate on the basis of false, implausible and nonexistent averages. Italy is a country with great disparities in industry and income across regions.

On the global economic map the lines that now matter are those defining what may be called "region states." The boundaries of the region state are not imposed by political fiat. They are drawn by the deft but invisible hand of the global market for goods and services. They follow, rather than precede, real flows of human activity, creating nothing new but ratifying existing patterns manifest in countless individual decisions. They represent no threat to the political borders of any nation, and they have no call on any taxpayer's money to finance military forces to defend such borders.

Region states are natural economic zones. They may or may not fall within the geographic limits of a particular nation—whether they do is an accident of history. Sometimes these distinct economic units are formed by parts of states, such as those in northern Italy, Wales, Catalonia, Alsace-Lorraine or Baden-Württemberg. At other times they may be formed by economic patterns that overlap existing national boundaries, such as those between San Diego and Tijuana, Hong Kong and southern China, or the "growth triangle" of Singapore and its neighboring Indonesian islands. In today's borderless world these are natural economic zones and what matters is that each possesses, in one or another combination, the key ingredients for successful participation in the global economy.

Look, for example, at what is happening in Southeast Asia. The Hong Kong economy has gradually extended its influence throughout the Pearl River Delta. The radiating effect of these linkages has made Hong Kong, where GNP per capita is $12,000, the driving force of economic life in Shenzhen, boosting the per capital GNP of that city's residents to $5,695, as compared to $317 for China as a whole. These links extend to Zhuhai, Amoy and Guangzhou as well. By the year 2000 this cross-border region state will have raised the living standard of more than 11 million people over the $5,000 level. Meanwhile, Guangdong province, with a population of more than 65 million and its capital at Hong Kong, will emerge as a newly industrialized economy in its own right, even though China's per capita GNP may still hover at about $1,000. Unlike in Eastern Europe, where nations try to convert entire socialist economies over to the market, the Asian model is first to convert limited economic zones—the region states—into free enterprise havens. So far the results have been reassuring.

These developments and others like them are coming just in time for Asia. As Europe perfects its single market and as the United States, Canada and Mexico begin to explore the benefits of the North American Free Trade Agreement (NAFTA), the combined economies of Asia and Japan lag behind those of the other parts of the globe's economic triad by about $2 trillion—roughly the aggregate size of some 20 additional region states. In other words, for Asia to keep pace existing regions must continue to grow at current rates throughout the next decade, giving birth to 20 additional Singapores.

Many of these new region states are already beginning to emerge. China has expanded to 14 other areas—many of them inland—the special economic zones that have worked so well for Shenzhen and Shanghai. One such project at Yunnan will become a cross-border economic zone encompassing parts of Laos and Vietnam. In Vietnam itself Ho Chi Minh City (Saigon) has launched a similar "sepzone" to attract foreign capital. Inspired in part by Singapore's "growth triangle," the governments of Indonesia, Malaysia and Thailand in 1992 unveiled a larger triangle across the Strait of Malacca to link Medan, Penang and Phuket. These developments are not, of course, limited to the developing economies in Asia. In economic terms the United States has never been a single nation. It is a collection of region states: northern and southern California, the "power corridor" along the East Coast between Boston and Washington, the Northeast, the Midwest, the Sun Belt, and so on.

What Makes a Region State

THE PRIMARY linkages of region states tend to be with the global economy and not with their host nations. Region states make such effective points of entry into the global economy because the very characteristics that define them are shaped by the demands of that economy. Region states tend to have between five million and 20 million people. The range is broad, but the extremes are clear: not half a million, not 50 or 100 million. A region state must be small enough for its citizens to share certain economic and consumer interests but of adequate size to justify the infrastructure—communication and transportation links and quality professional services—necessary to participate economically on a global scale.

It must, for example, have at least one international airport and, more than likely, one good harbor with international-class freight-handling facilities. A region state must also be large enough to provide an attractive market for the brand development of leading consumer products. In other words, region states are not defined by their economies of scale in production (which, after all, can be leveraged from a base of any size through exports to the rest of the world) but rather by their having reached efficient economies of scale in their consumption, infrastructure and professional services.

For example, as the reach of television networks expands, advertising becomes more efficient. Although trying to introduce a consumer brand throughout all of Japan or Indonesia may still prove prohibitively expensive, establishing it firmly in the Osaka or Jakarta region is far more affordable—and far more likely to generate handsome returns. Much the same is true with sales and service networks, customer satisfaction programs, market surveys and management information systems: efficient scale is at the regional, not national, level. This fact matters because, on balance, modern marketing techniques and technologies shape the economies of region states.

Where true economies of service exist, religious, ethnic and racial distinctions are not important—or, at least, only as important as human nature requires. Singapore is 70 percent ethnic Chinese, but its 30 percent minority is not much of a problem because commercial prosperity creates sufficient affluence for all. Nor are ethnic differences a source of concern for potential investors looking for consumers.

Indonesia—an archipelago with 500 or so different tribal groups, 18,000 islands and 170 million people—would logically seem to defy effective organization within a single mode of political government. Yet Jakarta has traditionally attempted to impose just such a central control by applying fictional averages to the entire nation. They do not work. If, however, economies of service allowed two or three Singapore-sized region states to be created within Indonesia, they could be managed. And they would ameliorate, rather than exacerbate, the country's internal social divisions. This holds as well for India and Brazil.

The New Multinational Corporation

WHEN VIEWING the globe through the lens of the region state, senior corporate managers think differently about the geographical expansion of their businesses. In the past the primary aspiration of multinational corporations was to create, in effect, clones of the parent organization in each of the dozens of countries in which they operated. The goal of this system was to stick yet another pin in the global map to mark an increasing number of subsidiaries around the world.

More recently, however, when Nestlé and Procter & Gamble wanted to expand their business in Japan from an already strong position, they did not view the effort as just another pin-sticking exercise. Nor did they treat the country as a single coherent market to be gained at once, or try as most Western companies do to establish a foothold first in the Tokyo area, Japan's most tumultuous and overcrowded market. Instead, they wisely focused on the Kansai region around Osaka and Kobe, whose 22 million residents are nearly as affluent as those in Tokyo but where competition is far less intense. Once they had on-the-ground experience on how best to reach the Japanese consumer, they branched out into other regions of the country.

Much of the difficulty Western companies face in trying to enter Japan stems directly from trying to shoulder their way in through Tokyo. This instinct often proves difficult and costly. Even if it works, it may also prove a trap; it is hard to "see" Japan once one is bottled up in the particular dynamics of the Tokyo marketplace. Moreover, entering the country through a different regional doorway has great economic appeal. Measured by aggregate GNP the Kansai region is the seventh-largest economy in the world, just behind the United Kingdom.

Given the variations among local markets and the value of learning through real-world experimentation, an incremental region-based approach to market entry makes excellent sense. And not just in Japan. Building an effective presence across a landmass the size of China is of course a daunting prospect. Serving the people in and around Nagoya City, however, is not.

If one wants a presence in Thailand, why start by building a network over the entire extended landmass? Instead focus, at least initially, on the region around Bangkok, which represents the lion's share of the total potential market. The same strategy applies to the United States. To introduce a new top-of-the-line car into the U.S. market, why replicate up front an exhaustive coast-to-coast dealership network? Of the country's 3,000 statistical metropolitan areas, 80 percent of luxury car buyers can be reached by establishing a presence in only 125 of these.

The Challenges for Government

TRADITIONAL ISSUES of foreign policy, security and defense remain the province of nation states. So, too, are macroeconomic and monetary policies—the taxation and public investment needed to provide the necessary infrastructure and incentives for region-based activities. The government will also remain responsible for the broad requirements of educating and training citizens so that they can participate fully in the global economy.

Governments are likely to resist giving up the power to intervene in the economic realm or to relinquish their impulses for protectionism. The illusion of control is soothing. Yet hard evidence proves the contrary. No manipulation of exchange rates by central bankers or political appointees has ever "corrected" the trade imbalances between the United States and Japan. Nor has any trade talk between the two governments. Whatever cosmetic actions these negotiations may have prompted, they rescued no industry and revived no economic sector. Textiles, semiconductors, autos, consumer electronics—the competitive situation in these industries did not develop according to the whims of policymakers but only in response to the deeper logic of the competitive marketplace. If U.S. market share has dwindled, it is not because government policy failed but because individual consumers decided to buy elsewhere. If U.S. capacity has migrated to Mexico or Asia, it is only because individual managers made decisions about cost and efficiency.

The implications of region states are not welcome news to established seats of political power, be they politicians or lobbyists. Nation states by definition require a domestic political focus, while region states are en-

sconced in the global economy. Region states that sit within the frontiers of a particular nation share its political goals and aspirations. However, region states welcome foreign investment and ownership—whatever allows them to employ people productively or to improve the quality of life. They want their people to have access to the best and cheapest products. And they want whatever surplus accrues from these activities to ratchet up the local quality of life still further and not to support distant regions or to prop up distressed industries elsewhere in the name of national interest or sovereignty.

When a region prospers, that prosperity spills over into the adjacent regions within the same political confederation. Industry in the area immediately in and around Bangkok has prompted investors to explore options elsewhere in Thailand. Much the same is true of Kuala Lumpur in Malaysia, Jakarta in Indonesia, or Singapore, which is rapidly becoming the unofficial capital of the Association of Southeast Asian Nations. São Paulo, too, could well emerge as a genuine region state, someday entering the ranks of the Organization of Economic Cooperation and Development. Yet if Brazil's central government does not allow the São Paulo region state finally to enter the global economy, the country as a whole may soon fall off the roster of the newly industrialized economies.

Unlike those at the political center, the leaders of region states—interested chief executive officers, heads of local unions, politicians at city and state levels—often welcome and encourage foreign capital investment. They do not go abroad to attract new plants and factories only to appear back home on television vowing to protect local companies at any cost. These leaders tend to possess an international outlook that can help defuse many of the usual kinds of social tensions arising over issues of "foreign" versus "domestic" inputs to production.

In the United States, for example, the Japanese have already established about 120 "transplant" auto factories throughout the Mississippi Valley. More are on the way. As their share of the U.S. auto industry's production grows, people in that region who look to these plants for their livelihoods and for the tax revenues needed to support local communities will stop caring whether the plants belong to U.S.- or Japanese-based companies. All they will care about are the regional economic benefits of having them there. In effect, as members of the Mississippi Valley region state, they will have leveraged the contribution of these plants to help their region become an active participant in the global economy.

Region states need not be the enemies of central governments. Handled gently, region states can provide the opportunity for eventual prosperity for all areas within a nation's traditional political control. When political

and industrial leaders accept and act on these realities, they help build prosperity. When they do not—falling back under the spell of the nationalist economic illusion—they may actually destroy it.

Consider the fate of Silicon Valley, that great early engine of much of America's microelectronics industry. In the beginning it was an extremely open and entrepreneurial environment. Of late, however, it has become notably protectionist—creating industry associations, establishing a polished lobbying presence in Washington and turning to "competitiveness" studies as a way to get more federal funding for research and development. It has also begun to discourage, and even to bar, foreign investment, let alone foreign takeovers. The result is that Boise and Denver now prosper in electronics; Japan is developing a Silicon Island on Kyushu; Taiwan is trying to create a Silicon Island of its own; and Korea is nurturing a Silicon Peninsula. This is the worst of all possible worlds: no new money in California and a host of newly energized and well-funded competitors.

Elsewhere in California, not far from Silicon Valley, the story is quite different. When Hollywood recognized that it faced a severe capital shortage, it did not throw up protectionist barriers against foreign money. Instead, it invited Rupert Murdoch into 20th Century Fox, C. Itoh and Toshiba into Time-Warner, Sony into Columbia, and Matsushita into MCA. The result: a $10 billion infusion of new capital and, equally important, $10 billion less for Japan or anyone else to set up a new Hollywood of their own.

Political leaders, however reluctantly, must adjust to the reality of economic regional entities if they are to nurture real economic flows. Resistant governments will be left to reign over traditional political territories as all meaningful participation in the global economy migrates beyond their well-preserved frontiers.

Canada, as an example, is wrongly focusing on Quebec and national language tensions as its core economic and even political issue. It does so to the point of still wrestling with the teaching of French and English in British Columbia, when that province's economic future is tied to Asia. Furthermore, as NAFTA takes shape the "vertical" relationships between Canadian and U.S. regions—Vancouver and Seattle (the Pacific Northwest region state); Toronto, Detroit and Cleveland (the Great Lakes region state)—will become increasingly important. How Canadian leaders deal with these new entities will be critical to the continuance of Canada as a political nation.

In developing economies, history suggests that when GNP per capita reaches about $5,000, discretionary income crosses an invisible threshold. Above that level people begin wondering whether they have reasonable access to the best and cheapest available products and whether they have an adequate quality of life. More troubling for those in political control, citizens also begin to consider whether their government is doing as well by them as it might.

Such a performance review is likely to be unpleasant. When governments control information—and in large measure because they do—it is all too easy for them to believe that they "own" their people. Governments begin restricting access to certain kinds of goods or services or pricing them far higher than pure economic logic would dictate. If market-driven levels of consumption conflict with a government's pet policy or general desire for control, the obvious response is to restrict consumption. So what if the people would choose otherwise if given the opportunity? Not only does the government withhold that opportunity but it also does not even let the people know that it is being withheld.

Regimes that exercise strong central control either fall on hard times or begin to decompose. In a borderless world the deck is stacked against them. The irony, of course, is that in the name of safeguarding the integrity and identity of the center, they often prove unwilling or unable to give up the illusion of power in order to seek a better quality of life for their people. There is at the center an understandable fear of letting go and losing control. As a result, the center often ends up protecting weak and unproductive industries and then passing along the high costs to its people—precisely the opposite of what a government should do.

The Goal is to Raise Living Standards

THE CLINTON administration faces a stark choice as it organizes itself to address the country's economic issues. It can develop policy within the framework of the badly dated assumption that success in the global economy means pitting one nation's industries against another's. Or it can define policy with the awareness that the economic dynamics of a borderless world do not flow from such contrived head-to-head confrontations, but rather from the participation of specific regions in a global nexus of information, skill, trade and investment.

If the goal is to raise living standards by promoting regional participation in the borderless economy, then the less Washington constrains these regions, the better off they will be. By contrast, the more Washington intervenes, the more citizens will pay for automobiles, steel, semiconductors, white wine, textiles or consumer electronics—all in the name of "protecting" America. Aggregating economic policy at the national level—or worse, at the continent-wide level as in Europe—inevitably results in special interest groups and vote-conscious governments putting their own interests first.

The less Washington interacts with specific regions, however, the less it perceives itself as "representing" them. It does not feel right. When learning to ski, one of the toughest and most counterintuitive principles to accept is that one gains better control by leaning down toward the valley, not back against the hill. Letting go is difficult. For governments region-based participation in the borderless economy is fine, except where it threatens current jobs, industries or interests. In Japan, a nation with plenty of farmers, food is far more expensive than in Hong Kong or Singapore, where there are no farmers. That is because Hong Kong and Singapore are open to what Australia and China can produce far more cheaply than they could themselves. They have opened themselves to the global economy, thrown their weight forward, as it were, and their people have reaped the benefits.

For the Clinton administration, the irony is that Washington today finds itself in the same relation to those region states that lie entirely or partially within its borders as was London with its North American colonies centuries ago. Neither central power could genuinely understand the shape or magnitude of the new flows of information, people and economic activity in the regions nominally under its control. Nor could it understand how counterproductive it would be to try to arrest or distort these flows in the service of nation-defined interests. Now as then, only relaxed central control can allow the flexibility needed to maintain the links to regions gripped by an inexorable drive for prosperity.

EUROPE at Century's End

The Challenge Ahead

BY RICHARD N. HAASS

The year 1999 marks the end of the first decade of the post–Cold War world. For Europe, a region central to the Cold War—where it both began and ended—1999 is also proving to be a year of historic import, the most important on the continent since 1989, when the wall came down and Europe's division came to an abrupt and for the most part unanticipated end.

This claim can be justified by pointing to many events and trends, including the fitful progress toward reconciliation in Northern Ireland and Germany's emergence as a more "normal" country, one able to use military force beyond its borders as part of a nondefensive NATO action. Still, four developments in 1999 stand out: European Monetary Union and the launch of the euro; the entry of Poland, Hungary, and the Czech Republic into NATO and the articulation of a new strategic concept for the Alliance

Richard N. Haass, vice president and director of the Brookings Foreign Policy Studies program, is the author of The Reluctant Sheriff: The United States after the Cold War *(Brookings, 1997) and editor of* Transatlantic Tensions: The United States, Europe, and Problem States *(Brookings, 1999).*

From *The Brookings Review,* Summer 1999, pp. 4-9. © 1999 by the Brookings Institution. Reprinted by permission.

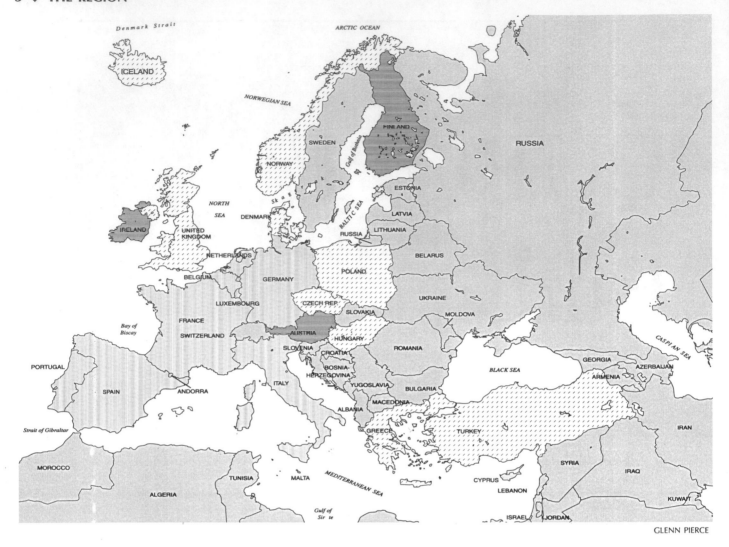

GLENN PIERCE

as it passed the half-century mark; the continuing deterioration of conditions within Russia and in U.S.-Russian relations; and the Kosovo conflict, the largest military clash on the continent since the Second World War.

Bearish on Russia

Russia is in many ways the most vexing, significant, and unexpected problem to cloud Europe's horizon. Russian weakness is proving a more complex challenge to Europe and the United States than did Soviet strength. The stakes are great, not so much in the economic sphere, where the actual or potential global impact is limited given Russia's small and increasingly barter-driven economy, but rather strategically, given Russia's ability to influence events in Europe through its political and military power and its nuclear arsenal.

But Russia's economic situation has grave strategic implications. An economically weak Russia is much more likely to pose a strategic threat, be it through arms sales to problem countries, authorized (or unauthorized) provision of technology to the

unconventional weapons programs of so-called rogue states, a greater reliance on nuclear weapons to offset a growing inferiority in conventional arms, and, most worrisome, a loss of control over its own nuclear arsenal.

There is no single answer to the challenge posed by a weak Russia. It is, however, possible to rule out some alleged answers that are

clearly misguided, including neglect. Russia needs help from the outside, but only on a conditional basis. Conditionality is necessary if economic assistance is to gain the required political support within donor societies and if funds are to do anything more than enrich corrupt individuals within Russia. Among the other lessons of the past few years is that market economic reform cannot occur (much less be sustained) in a vacuum if it is to be something other than crony capitalism;

basic infrastructure improvements—property rights, securities regulation, generally accepted accounting practices, bankruptcy proceedings, and so on—are essential.

All this will take time, though, something that does not square with the urgent challenges posed by Russia's enormous but unsafe nuclear arsenal. Bruce Blair and

The United States cannot have it both ways, urging that Europe do more, but do America's bidding and no more.

Clifford Gaddy suggest a range of measures, from the relatively modest—a U.S.-Russian accord to reduce the alert status of their weapons to lessen the chance of crisis instability, a cut in the number of deployed strategic nuclear systems to well below START II levels—to the ambitious, including a grand bargain in which the United States and Europe would buy much of the Russian nuclear stockpile. The latter would be costly but so would the alternatives—modernizing

U.S. nuclear forces or building large defensive systems to protect against a Russian attack. Moreover, trading bombs for dollars would improve American and European security and bolster Russia's economy. Such thinking may be too "outside the box" for many but it is hard to see where staying the course—and seeing conditions within Russia or in U.S.-Russian relations continue to worsen—would serve any American or European interest.

Working out a constructive relationship with Russia over the past decade has been complicated by differences over how best to promote peace and security throughout Europe. Until 1999, much of the disagreement centered on Western plans to enlarge NATO by offering membership to former members of the Warsaw Pact, among others.

These differences caused strains but not a breach in relations with Russia, as evidenced by Russian willingness to sign an agreement—the so-called Founding Act—that institutionalized NATO-Russian consultations. Russia's participation in the Bosnia peacekeeping effort was another sign that it was prepared to contribute to European security despite its unhappiness over NATO's growth.

Whether such a restrained "agreement to disagree" can continue if NATO enlargement proceeds is less clear. Bringing Russia into NATO at some point is an option, but one that would require the emergence of a very different Russia—and result in the emergence of a very different NATO. But regardless of whether Russia ever formally joins NATO, James Goodby is correct when he argues that European security will benefit directly and dramatically from a Russia that is democratic, prosperous, and stable—and not alienated from the United States and the countries of Western Europe. Avoiding such alienation, Goodby suggests, will require sensitivity and compromise on both sides: the West will have to keep Russian concerns in mind as it uses force in Europe and proceeds with NATO enlargement; Russia for its part must work with the United States and Europe in combating the proliferation of weapons of mass destruction, terrorism, and threats to innocent peoples from their governments. The only thing that is certain is that making Europe "whole and free"—not to mention peaceful and prosperous—promises to be a long-term, difficult proposition. For the immediate future, James Goldgeier puts forward the logical and, to many, persuasive argument that it makes no sense to stop NATO enlargement now. To do so would only redraw the line dividing Europe and deny the advantages of NATO membership to other deserving states. But further enlargement is hardly problem free. Beyond its economic costs and the risk that it would further complicate Alliance decisionmaking is the risk that adding Baltic states to the

Alliance could trigger a crisis with Russia—and within Russia.

Alas, one need not wait for a decision on NATO enlargement to set off a crisis with Russia; the war in Kosovo has done that. Russians are unhappy and angry over the attacks on their fellow Slavs. That these attacks were carried out without UN Security Council authorization and by an enlarged NATO from which they are thus far excluded makes matters that much worse. The Russian reaction is both deep and wide, reflecting widespread frustration in that country over its reduced circumstances at home and its reduced status abroad.

The war in Kosovo is also a crisis for NATO. For the 40 years of the Cold War, NATO unity was forged around a common threat; under Article 5 of the NATO treaty, Alliance members pledged to defend one another against external attack. With the demise of the Soviet Union and the end of the Cold War, many analysts and politicians believe that NATO must evolve or risk fading away. They argue that NATO remains valuable—especially in keeping the United States involved in Europe's security, given the limits on what Europeans are able and willing to do in the defense realm, the potential primacy of a unified Germany and uncertainty over how Russia will evolve—but that a continued focus on defending against a nonexistent threat is not enough.

Alliance's cohesion just at a time it is seeking to define a new purpose.

Economic Europe

Few if any observers predicted at the outset of 1999 that NATO would be the European or transatlantic institution that would most come under strain during the year. The European Union (EU) was the more obvious candidate. Indeed, the year began with the launch of the 11-member European Monetary Union and its common currency, the euro. This is the latest and one of the most important of the many steps Europe has taken toward integration over the past half century—and, as Robert Solomon points out, the first time since the fall of the Roman Empire that much of Western Europe has had a single currency

Creation of the euro effectively brings about a single European financial market, which should facilitate economic activity and, with it, growth. Still, the euro is not without its skeptics. Many question whether member governments can and will maintain the necessary fiscal discipline if they are faced with high unemployment or recession. The euro and EMU are the products of a continent that is increasingly controlled collectively in the economic sphere—while governments and politicians are still mostly

> The Russian reaction to the NATO attacks on Yugoslavia is both deep and wide, reflecting widespread frustration in that country over its reduced circumstances at home and its reduced status abroad.

The principal alternative mission for NATO, as set forth by Ivo Daalder, is to broaden its mandate from combating external threats to countries within the treaty area to promoting peace and security throughout all of Europe. The change could come about in two ways: through continued enlargement and through taking on the sorts of challenges that developed in Bosnia and Kosovo.

When leaders of NATO's 19 member states convened in Washington last April, they agreed to such a new strategic concept. That they did so amidst the war in Kosovo, however, highlighted the gap between rhetoric and reality. It is one thing to talk about meeting challenges to European peace, quite another to do it. The risk for NATO is that the experience of Kosovo will jeopardize the

elected in national and local contexts. Whether this tension can be managed remains to be seen.

On the American side of the Atlantic, there are those who doubt whether the euro is good for the United States. Some fear it will make Europe more of a closed market or weaken the role and value of the dollar. Others predict it will lead to greater European political and military unity—and greater independence from the United States. Elizabeth Pond takes a relatively sanguine view of these possibilities—but points out that the euro is likely to promote economic reform within Europe and make Europe more like America, benefiting some American investors and exporters but posing severe hardships for others who lose out to new competitors. As the U.S.-European skirmish over bananas indicates, trade frictions and protectionist pressures are never less than dormant and can surface with a vengeance, bearing adverse consequences for transatlantic political and economic relations.

In trade as in other matters, it is obvious that the United States and Europe, though drawn together in many ways, do not always or automatically see eye to eye. One specific disagreement involves Turkey a country the EU has kept at arm's length for a host of reasons, including dissatisfaction with Turkey's handling of its Kurdish problem and human rights more generally, tensions between Turkey and EU member Greece, and concern that Turkish entry into the EU could have adverse consequences for agricultural policy and labor markets. The United States has for the most part taken a more sympathetic approach to Turkey, a country viewed as vital not only for its contributions to European security but also for its links to the greater Middle East. As Heinz Kramer explains, bridging these transatlantic differences will require new flexibility in European thinking—and internal reform in Turkey itself, something that its domestic politics threaten to make more not less difficult.

The Way Ahead

The challenges facing Europe in 1999 and beyond are enormous. Indeed, the agenda could hardly be more crowded or daunting: Kosovo and the broader problem of bringing stability to the former Yugoslavia; ensuring the success of EMU; enlarging and deepening both the EU and NATO; and promoting Russian democracy, economic revival, and integration with the rest of the continent.

Many of these challenges are as much transatlantic as European in nature and will require close collaboration between the United States and Europe—whether as individual states, within NATO, within the EU, or within some other forum. Consultations using all these frameworks will remain essential—and would be more useful if Europe begins to speak with one voice on matters of security. Also useful would be a transatlantic commitment to open trade, to meet the norms and adhere to the decisions of the World Trade Organization; trade can be a source of prosperity or discord but not both.

But consultation and even compromise will not be sufficient. For the transatlantic tie to work and for Europe to prosper, Europeans must be prepared to assume a greater share of the burden of action, especially in the military domain. They must be ready not just to spend somewhat more on their military capacity, but to spend it on forces that are relevant for the post–Cold War world rather than its predecessor. Europeans will need, too, to resist the temptation to oppose American leadership simply out of resentment.

At the same time, the United States will have to accept that a greater European willingness and capacity to share the burdens of European and global security will translate into enhanced European influence, especially if Europe is prepared to act politically and militarily under EU rather than NATO auspices. The United States cannot have it both ways, urging that Europe do more, but do America's bidding and no more. In addition, the United States will need to curb its enthusiasm for economic sanctions in general and for secondary sanctions—those imposed on countries (often European) who do not support U.S. sanctions against such countries as Iran Libya, or Cuba—in particular.

But whatever the specific changes and compromises, it is critical that the two sides get it right. Europe and the United States remain essential to one another. It is not simply a matter of economics, although transatlantic trade and investment count for a lot. Nor is it simply a shared interest in Europe's stability, although this too obviously matters a great deal to both Europeans and Americans. Rather, it is that both Americans and Europeans have a major stake in what sort of a world emerges in the aftermath of the Cold War. Their ability to help bring about a world to their liking—one that promotes their common interests and values alike—depends on their ability to agree on a set of common priorities and work together on their behalf. Transatlantic cooperation proved central to the successful outcome of the Cold War; continued partnership is likely to prove no less central to the course of 1999 and the years that follow.

RUSSIA'S FRAGILE UNION

by Matthew Evangelista

On November 20, 1998, Galina Starovoitova, one of Russia's few democratic politicians worthy of the name, was murdered in the entryway of her St. Petersburg apartment. Her death attracted brief notice in the Western press at first. But following an outpouring of anger and sympathy among the Russian public, the media paid attention for a few more days—long enough to cover the demonstrations that coincided with her funeral and the initial investigation into her murder.

Starovoitova, it seems, was the victim of a contract killing of the kind that Russian bankers and businessmen, in the absence of a functioning legal system, typically use to settle economic disputes. But unlike other politicians who had been murdered in similar circumstances, Starovoitova was not known to be involved in any shady deals. She was widely considered honest and incorruptible. A human rights activist and colleague of Andrei Sakharov's during the Soviet period, her work as a deputy of the Russian State Duma had been driven by principle.

Much as Sakharov's death in 1989 demoralized the supporters of the democratic reforms triggered by Mikhail Gorbachev's perestroika, Starovoitova's murder signaled the demise of hope for any way out of the greed, violence, and political bankruptcy of Boris Yeltsin's Russia.

Another death last November went unnoticed by the Western media, but it also served to highlight an important dimension of the Russian crisis. Nearly 12,000 kilometers east of Moscow, in the city of Petropavlovsk-Kamchatsky, 13-year-old Zoya Korshunova was killed when the batteries in a flashlight exploded in her mouth. She had been trying to do her homework in the dark, in a region suffering such an acute shortage of energy that electricity was available only (and briefly) three times a day. Zoya's death received some attention in the Russian media. So did a plea to the United Nations from the local authorities on the Kamchatka peninsula for emergency humanitarian aid.

Russian Prime Minister Yevgeny Primakov, deeply embarrassed, sent a special team to help resolve Kamchatka's energy crisis. Meetings with

AP/WIDE WORLD

November 1998: The funeral of Galina Starovoitova, gunned down at her St. Petersburg home.

Matthew Evangelista teaches international relations at Cornell University. He is the author of the recently published Unarmed Forces: The Transnational Movement to End the Cold War. *Research support for this article was provided by the Smith Richardson Foundation.*

Reprinted by permission of *The Bulletin of the Atomic Scientists,* May/June 1999, pp. 50-55. © 1999 by the Educational Foundation for Nuclear Science, 6042 South Kimbark, Chicago, IL 60637. A one-year subscription is $28.

officials from the ministries of Energy, Finance, and Defense came up with a temporary solution: The navy would provide oil from its stockpile.

The circumstances surrounding the two deaths tell us much about today's Russia. Starovoitova's murder is still unsolved, but speculation about it abounds. She had many enemies. For instance, when I met her in Moscow in early November, three weeks before she was killed, Starovoitova had just returned from a Duma session where she had unsuccessfully promoted a motion to censure Gen. Albert Makashov, a notoriously anti-Semitic communist deputy, for remarks inciting violence against Jews. She was already receiving death threats from Makashov's supporters.

It is perhaps more likely, though, that her murder was a local affair unrelated to her role in the Duma. Her concern for her St. Petersburg constituents had led Starovoitova to try to clean up corruption in the municipal elections by running a slate of like-minded candidates. Her efforts annoyed local organized crime figures.

Nevertheless, according to the *St. Petersburg Times*, the investigation of Starovoitova's murder—led by former KGB officials—has focused on digging up dirt on her democratic allies rather than actually solving her murder. "We are going to solve this case in such a way that it buries your democratic movement," one of the investigators told a *Times* journalist who was close to Starovoitova.

A Potemkin village

In the wake of Starovoitova's death, some of the leading "reformers" on the Russian political scene—including Yegor Gaidar, Boris Nemtsov, Anatoly Chubais, and Sergei Kirienko—put aside their internecine squabbles to form a center-right

Russia may not break apart, but the regions will surely take more power from the center.

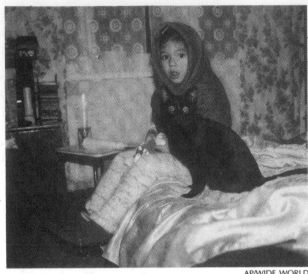

AP/WIDE WORLD

A three-year-old and her cat in an unheated apartment in the far eastern city of Petropavlovsk, Kamchatka peninsula.

bloc, dubbed "Just Cause," ostensibly under Starovoitova's banner. The efforts of these four politicians—all former prime ministers or deputy prime ministers in Yeltsin's cabinets—only highlighted the advanced state of Russia's political bankruptcy.

In the Gorbachev era, common political discourse identified the orthodox communists as the "right," and supporters of perestroika and democratic socialism as the "left." Now the communists are back on the left and the self-styled reformers are on the right. The words "reform" and "democrat" have become epithets in today's Russia, thanks to Chubais and the others.

During their years in power, many of the reformers preached the virtues of the market while engaging in privatization schemes in which the most valuable state assets were sold at fire-sale prices to their cronies. Just Cause is widely considered the party of the nouveaux riches, a

party that follows the dictates of the International Monetary Fund and the U.S. Treasury Department, but which is out of touch with the needs of ordinary Russians.

The reformers are blamed for the devaluation of the Russian ruble last August, the default on Russian loans, and the subsequent collapse of the Russian banking system, when millions lost their savings. That financial collapse precipitated a crisis that continues today. Most important, perhaps, the crisis revealed that the vaunted transition from a command economy to a market economy was a mere Potemkin village—an overused but sadly appropriate metaphor for Russia's reforms.

Much of Russia's economic activity before the August meltdown took place not through normal market mechanisms, but on the basis of sometimes extraordinarily complex systems of barter. The August collapse undermined the Rube Goldberg economy and put the survival of the regions beyond Moscow at risk.

Bartering for heat and light

The story of how Zoya's hometown of Petropavlovsk ended up without fuel oil illustrates how the barter system works—and why it failed in this case. Kamchatskenergo, a partly private, partly state-owned company, supplies most of the energy for the Kamchatka peninsula, which was a key military outpost on the Pacific during the Soviet era. Most of its customers, including the local military base, are agencies of the Russian government.

But the government has been notoriously negligent in paying its bills, as many Russian workers know from the personal experience of going months without receiving wages. In this case, Kamchatskenergo was able to continue supplying heat and energy to government-run schools, hospitals, and other institutions, without payment, in return for tax credits.

As Colin McMahon of the *Chicago Tribune* figured out when he interviewed the director of Kamchatskenergo, the energy company was allowed to trade its tax credits to a profitable Moscow-based oil firm. In turn, the Moscow firm was excused from paying a corresponding amount of taxes to the federal government.

The Moscow company provided crude oil that was sold or bartered on the international market in return for processed fuel oil. In turn, Kamchatskenergo used the fuel oil, acquired at lower-than-market prices, to power its two main energy plants. But when the ruble's value dropped by a third last August, Kamchatskenergo could no longer afford to buy imported oil, which was denominated in dollars.

Meanwhile, Kamchatskenergo's Moscow partner was having trouble selling its crude oil on the glutted international market. All of these events conspired to leave much of the Kamchatka peninsula without heat or electricity as temperatures dropped below zero by mid-November.

The Chechnya debacle

The Russian Federation is formally made up of 89 "subjects," including 21 ethnically identified republics such as Tatarstan, Chechnya, and Bashkortostan, and 68 non-ethnic units of various kinds. All 89 are commonly known as "the regions."

The central government's reaction to the crisis in Kamchatka—and in the regions in general—was revealing. By sending in troubleshooters from Moscow to knock heads together, Prime Minister Primakov reverted to the tried-and-true methods of the Soviet period. But Moscow no longer wields as much power as it once did vis-a-vis the regions. The deal to supply fuel from the navy depot, for instance, unraveled within weeks and Petropavlovsk again ran short of energy.

AP/WIDE WORLD

September 1, 1996: The center of Grozny, a day after Russia's Gen. Alexander Lebed and separatist commander Aslan Maskhadov signed a pact ending the conflict in Chechnya.

Moscow's inability to cope with regional crises has led to speculation that the federation might break apart, much as the Soviet Union did at the end of 1991. This is not the first time such concerns have been raised. When the Soviet Union split into its 15 constituent ethnic republics, many feared that Russia would go the same way.

Boris Yeltsin himself fueled the earliest concerns about a Russian breakup in 1991. As part of his campaign to undermine Soviet President Gorbachev, Yeltsin, then president of Russia, went to Tatarstan's capital city of Kazan and issued a challenge for the regions to "take as much autonomy as you can swallow."

Chechnya and Tatarstan took him at his word and declared formal independence. In 1992 both refused to sign the treaty that the Yeltsin administration promoted as the foundation for relations between the center and the regions. Moscow eventually worked out an uneasy *modus vivendi* with Tatarstan; the republic, acting as if it were a sovereign state, negotiated "treaties" with the federal government to regulate essentially internal matters, such as taxation.

Chechnya did not fare as well. Under the erratic leadership of Dzhokar Dudaev, the republic insisted on a special independent status. In retrospect, many observers believe that if Yeltsin had been willing to meet directly with Dudaev in 1992, the conflict could have been resolved peacefully. Instead Yeltsin thwarted efforts to open negotiations with Dudaev.

In 1992 Galina Starovoitova, then serving as Yeltsin's adviser on ethnic politics, contacted Dudaev in the Chechen capital of Grozny by telephone. Russian hardliners, with Yeltsin's blessing, attacked her efforts at reconciliation.

The next time Starovoitova tried to contact Dudaev, she found that the government phone service to Grozny had been cut. Seeing the writing on the wall in the increasing militarization of Moscow's relations with Chechnya, Starovoitova resigned in protest—well before the Russian armed forces launched an invasion in December 1994.

The war against Chechnya was a low point in Yeltsin's already sagging popularity and a high point in concern about the future of the Russian Federation. At first many of the neighboring regions in the Caucasus expressed solidarity with the Chechens, as citizens from Dagestan and Ingushetia, for example, tried to prevent Russian military forces from crossing their territory.

In the wake of the Chechen war, Andrei Shumikhin, a Russian analyst, expressed a fear shared by many that "a 'brush fire' of drives for independence may pick up elsewhere across Russia, leading to the eventual destruction of Russian territorial integrity."

Starovoitova criticized Moscow's "crude use" of "notorious tools of imperial policy" and predicted that the military intervention in Chechnya would "produce mistrust of the center's policy and centrifugal tendencies in Tatarstan, Bashkortostan," and elsewhere in the Russian Federation.

Gen. Aleksandr Lebed, who led Soviet forces in the disastrous intervention in Afghanistan, warned that an invasion of Chechnya would lead to a similar, protracted guerrilla war, resulting in "a Pyrrhic victory."

Lebed was right. The war in Chechnya dragged on for nearly two years, claimed tens of thousands of casualties, and produced hundreds of thousands of refugees.

But Starovoitova and others who feared that the war would tear apart the federation were wrong. The war ended with a cease-fire brokered by Lebed himself in 1996 and a treaty on future relations signed by Yeltsin and Dudaev's successor Aslan Maskhadov in May 1997.

Creeping autonomy

After the end of the Chechen war, a new—and somewhat contradictory—consensus emerged in the West and in Russia:

■ The Chechen victory would not inspire other regions to attempt to achieve independence. The cost of gaining independence was too high—tens of thousands of lives, mainly civilian, and the total destruction of Grozny, the capital city,

as well as many regional centers and villages.

■ The war was also disastrous for Russia, undermining Yeltsin's popularity, demoralizing the armed forces, and draining the federal budget. It was unlikely that Moscow would ever again resort to such a brutal and foolhardy use of violence to resolve a conflict with one of the regions.

■ Finally, common interests would maintain the federation, particularly economic interests, such as trade, distribution of taxes and subsidies, and a common currency; a common transportation network; and a common energy grid.

But the financial crisis of last August undermined such arguments. Perhaps it is too early to speak of a new consensus, but many Russian and Western observers believe a major change in center-regional relations is leading, if not to the breakup of the federation, at least to a substantial decentralization of power and more autonomy for the regions.

A conference in Washington in December, sponsored by the Harvard-based Program on New Approaches to Russian Security, captured the mood among U.S. specialists. Kathryn Stoner-Weiss, a political scientist at Princeton, spoke of "creeping regional autonomy," and Henry Hale, from Harvard's Davis Center for Russian Studies, described the "consolidation of ethnocracy" within the republics and "regional autocratization" throughout the country.

Eva Busza, from the College of William and Mary, focused on the implications of the economic crisis for regionally based military forces. As resources from the center dry up and troops become "reliant on the goodwill of regional leaders for their day-to-day survival," what might those leaders demand in return, particularly as their policies come into conflict with those of the central government?

In many respects the concerns of Western analysts simply reflect fears in Moscow. Primakov put it starkly in a speech to the Duma last August,

shortly after being named prime minister: "We are facing a very serious threat of our country being split up." And on Russian television last September, he argued that "89 constituent parts of the federation is too many." He supported plans to consolidate the regions into as few as eight regional economic groups.

In fact, several interregional economic associations already exist, and their horizontal economic links may serve as a "bulwark against disintegration," as Graeme Herd of Aberdeen University has argued.

But Prime Minister Primakov is not satisfied with the status quo. In January, at a meeting of Siberian governors representing the Siberian Accord, an interregional association, Primakov called for the "restoration of the vertical state power structure, where all matters would be solved jointly by the center and local authorities." He also criticized "talk of conflict between the center and the regions" and insisted that separatist trends "must be quelled, liquidated, and uprooted."

The concern of Western analysts about "regional autocratization"— that is, the demise of democratic procedures within the regions—is also heard in Moscow. Many regional governors have manipulated electoral laws to eliminate any challenges to their rule. Galina Starovoitova told me of a particularly blatant, but not unusual, technique employed by the leader of the small ethnic republic of Marii El, located just north of Tatarstan. He sponsored a law requiring that any candidates for the presidency of the republic be fluent in Russian as well as two dialects of the Marii language.

With ethnic Marii people a minority in the republic (43 percent of the population), the law would have eliminated candidates from other groups— such as Russians who make up 48 percent—as well as the many Marii who have lost their native language. In effect the law would have limited the field to the incumbent himself.

Starovoitova made her political reputation in the late 1980s by supporting autonomy for the Armenian enclave of Nagorno-Karabakh in Azerbaijan. But she was not one to offer blind support for ethnic chauvinism at the expense of democratic principles. In one of her last achievements in the service of democracy before her murder, she challenged the Marii El electoral law as unconstitutional and the Russian Constitutional Court ruled in her favor.

Unfortunately, widespread manipulation of local elections often goes unremarked by the Moscow authorities, especially if the regional leaders have supported Yeltsin's own electoral efforts in the past. Primakov favors doing away with regional gubernatorial elections altogether in favor of appointing local rulers directly from Moscow.

More talk; less defiance

Although the concern about anti-democratic tendencies in the regions is well founded—some local politicians have even targeted critical journalists for assassination—the prospects for either successful regional political consolidation or more centralized control from Moscow seem dubious.

Nevertheless, outright disintegration does not appear likely. That impression comes in part from my recent discussions with representatives from some of the regions viewed as most likely to seek greater autonomy—Tatarstan, Bashkortostan, and Dagestan—and one, Chechnya, that has already achieved de facto independence.

Despite the undeniable crisis atmosphere in the regions, there seems to be a strong commitment to working out compromises with the central government. Few regional leaders really want to see Russia go the way of the Soviet Union.

All of the regions have permanent representatives stationed in Moscow who are in daily contact with federal officials to discuss issues ranging from the budget to foreign investment to constitutional disputes. (The constitutions of many regions con-tradict the federal constitution on key points.)

Even Chechnya, which claims to be independent, has kept its Moscow office open. The most noticeable change is that the staff now answers the phone with a hopeful "Embassy," even though no major country has recognized Chechen sovereignty.

The Chechen office is headed by Vakha Khasanov, a former factory manager from the Nadterechny region of Chechyna, an area that, as Khasanov put it, tried to maintain "neutrality" during the war. Khasanov explained that he was not trained to be either a diplomat or a politician.

His refreshingly outspoken views suggested as much. He criticized the late Chechen leader Dzhokar Dudaev, whose eccentric behavior and megalomania contributed much to the outbreak of the war. He was forth-right in admitting that the crime rate in the republic is so high as to discourage any foreign investment. (He told a recent commercial delegation from Malaysia that they would need a small army of security guards even to consider a potential site visit to the republic.) The prerequisite for solving Chechnya's problems, in Khasanov's view, lies in improving relations with Russia.

Further, relations with Moscow are not the only thing that preoccupies the Chechen government. Chechnya teeters on the brink of civil war as the relatively moderate, popularly elected president Aslan Maskhadov—the top military commander during the war—faces challenges from his erstwhile subordinates, including Salman Raduev, whose contributions to the Chechen victory included notorious terrorist attacks and the taking of civilian hostages. Raduev has threatened cross-border attacks on Russian military bases in neighboring Dagestan, putting at risk President Maskhadov's attempts at working out a *modus vivendi* with Moscow.

Tatarstan offers a sharp contrast to Chechnya, in that it was able to gain substantial independence from Moscow without provoking violence. Indeed, the "Tatarstan solution" is often cited as a lost opportunity that Dudaev and Yeltsin should have pursued.

Given its level of de facto independence, one might expect the recent economic crisis to have driven Tatarstan even further away from the central government. To be sure, its authoritarian leader Mintmir Shaimiev has implemented several measures that isolate the republic, including a ban on the export of food stuffs or the import of Russian vodka. At the same time, Tatar officials in the Moscow office are busy trying to reconcile differences with the central government.

Some of their reasons are obvious. "It's too big a price what was paid in Chechnya, and for the Russian Federation," says Mikhail Stoliarov, a deputy head of the Tatarstan delegation in Moscow.

Tatarstan has come a long way since its defiant claims of independence from Russia in the early 1990s. Stoliarov believes the situation then was fraught with the danger of a Chechen-style military conflict, but not anymore. "I'm talking about lessons of Chechnya, which should be a warning to everybody, to all the radicals in Tatarstan as well as in any other republic—and to radicals in Russia, whatever, governmental or non-governmental radicals. So it's a warning that's clear, the Chechen conflict should not be repeated in any form."

Tatarstan and neighboring Bashkortostan are among the richest regions in the Russian Federation, thanks to substantial oil and natural gas deposits and high-technology industries inherited from the Soviet military-industrial sector. They are among the small number of "donor" regions—regions that receive no subsidies from the central budget, but instead contribute through their taxes to the redistribution of income to poorer, "recipient" regions.

In my discussions I probed for some signs of resentment toward the poorer regions, but the reactions of the Tatar and Bashkiri representatives were more mixed. They spoke of pride in their own regions and a desire to protect their achievements. They also seemed to view their economic development as a product of Soviet-era policies, and their abundant resources as a matter of good fortune.

In describing the sources of his republic's wealth, Erik Yumabaevich Ablaev, the representative from Bashkortostan, said: "We understand very well that these were not the achievements of our republic alone. They were developed in the course of 70 years by the whole Soviet people." The vast investments in oil processing were "created by the efforts of the entire Soviet Union."

Ablaev was sensitive to the charge that Bashkortostan's wealth accords it a privileged position vis-a-vis Moscow. But he also stressed the negative side of the republic's development. Industrial pollution poisons the land, water, and air. According to Ablaev, every second child in Bashkortostan is born unhealthy, the victim of an environmental disaster that the central government in Moscow ignores. "They speak of 'privileges,'" he complained. "If we had healthy children, that would be a privilege."

At the other end of the spectrum is Dagestan, the poorest of the 89 regions, except for Chechnya. On economic grounds the republic has no reason to secede, as one of its Moscow representatives readily acknowledged.

Although Chechnya was also poor, it had something Dagestan lacks—a relatively homogeneous population with a strong tradition of opposition to Moscow. Dagestan's two million residents comprise 32 ethnic groups and speak 33 languages. The republic still reflects much of Soviet "internationalist" ideology. Robert Chenciner, the author of a recent book on Dagestan, calls it "a microcosm of the ethnic mosaic of the Soviet Union."

That is pretty much how Mamay Mamayev, the Moscow representative, described his region, as he pointed to it on an old map of the Soviet Union tacked on the wall behind his desk. Most likely, he is not

AP/WIDE WORLD

January 1996: Magamed Abdurazakov, Dagestan's internal affairs minister, listens to Chechen rebels engaged in a civil war with Russian troops.

the only Dagestani nostalgic for the stability of Soviet rule. The Chechen war wrought havoc on Dagestan, where some 100,000 refugees fled from the destruction of Grozny and surrounding villages. "Some of them are still living at our house," Mamayev said.

Further, the republic is now awash with weapons. In the capital city of Makhachkala, competing political factions command their own armed militias. Mamayev played down the violence, inviting me to visit the capital, and insisting that only the border area with Chechnya posed any risks. The day after my interview, an American teacher was kidnapped in Makhachkala in broad daylight.

The deeper threat

Despite the crisis atmosphere in much of the country, concern about the imminent disintegration of the Russian Federation seems unwarranted. Relations between the regions and Moscow will not go smoothly. But they will probably continue to be worked out on an ad hoc basis, as they have been for the last several years.

And despite Prime Minister Primakov's ambitious plans for consolidation of the regions, there are few governors or presidents of republics who would willingly relinquish political and economic power to be swallowed up by a larger agglomeration more closely controlled from Moscow.

The real story of the regions is the one represented by the deaths of Galina Starovoitova and Zoya Korshunova. Corruption and political violence are undermining the fragile foundations of Russian democracy. Lack of food, lack of medicine, and lack of heat over the winter months have produced humanitarian catastrophes throughout the country. The Russian government, bulging with more bureaucrats than even Leonid Brezhnev employed, seems utterly incapable of dealing with the various crises, whether large or small.

An official at Radio Free Europe put it just right in January: "It is the collapse of the Russian state, not the breakup of the federation or economic depression, that may in the long run prove the greatest threat to Russian democratic development and international stability."

> "The Kosovo problem is as much about the resolve of American leadership and the credibility of NATO as it is about the complexities of Balkan politics . . . Neither the crisis in Kosovo nor the broader problems of instability in the Balkans will be resolved until the United States exercises political and military leadership to do so; this will require sustained presidential attention."

Kosovo: America's Balkan Problem

JAMES HOOPER

Serbia is a revolutionary ultranationalist state whose leaders are committed to destabilizing the Balkans. Since 1991 the Serb leadership has launched wars against Slovenia, Croatia, Bosnia, and now Kosovo. But unlike the other conflicts, the Serb war against Kosovo's Albanian population raises an especially pointed geopolitical problem: elements of the guerrilla force fighting the Serbs hope to create a common Albanian homeland by uniting the Albanian people living in the Balkan states of Serbia, Albania, Macedonia, and Montenegro. The inherently destabilizing rise of pan-Albanian nationalism as a consequence of the Kosovo Albanians' subjection to Serbia's ultranationalist policies has imparted a sense of urgency to international diplomatic efforts to resolve the crisis.

The Kosovo conflict has also challenged Washington's resolve to handle its most difficult post–cold war transatlantic security problem. The future of America's collective defense strategy in Europe depends on the Clinton administration's determination to reverse the recent failures of its Kosovo policy and establish the foundation for stability in southeastern Europe.

WHEN MYTH CONFRONTS REALITY

The Serbs regard Kosovo as the birthplace of Serbian culture. Most of the important monasteries of the Serbian Orthodox Church are located in Kosovo. Indeed, the national myth of Serbia as the tragic sentinel of Western civilization astride the lands of the Ottoman Empire stems from the Ottoman victory over the Serbs at the Battle of Kosovo in 1389. The annual celebrations of that defeat have reinforced the Serbs' vision of

their own victimhood. According to the myth, Serbian blood has consecrated the soil of Kosovo; the integrity of the Serbian nation would be inconceivable without Kosovo. It was the calculated reshaping of that myth into a political justification for dictatorship, aggression, and genocide that defined and fueled the rise of Slobodan Milosevic.

Myth, however, collides with inconvenient realities in Kosovo. Of the province's approximately 2 million inhabitants, over 90 percent are ethnic Albanians. The Serbs constitute only the largest of several tiny minorities.

Kosovo's Albanians have shared a turbulent history with the Serbs this century. The Kosovars' post–World War I rebellion against the Serbs lasted for almost two decades until it was ruthlessly eradicated. The two peoples did enjoy stable, if uneasy, relations during the latter years of President Joseph Broz Tito's rule after Tito provided them with a considerable degree of autonomy through a new constitution in 1974. This period of self-rule lasted until 1989. It was then, on the 600th anniversary of the Battle of Kosovo, that Milosevic removed Kosovo's autonomy, established direct Serbian rule over the province, expelled the Albanians from the Kosovo parliament, the state bureaucracy, and state-owned industries, and closed the state-run school system and most of the medical system to them. It was a Serbian version of apartheid, which enabled Milosevic to use the power of the state to enforce the rule of the small Serbian minority over the ethnic Albanians.

Most local diplomats and Balkan experts believed that war was inevitable and would begin soon, a belief that made the war in Kosovo that began in February 1998 perhaps the most frequently predicted conflict in recent memory. Yet its timing took most observers by surprise. After 1989 the Kosovars confounded the international community by eschewing a war of national liberation, embracing instead the nonviolent approach espoused by leading Kosovo intellectual Ibrahim Rugova and constructing a parallel civil society. They elected moderate political leaders (these elections were not officially recognized by Serbia or the international community, and Rugova received

JAMES HOOPER *is the executive director of the Balkan Action Council in Washington, D.C. A former Foreign Service officer, he served as deputy director for Eastern Europe responsible for the Balkans and Baltics between 1989 and 1991 and deputy chief of mission in Warsaw between 1994 and 1996.*

Reprinted with permission from *Current History* magazine, April 1999, pp. 159–164. © 1999 by Current History, Inc.

little more than polite audiences and rhetorical encouragement from Western governments). The Kosovars also established their own educational system through classes operating in private homes and community buildings, maintained a separate health care system, and revived their economy through revenues derived from a voluntary tax levied on ethnic Albanians living abroad and remittances from Albanians temporarily working in Europe.

THREATS AS STRATEGY

When the war in Bosnia broke out, the United States, fearing a spillover of genocide from the raging conflict in that newly independent country, delivered a warning to Milosevic in December 1992 that if he cracked down on the ethnic Albanians, he would face unilateral American military intervention in Kosovo and Serbia proper. That threat, reaffirmed by President Bill Clinton's administration after it took office early in 1993, helped keep the peace in Kosovo for five years, but did not relieve the apartheid-like repression endured by the ethnic Albanians.

The Kosovo Albanians had been led by the United States to expect that their concerns would be addressed at a peace conference on Bosnia. When United States negotiators, in deference to Milosevic, excluded Kosovo Albanian delegates from those peace talks in Dayton, Ohio, in November 1995 and avoided discussion of the Kosovo problem, Rugova's nonvio-

lent strategy lost its credibility. The rise of the guerrilla Kosovo Liberation Army (KLA) and expansion of popular support for an armed independence struggle dates from the post-Dayton period, when many Kosovo Albanians concluded that the reward for nonviolence was international neglect.

The KLA has become such a prominent feature of the Kosovo landscape that it is easy to forget that only one year ago it was a tiny force with little public visibility. Jonathan Landay, one of the first American reporters to explore the origins of the KLA, noted in several articles for the *Christian Science Monitor* last year that the KLA had been founded in 1993 by former political prisoners and young activists disillusioned with Rugova's strategy of nonviolence. He reported that several KLA leaders had been beaten and jailed in 1982 for belonging to the Kosovo People's Movement, which campaigned for greater ethnic Albanian political rights. The head of the KLA political directorate, Hashim Thaci, who played a key role at the recent peace conference in Rambouillet, France, was also a former political prisoner.

The actions of Serbia and the inaction of the United States made possible the expansion of the KLA from a minuscule force conducting pinprick attacks against isolated Serbian police stations into a serious insurgent force. The Serbian crackdown began in February 1998, when Serb security forces, in response to the intermittent KLA attacks on the police, used armor and artillery force to destroy several villages in the Drenica region. Milosevic then halted his attacks and watched for Washington's reaction to determine whether the Clinton administration would follow through on its threat to use force.

When the United States limited its response to tough rhetoric and more meetings of the six-nation Contact Group on the former Yugoslavia—the United States, Britain, France, Germany, Italy, and Russia—Milosevic and the Kosovo Albanians each concluded that the West had abandoned Kosovo, and acted accordingly. As Milosevic escalated the conflict, backing for moderate Kosovo Albanian political leaders evaporated while support for the KLA grew rapidly. Almost overnight, the KLA moved from the margins of ethnic Albanian society into the mainstream; the insurgency mushroomed as every village under attack identified itself with the KLA.

Washington's own unwillingness to follow through on its threat of air strikes against Serbian military targets played directly into the hands of the KLA as well as Belgrade. All of the fighting that has taken place since February 1998 and its many consequences for Kosovar society—the killing of approximately 2,000 people, the displacement of over 500,000 from their homes, the destruction of more than 500 villages

> *Many Kosovo Albanians concluded that the reward*
> *for nonviolence was international neglect.*

and 19,000 homes—have happened as a result of United States inaction.

Fighting between the KLA raged until October 13, when United States special envoy Richard Holbrooke, backed by the threat of NATO air strikes, concluded a cease-fire deal with Milosevic. The agreement put NATO on the front line of the conflict. Under the terms of the agreement worked out by Holbrooke, the alliance was responsible for enforcing the agreement through the continuing threat of air strikes and verifying its provisions through reconnaissance overflights by unarmed aircraft. The cease-fire also provided for the Organization for Security and Cooperation in Europe (OSCE) to deploy 2,000 unarmed monitors to verify the agreement on the ground.

The deal was fatally flawed, however, by Holbrooke's decision to allow Milosevic to retain nearly 20,000 military, paramilitary, and special police personnel in Kosovo. Milosevic never complied with signed commitments to reduce Serbian forces even to this level, nor did NATO compel him to meet his obligations. In effect, Holbrooke traded NATO's fragile and hard-won consensus to use airpower against Serbian targets for a piece of paper from Milosevic. The failure of the Holbrooke-Milosevic deal generated renewed fighting in late December and a massacre of ethnic Albanian civilians in the village of Racak in January that hardened KLA attitudes, left moderate Kosovo Albanian political leaders even more isolated, and spurred the Contact Group to hurriedly organize the Rambouillet peace conference to avert a complete breakdown on the cease-fire.

RENDEZVOUS AT RAMBOUILLET

The Kosovo peace conference held at Rambouillet from February 6 to 23 was sponsored by the Contact Group, hosted by France, chaired by the French and British foreign ministers, and essentially led by the United States. The European allies had deeply resented their treatment at the 1995 Dayton peace conference, where they were present but largely ignored by American negotiators. At Rambouillet they took a more visible public role in the negotiations. They also sought to project a more equal partnership with the United States in addressing Balkan security issues.

Washington's strategy at Rambouillet was to offer the Kosovo Albanians interim self-government for three years with no guarantee of independence, a NATO peacekeeping force to protect them from the Serbs, and the threat of NATO air strikes to induce Serbian cooperation.[1] The self-government plan

would establish democracy in Kosovo through a unicameral parliament, a president chosen by the legislature and given significant powers to govern, and an independent judiciary. Various international organizations were invited to help reinforce implementation of the political agreement.

The draft agreement did not include any mechanism to determine the final status of Kosovo. Instead, the Kosovo Albanians were expected to put aside their goal of independence for at least three years while implementing self-government. Because the Kosovars demanded language on a referendum on the province's status, the United States delegation ultimately provided them with a letter promising a referendum after three years, although the Kosovars complained that there were no guarantees that its results would be binding.

The security annex to the agreement would have NATO peacekeepers ensure the withdrawal from Kosovo of most Serbian security forces and supervise those forces allowed to remain. Belgrade could retain 2,500 army troops indefinitely for border patrolling. These forces would be assigned to specific cantonments and placed under the control of the NATO peacekeepers. The Serbs would also be permitted to keep up to 2,500 Interior Ministry special police in Kosovo for up to two years. This police force was to be placed under OSCE supervision, and OSCE was expected to consult with NATO peacekeepers on maintaining effective supervision over the police. A 3,000-person multiethnic police force reflecting the ethnic percentages of Kosovo's population was also outlined in the agreement.

Serbian and Kosovo Albanian paramilitaries were to be disarmed and disbanded by the NATO peacekeepers. By treating the KLA on a par with Serbian paramilitaries rather than the Serbian army—which would be allowed to remain in Kosovo in a sizable deployment—the Contact Group signaled its intention of dissolving the insurgent army. KLA delegates at the conference firmly conveyed their unwillingness to accept dissolution, especially in view of the 5,000 Serbian police and military forces permitted to remain in Kosovo.

The Rambouillet strategy depended on Kosovar acceptance of the package as the first step toward a NATO showdown with Serbia. It was assumed that the Kosovo Albanian delegation would be prepared to embrace the agreement, despite its shortcomings, in order to obtain self-government, NATO ground troops, and possible air strikes against Serbia if Serbia balked at the agreement's conditions. The real challenge was presumed to be Milosevic, who firmly opposed the stationing of NATO troops on Serbian territory.

When the conference opened on February 6, several problems immediately became apparent. The Kosovars protested the absence of NATO representatives at Rambouillet. They learned that the French, with the acquiescence of the Ameri-

---

[1]The full text and annexes of the proposed agreement can be found on the Balkan Action Council website at www.balkanaction.org

cans, had insisted on NATO's nonparticipation to accentuate the leading role of the Europeans at the talks. This afforded no opportunity for NATO officials to establish relationships of trust with the Kosovo Albanians. The Kosovars immediately requested that NATO Supreme Commander General Wesley Clark meet with the delegation at Rambouillet, but the chief United States negotiator, Ambassador Christopher Hill, ignored their request.

Another warning sign was the selection of the leadership of the 16-man Kosovo Albanian delegation. With no protest from the United States or other conference organizers, the five KLA members in the delegation secured the chairmanship for their 29-year-old political director, Hashim Thaci. Kosovo's powerless elected president, Ibrahim Rugova, head of the majority Democratic League of Kosovo, and Rexhep Qosja, head of the small opposition United Democratic Movement, rounded out the leadership triumvirate, with Veton Surroi, the influential editor of Kosovo's largest Albanian-language newspaper, *Koha Ditore,* serving as the delegation's spokesman. Thaci, however, dominated the delegation.

Throughout the first week of the conference the Serbian and Kosovo Albanian delegations reviewed the political agreement, the draft constitution for Kosovo, and several annexes related to economic policy, a human rights ombudsman, humanitarian assistance, and other issues; they did not receive the security annex until the thirteenth day of talks. The Kosovars prepared written responses to these drafts and conveyed them to the organizers. By the second week of the conference the Kosovars had largely finished their work and sought feedback from the Contact Group. But the organizers, taking the Kosovo Albanians for granted, were uncommunicative and reluctant to engage in negotiations with them.

When United States Secretary of State Madeleine Albright arrived in Rambouillet on February 20 for final negotiations, she found that the conference was on the verge of collapse. As anticipated, the Serbs were unwilling to accept the agreement. Milosevic had even snubbed the American delegation chief when he visited Belgrade during the talks.

Unexpectedly, Kosovo Albanian delegation chairman Thaci was also refusing to sign on. Thaci had flown the previous day to Slovenia to meet with Adem Demaci, the KLA's political representative who had spent 28 years in Serbian jails for advocating Kosovo's independence. Demaci, an avowed hardliner, had refused to attend the talks, positioning himself to lead a Kosovar rejectionist front, or at least ensure that he would become a kingmaker in the new government, if not its president. Demaci, who reportedly is related to Thaci and influential with several KLA regional commanders, advised the delegation chairman to reject the agreement, and implied that he was speaking for key KLA leaders not attending the conference as well as himself. (The decision by the KLA in early March to dismiss Demaci lends credence to the view that he had overstepped his authority or misrepresented KLA opinions.)

Despite pleas from most other members of the delegation to sign, Thaci proved adamant, and the conference was extended for three additional days to February 23. He claimed the proposed agreement contained insufficient guarantees for independence.

Others present at the talks have told the author that Thaci's real concern was the desire to maintain the KLA intact rather than accepting its dissolution as called for by the agreement.

Finally, on February 23, as the result of a last-minute compromise proposal, the Kosovo Albanians said they would conditionally accept the agreement pending their return to Kosovo to obtain the reaction of the KLA and the Kosovo Albanian people during the next two weeks. The Serbian delegation refused to sign the agreement; plans to pressure Milosevic to accept it were suspended until after the conference reconvened on March 15, if the Kosovars conveyed their unconditional acceptance. The conferees departed Rambouillet amid ominous signs of a major Serbian military buildup in and around Kosovo and reports that Belgrade intended to resume full-scale war in Kosovo.

FAILURE BECOMES THEM?

The conditional acceptance of the proposed agreement by the Kosovars enabled the United States and the Contact Group to claim a partial success at Rambouillet. But the outcome was a major failure for United States leadership, the prestige of Secretary of State Albright, the credibility of the NATO alliance, and the standing of the Kosovo Albanians as potential security partners of the West. The international consensus in support of forceful NATO intervention, which had begun to erode when the Kosovars missed the February 20 deadline, had virtually evaporated by the end of the conference on February 23. The Serbians emerged from Rambouillet as the major beneficiaries of its breakdown.

Four principal reasons account for the failure at Rambouillet, most of which reflect an amateurish approach by the American delegation. First, United States diplomats assumed that the Kosovars would be compliant and thus made little effort to engage them in serious negotiations or ascertain the depth of their concerns about some provisions of the proposed agreement. This tactic changed with the arrival of Secretary Albright, but by then it was too late.

Second, the Americans were insistent on keeping NATO representatives, including Secretary General Javier Solana and Supreme Commander Clark, away from the proceedings. Had NATO officials been present from the beginning, they would have been able to identify and address the real concerns of the Kosovars about security matters and suggest measures to win their confidence. A brief off-site meeting with General Clark was hastily arranged for the five KLA delegates on the penultimate day of the conference, but it came too late to sway the KLA.

Third, United States diplomats should have been more wary of allowing KLA political director Thaci to act as the chairman of the Kosovar delegation. In effect, they condoned a military coup against Kosovo's moderate civilian politicians. The Americans then gave Thaci and the delegation the security annex that called for the KLA to disband. Thus, United States diplomats helped empower the KLA only to antagonize them on a core issue without offering them access to NATO officials who might have been able to assuage their concerns.

Finally, the most important policy reason for failure was the assumption that the KLA would accept its own dissolution as part of the price for obtaining a settlement. After one year of war, in which it suffered military defeat by the Serbs and then displayed the resilience to return to the battlefield stronger than ever, the KLA has become a permanent fixture in Kosovo. Its regional commanders will have the decisive voice among Kosovars in determining the final outcome of the negotiating process and implementing the settlement. The unwillingness of the Americans to accept this at Rambouillet proved their undoing.

AMERICA'S CHOICES

The failure at Rambouillet presented the United States with a nightmare scenario. Until the Kosovo Albanians established themselves as reliable security partners of the West by definitively accepting the agreement, NATO would be unwilling to deploy ground troops to Kosovo. The collapse of the international consensus in support of air strikes against the recalcitrant Serbs, meanwhile, emboldened Milosevic to build up his forces in and around Kosovo and threaten total war against the Kosovo Albanians. The United States was left with no committed partner, one very hostile adversary, and no policy to deal with the rapidly escalating tensions in Kosovo.

As of early March, the future of Kosovo hangs in the balance. If the Kosovo Albanian delegation accepts the document negotiated in February and returns to Rambouillet on March 15, Washington policymakers will confront inherently difficult problems. They need the Kosovo Albanians as partners to implement a settlement that ends the conflict, establishes democratic self-rule in Kosovo, and transforms Kosovo into an exporter of regional stability. This will require coming to terms with the Kosovo Liberation Army and transforming it into a pillar of support for democratic civilian moderates. Yet the KLA contains elements that are antidemocratic and seek to create a revolutionary state to accomplish the union of all Albanian peoples of southeastern Europe. Kosovo Albanian moderates, the natural allies of the United States, have been eclipsed by Washington's neglect and the rise of the KLA.

The starting point for implementing a stabilizing settlement in Kosovo is the acceptance of the KLA as a permanent feature of Kosovo politics. Washington's objective should be the successful integration of a reduced and circumscribed KLA into the democratic political structure that negotiators at Rambouillet tried to create. That can only take place in the context of a NATO military intervention. NATO will need to shield Kosovo from further Serbian assaults while protecting Kosovo's political moderates from extremist elements in the KLA. If the United States fails, or policymakers wash their hands of the problem, the consequence will be renewed fighting that spills over borders and creates the conditions for broadened regional conflict.

The United States should adopt earned independence as its political goal in Kosovo. Milosevic's repression and crackdown have effectively abrogated Serbia's claim on Kosovo, but the Kosovo Albanians are not yet ready for independence. They should not receive it automatically, which is why autonomy, or democratic self-government, makes sense as an intermediate step. The international community, led by the United States and relying on NATO, should put in place and help sustain Kosovo's self-governance until the Kosovo Albanians demonstrate that they have established a durable and stable democracy.

The most immediate problem, however, is that Milosevic may seek to exploit the post-Rambouillet drift, confusion, and indecision on the part of NATO governments with a renewed military offensive. The credibility of NATO's air strike threat needs to be restored urgently. That can best be accomplished by compelling Serbian compliance with the force levels and activities detailed in the October cease-fire agreement.

Milosevic will continue to destabilize the region and generate crises that drain NATO's credibility and challenge American leadership until he is stopped and replaced by a democratic government. Formulating a strategy to achieve that objective should be the first order of business for the administration's Balkan policymakers after the immediate hurdles are cleared in Kosovo. They would be well advised to use the leverage at their disposal by playing upon Milosevic's fear of indictment by the UN war crimes tribunal.

The Kosovo problem is as much about the resolve of American leadership and the credibility of NATO as it is about the complexities of Balkan politics. The Europeans lack the cohesion and political will to carry the burdens of restoring security to the region. The Contact Group, moreover, is too unwieldy to rely on as a vehicle for effective diplomatic action. Originally created as an institutional alibi for Western inaction during the Bosnian war, it remains an impediment to serious policymaking. Neither the crisis in Kosovo nor the broader problems of instability in the Balkans will be resolved until the United States exercises political and military leadership to do so; this will require sustained presidential attention.

The stability of the Balkans has become a strategic interest for the United States, but Washington has no coherent policy to protect United States interests in the region. Instead, Washington is letting Milosevic, rather than NATO, with the help of Balkan moderates, establish the ground rules for post–cold war security in Europe. Since 1991 United States policy has been reactive, lurching from war to war and backing into one military commitment after another. There are American troops in Bosnia and a commitment to send them to Kosovo. This is unlikely to be the last time that the United States military is called in to respond to a Balkan crisis. The sooner Washington acknowledges that burdensome reality, the more likely it will be to take effective steps to deter further crises.

Sea Change in the Arctic

An oceanful of clues points to climatic warming in the far North

By RICHARD MONASTERSKY

The first sign that something was wrong came just a few days after the icebreaker Des Groseilliers left Tuktoyaktuk, a port on the Arctic coast of western Canada. Steaming north with a full load of 20 scientists and provisions for 18 months, the Canadian coast guard ship was heading for a date with the Arctic pack ice—the perennially frozen layer that covers the top of the globe like a white skullcap.

Old hands at polar research, such as oceanographer Miles G. McPhee, expected to meet the ice between 71° and 72°N. That's where the southern edge of the pack showed up in the 1970s, when McPhee made several trips far into the Arctic. Two decades later, the Canadian icebreaker cruised past 72°N at full speed, with no ice in the water to slow its progress. McPhee, who runs a research company out of Naches, Wash., wondered silently, "Oh my God, where did all the ice go?"

The case of the shrinking pack ice is only one of many climatic conundrums troubling scientists who study the Arctic. In recent years, researchers have discovered myriad signs of changes in the far North, affecting everything from the ocean currents flowing 1,000 meters beneath the ice pack to the howling winds

at the top of Earth's atmosphere. "I think the changes are verging on what could be called dramatic," says John M. Wallace, an atmospheric scientist at the University of Washington in Seattle.

Some aspects of the shifting Arctic weather resemble the patterns expected from greenhouse warming, leading scientists to wonder whether they are witnessing early warning signs of conditions to come during the next century. The polar regions, especially the Arctic, are regarded as the climatic equivalent of canaries in a coal mine. The poles are the places on Earth most sensitive to global warming. With the canaries looking distinctly wan, the search is now on to determine what has caused their decline.

These are some of the scientific issues that propelled McPhee and his colleagues northward on the *Des Groseilliers* in the fall of 1997. Following the plans of the $19.5 million international research project, the Canadian icebreaker finally plowed into the pack ice, cut its way to the middle of a large floe, and then stopped its engines at 75°N.

For an entire year, the ship remained frozen amid the ice, dragged 2,800 kilometers on a roundabout course northwest across the Arctic Ocean, a landless expanse of water and ice. Over that

time, planes ferried a total of 170 researchers out to the ship in shifts of several weeks to months. The scientific teams fanned out across the nearby ice to collect measurements on how heat shuttles among the ocean, the sea ice, and the atmosphere through the different seasons. They called the project SHEBA, an acronym for Surface Heat Budget of the Arctic.

From the beginning, SHEBA taught investigators to question the prevailing ideas about the Arctic. Project planners had expected to park the icebreaker amid floes measuring 3 m thick, the kind of ice seen in the 1970s during the last major U.S. Arctic initiative. The SHEBA crew, however, was dismayed by what it encountered in 1997.

"When we went up there, the first problem we had was trying to find a floe that was thick enough. The thickest ice we could find was 1.5 to 2 m," says SHEBA chief scientist Donald K. Perovich of the U.S. Army Cold Regions Research and Engineering Laboratory in Hanover, N.H.

Once the crew set up the ice station, another startling fact came to light. During the first few weeks of work, McPhee and his colleagues found that the shallow waters of the Arctic Ocean were warmer and less salty than they had been 22 years earlier. This observation implies that a significant quantity of ice had melted during the previous summer, says McPhee's team, which published the SHEBA finding in the May 15, 1998 GEOPHYSICAL RESEARCH LETTERS.

More shocks came toward the end of the project in October 1998. "The big surprise of our work was that the ice ended up at the end of the year thinner than when we had started," says Perovich. The warm winter and long summer of 1988 had shaved off about one-third of a meter from the already thin ice.

SHEBA investigators suggest that some of the changes they saw could have stemmed from the 1997–1998 El Niño in the equatorial Pacific, although this temporary warming can't explain all the findings. Arctic sea ice has been declining for many years.

In 1997, researchers from NASA's Goddard Space Flight Center in Greenbelt, Md., reported that the area covered by sea ice decreased by more than 5 percent between 1978 and 1996. Other satellite measurements of the sea ice have revealed evidence of increased summertime melting since 1979 (SN:2/21/98, p. 116).

Sea ice is only a thin skin over the Arctic Ocean, but in many ways it serves as the linchpin of the region's climate and perhaps that of the whole globe. If the pack diminished substantially, climate models predict big changes in the ocean and atmosphere. The reason stems mainly from something scientists call the ice-albedo feedback.

Because ice is so bright, it reflects more than half the sunlight that hits it during summer. The dark water, by contrast, absorbs 90 percent of the incident sunlight, says Perovich. The existence of the sea ice, therefore, keeps the Arctic Ocean cool by shielding it from solar energy.

If the ice starts to disappear, according to theory, the ocean will rapidly warm and melt more ice—a potentially runaway process that could strip the

Arctic of its protective cap and allow solar energy to stream into the ocean unhindered. This ice-albedo feedback would greatly amplify the effects of greenhouse gas polution, causing the Arctic to warm much more than the rest of the globe, according to climate models. The Arctic shift would have a domino effect, rerouting ocean currents and weather patterns further south.

Until recently, scientists had little access to the Arctic Ocean, which served as the sparring ring for superpower submarines. With

Three sets of submerged mountain ridges (black) split the Arctic Ocean into separate basins. The SHEBA project studied conditions on the Alaskan side of the Arctic.

the end of the Cold War and the withdrawal of military forces, scientists started invading the region. Researchers who once sat on opposite sides of the Iron Curtain began sharing data and collaborating on projects. Since 1993, the U.S. Navy has invited oceanographers along on several submarine cruises beneath the ice pack.

In a strange confluence of climate and current events, the thaw in political tensions a decade ago coincided with an apparent shift toward warmer conditions in the North. Some of the biggest changes have occurred within the Arctic

Ocean, an oblong body of water nearly one and a half times the size of the United States. The ocean is divided into four unequal basins separated by three roughly parallel mountain ranges on the seafloor—an arrangement that looks somewhat like an empty TV dinner tray.

Much of the water in the Arctic comes from the Atlantic Ocean, coursing north as a vestigial extension of the Gulf Stream. On the opposite side of the pole, water from the Pacific Ocean enters the Arctic via the Bering Strait. As the Pacific water passes through the shallow strait, it loses much of its heat to the atmosphere and consequently grows colder than the Atlantic water. The two distinct types of water meet midway in the Arctic, creating a front much like the kind that separates warm and cold air masses in the atmosphere.

Water measurements made by Russian and Western scientists indicate that the front between these two types of water bisected the Arctic Ocean for most of the past half-century. From 1949 through the late 1980s, the front generally lined up with a submerged mountain range called the Lomonosov Ridge, which runs from northern Greenland toward eastern Siberia, passing close to the North Pole.

During a scientific submarine cruise in 1993, researchers on board the USS *Pargo* found that the front had shifted substantially. The warmer, saltier Atlantic waters had pushed further into the Arctic, causing a retreat of the Pacific waters, says James H. Morison of the University of Washington, who participated in the *Pargo* cruise.

At the same time, the Atlantic-water layer over the Lomonosov Ridge warmed by 1°C, reaching a temperature not seen in the data going back to 1949, according to a January 1998 report by Morison and his colleagues in DEEP-SEA RESEARCH PART I.

Other shifts have altered the layered arrangement of water in the Arctic. To understand these divisions, imagine lowering a thermometer and a salt meter off a ship in the middle of the Arctic. The uppermost layer of the ocean would be extremely fresh with a temperature near the freezing point. Below that, the instruments would pass through a layer called the halocline, where the water remains cold but grows saltier with depth. Lower still would be a thick region of relatively warm water, so dense with salt that it remains stuck on the bottom, trapped below the fresher, colder layer above.

In a series of three submarine cruises during the 1990s, oceanographers witnessed the halocline weakening in the central part of the Arctic. Early in the decade, a particularly cold sheet of halocline water straddled the Lomonosov Ridge. By 1995, however, the sheet had disappeared from the European side of this range, according to Michael Steele of the University of Washington. He and colleague Timothy Boyd of Oregon State University in Corvallis reported their finding in the May 15, 1998 JOURNAL OF GEOPHYSICAL RESEARCH.

The retreat of this layer could have profound effects, says Steele, because the halocline helps stratify the Arctic Ocean, keeping the warm, deep waters from reaching the surface. "Because there is decreased stratification, there should be enhanced heat transfer from the warm layer up to the surface, and hence thinner ice," says Steele.

Data collected by British submarines show that ice has been thinning in this region for some time. In 1990, researchers with the Scott Polar Research Institute at Cambridge University in England reported that ice thickness on the European side of the Lomonsov Ridge had decreased by 15 percent between 1976 and 1987. Preliminary analysis of measurements made in 1996 now suggest that the downward trend has continued, says Cambridge's Norman Davis.

With so many signs of disruptions in the Arctic climate, scientists are looking skyward for answers. The ice pack and even the water beneath it, they suspect, are taking their cues from the pattern of winds that swirl around the edges of the

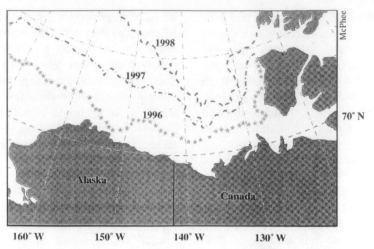

Frozen retreat: The edge of summertime sea ice in the Beaufort Sea pulled northward between 1996 and 1998, according to satellite data. Oceanographers are watching closely to see whether the ice withdrawal continues this summer.

Arctic. "Really, that is where it all starts," says Steele.

What's going on in the Arctic may fit into a larger picture that includes much of the globe's northern half. According to a theory proposed last year, atmospheric pressure tends to fluctuate in seesaw fashion over large parts of the Northern Hemisphere. When pressure increases over the Arctic, it decreases in a donut-shaped ring at the latitude of Washington, D.C. During the reverse part of the irregular cycle, air pressure drops in the Arctic and rebounds in the midlatitudes. Wallace and David W. J. Thompson of the University of Washington called this pattern the Arctic oscillation in the May 1, 1998 GEOPHYSICAL RESEARCH LETTERS.

Atmospheric temperatures in northern Europe and Asia dance in step with this Arctic oscillation. When the atmospheric pressure sags, air temperatures tend to rise and vice versa, says Wallace, who has tracked data going back to 1900.

For most of the century, the Arctic oscillation shifted randomly from month to month and year to year, with no distinct preference for either extreme. Recently, however, the oscillation has leaned strongly toward low Arctic air pressure and high temperatures in northern Eurasia. The shift started in the early 1970s and intensified in the late 1980s, bringing atmospheric conditions unseen in the century of measurements.

"That's the link to perhaps some of the other things going on in the Arctic,"

says Wallace. As the air pressure drops, the westerly winds encircling the Arctic grow stronger. The winds weaken the ice pack by pulling it apart and opening up watery channels between individual ice floes, says Wallace.

The souped-up westerlies also alter conditions in heavily populated parts of the globe. As they scream over the North Atlantic Ocean, the winds carry more heat downwind onto land. This translates into warmer winters in northern Europe and drier winters in southern Europe. "We've seen a strong warming in high latitudes over land in large areas of Russia," says Wallace. "We're seeing changes in the circulation that are big enough in winter that people in Europe are noticing them."

The key question is, What has caused all this change? In a study using computer models of the atmosphere, a team of Russian researchers has found evidence that thinning of the Arctic ozone layer could be driving the shift in the atmosphere. As chlorofluorocarbons and other chemicals eat away ozone, the polar stratosphere cools off. In the Russian model, this cooling effect enhances the westerlies that circle the Arctic and hence reduces the atmospheric pressure there.

Greenhouse gases could also have a hand in the Arctic shift. Carbon dioxide and other heat-trapping pollutants tend to warm the lowest layer of the atmosphere—the troposphere—while cooling off the stratosphere above. Researchers at NASA's Goddard Institute for Space Studies in New York City used a computer model to investigate this split effect of greenhouse gases. They found that the combination speeds up the westerly polar winds, pushing the Arctic oscillation in the direction seen during the past 30 years, says NASA's Drew T. Shindell.

Wallace thinks there is a good chance that one or both of these human influences could have precipitated the Arctic changes. Yet he cautions against drawing any conclusions. "We have to allow some possibility that this is of natural origin," as part of a long-term cycle that will eventually reverse, he says. To rule out a natural cause, the Arctic oscillation would have to keep on its present course for another 5 to 10 years, he says.

If that indeed transpires, researchers may not be able to repeat an experiment like SHEBA. With sustained warming and stronger westerlies, the ice-albedo feedback could kick in and rapidly rid the Arctic of its white cover during summer. McPhee and his colleagues raised this possibility last year in their paper in GEOPHYSICAL RESEARCH LETTERS. In the title, they asked, "Is perennial sea ice disappearing?"

Early in his career, McPhee would have found such a question unthinkable. Now, he says, "I'm starting to wonder whether we're not going to see it happen."

GREENVILLE
FROM BACK COUNTRY
to FOREFRONT

Eugene A. Kennedy

What factors are crucial in determining the success or failure of an area? This article explores the past and present and glimpses what may be the future of one area which is experiencing great success. The success story of Greenville County, S.C. is no longer a secret. This article seeks to find the factors which led to its success and whether they will provide a type of yardstick to measure the future.

The physical geography of this area is explored, as well as the economic factors, history, transportation, energy costs, labor costs and new incentive packages designed to lure new industries and company headquarters to the area.

Physical geography: advantageous

Greenville County is situated in the northwest corner of South Carolina on the upper edge of the Piedmont region. The land consists of a rolling landscape butted against the foothills of the Appalachian Mountains. Monadnocks, extremely hard rock structures which have resisted millions of years of erosion, rise above the surface in many places indicating that the surface level was once much higher than today. Rivers run across

An open courtyard off Main Street, downtown Greenville, S.C.

the Piedmont carving valleys between the plateaus. The cities, farms, highways and rail lines are located on the broad, flat tops of the rolling hills.

Climatologically, the area is in a transition zone between the humid coastal plains and the cooler temperatures of the mountains, resulting in a relatively mild climate with a long agricultural growing season. The average annual precipitation for Greenville County is 50.53 inches at an altitude of 1040 feet above sea level. The soil is classified as being a Utisoil. This type of soil has a high clay base and is usually found to be a reddish color due to the thousands of years of erosion which has leached many of the minerals out of the soil, leaving a reddish residue of iron oxide. This soil will produce good crops if lime and fertilizer containing the eroded minerals are added. Without fertilizers, these soils could sustain crops on freshly cleared areas for only two to three years before the nutrients were exhausted and new fields were needed. This kept large plantations from being created in the Greenville area. The climate and land are such that nearly anything could be cultivated with the proper soil modification. Physical potential, although a limiting factor, is not the only determining factor in the success of an area.

 From *Focus*, Spring 1998, pp. 1-6. © 1998 by the American Geographical Society. Reprinted by permission.

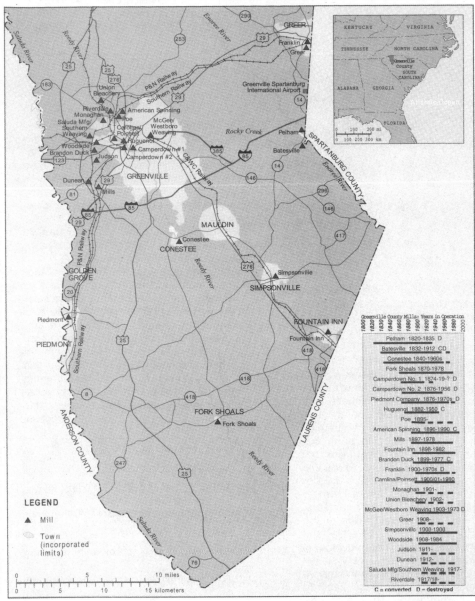

Greenville County Mills: Years in Operation

Mill	1800	1820	1830	1840	1860	1880	1900	1920	1940	1960	2000
Pelham 1820-1835 D											
Batesville 1832-1912 CD											
Conestee 1840-1960s											
Fork Shoals 1870-1978											
Camperdown No. 1 1874-19-? D											
Camperdown No. 2 1876-1956 D											
Piedmont Company 1876-1970s D											
Huguenot 1882-1950 C											
Poe 1895-											
American Spinning 1896-1990 C											
Mills 1897-1978											
Fountain Inn 1898-1982											
Brandon Duck 1899-1977 C											
Franklin 1900-1970s D											
Carolina/Poinsett 1900/01-1990											
Monaghan 1901-											
Union Bleachery 1902-											
McGee/Westboro Weaving 1903-1973 D											
Greer 1908-											
Simpsonville 1908-1900											
Woodside 1908-1984											
Judson 1911-											
Dunean 1912-											
Saluda Mfg/Southern Weaving 1917-											
Riverdale 1917/18-											

C = converted   D = destroyed

LEGEND

▲ Mill

◻ Town (incorporated limits)

KAREN SEVERUD COOK, UNIVERSITY OF KANSAS MAP ASSOCIATES

As Preston James, one of the fathers of modern geography pointed out, the culture of the population which comes to inhabit the area greatly determines the response to that particular physical environment.

From European settlement through the textile era

An Englishman named Richard Pearis was the first to begin to recognize the potential of what was then known as the "Back Country." The area was off limits to white settlers through a treaty between the British and the Indians. In order to get around the law, Pearis married a native American and opened a trading post in 1768 at the falls of the Reedy River.

He soon built a grist mill and used the waterfalls for power. Pearis prospered until the end of the Revolutionary War. He had remained loyal to the British, lost all his property when the new nation was established and left the country.

Others soon realized the potential of what was to become Greenville County. Isaac Green built a grist mill and became the area's most prominent citizen. In 1786, the area became a county in South Carolina and was named for Isaac Green. The Saluda, Reedy and Enoree Rivers along with several smaller streams had many waterfalls, making them excellent locations for mills during the era of water power. During the Antebellum period, Greenville County's 789 square miles was inhabited by immigrants with small farms and also served

as a resort area for Low Country planters who sought to escape the intense summer heat and the disease carrying mosquitoes which flourished in the flooded rice fields and swampy low country of the coastal plains. Most of the permanent residents farmed and a few mills were built to process the grains grown in the area.

Although very little cotton was ever actually grown in Greenville County, cotton became the driving force behind its early industrialization. Beginning with William Bates in 1820, entrepreneurs saw this area's plentiful rivers and waterfalls as a potential source of energy to harness. William Bates built the first textile mill in the county sometime between 1830 and 1832 on Rocky Creek near the Enoree River. This was known as the Batesville Mill. Water power dictated the location of the early southern textile mills, patterned after the mills built in New England. Mill owners purchased the cotton from farms and hauled it to their mills, but lack of easy transportation severely limited their efforts until 1852. That year, the Columbia and Greenville Railroad finally reached Greenville County. Only the interruption of the Civil War kept the local textile industry from becoming a national force during the 1850s and 1860s.

The area missed most of the fighting of the Civil War and escaped relatively unscathed. This provided the area with an advantage over those whose mills and facilities had been destroyed during the war. The 1870 census reported a total $351,875 in textiles produced in the county. This success encouraged others to locate in Greenville. Ten years later, with numerous mills being added each year, the total reached $1,413,556.11. William Bates' son-in-law, Colonel H. P. Hammett, was owner of the Piedmont Company which was the county's largest producer. Shortly after the construction of the water powered Huguenot Mill in the downtown area of the City of Greenville County in 1882, the manufacture of cotton yarn would no longer be controlled by the geography of water power.

The development of the steam engine created a revolution in the textile industry. No longer was the location of the mill tied to a fast moving stream, to turn a wheel that moved machinery. Large amounts of water were still needed but the dependence upon the waterfalls was severed. Between 1890 and 1920, four textile plants were built in the county outside the current city limits of the City of Greenville. At least thirteen large mills were built near the city to take advantage of the rail system, as shown on the map, "Greenville County Mills." Thus, with cheaper and more efficient steam

F.W. Poe Manufacturing Co., Old Buncombe Road, Greenville, S.C. Built in 1895, purchased by Burlington Industries and closed in 1997. Palmetto State Dyeing and Finishing Co. opened in 1987. The company employs approximately 110 people.

power, transportation costs became a deciding factor. These mills built large boiler rooms adjacent to their plants and dug holding ponds for water.

Another drastic change took place in the textile industry around 1900. This change would provide even greater flexibility for the mill owners. A hydroelectric dam was constructed on the Saluda River, five miles west of the city of Greenville. It was completed in 1902 and would provide cheap electricity for the county. John Woodside, a local mill owner who foresaw electricity as the next step in the evolution of the industry, built what was then the largest textile mill in the world in the city of Greenville that same year. He located it further from a water source than previously thought acceptable. However, John had done some primitive locational analysis and chose the new site well. It was located just beyond the city boundary to limit his tax liability and directly between the lines of two competing rail companies—the Piedmont Railroad and the Norfolk and Western (now known as Norfolk and Southern). John Woodside's mill proved to be a tremendous success. With water no longer a key factor of location, the owners identified transportation as the key factor of location. Others began to build near rail lines.

The textile industry made Greenville County very prosperous. The mills needed workers and shortly outstripped the area's available labor supply. Also, many did not want to work in the hot, poorly ventilated, dangerous conditions found in the mills. When most of the mills were still built of wood, the cotton fibers

floating in the air made fire a very real danger. Many businesses sprang up to service the needs of the workers and the textile mill owners. Farmers, sharecroppers, former slaves and children of former slaves were recruited to work in the mills. Housing soon became scarce and the infrastructure wasn't equipped to handle the influx of new workers. To alleviate the problem, the mill owners built housing for their workers. These were very similar to the coal camps of Appalachia and other factory owned housing in the north. They were very simple dwellings built close to the mill so the workers could easily walk to and from work. They also provided company-owned stores, doctors and organized recreational activities for their employees, creating mill communities. Many people who worked for the mills would have told you they lived at Poe Mill or Woodside, the names of their mill communities, rather than Greenville.

In the 1960s, rail transportation of textiles was a cost the owners wished to lower. They found a cheaper, more versatile form of transportation in the trucking industry. The interstate highway system was now well developed and provided a means of keeping costs down for the operators. In the 1970s, owners began to identify wages and benefits as a major factor in their cost of operation and many firms relocated in for-

eign countries, which offered workers at a fraction of the wages paid in the United States and requiring few if any benefits.

Meeting the challenge of economic diversification

Greenville County used its natural physical advantage to become the "Textile Capital of the World." Many of the other businesses were tied directly or indirectly to the textile industry. These ranged from engineering companies who designed and built textile machinery to companies which cleaned or repaired textile machines. Employment in the textile industry in Greenville County peaked in 1954 with 18,964 workers directly employed in the mills. As the industry began to decline, the leaders of the industry along with local and state leaders showed great foresight by combining their efforts into an aggressive move to transform Greenville County into a production and headquarters oriented economy. A state sponsored system of technical schools greatly facilitated this effort. Workers could get the training they needed to pursue almost any vocation at these centers. This system still is a factor in Greenville County's success.

Table 1

CORPORATE HEADQUARTERS IN GREENVILLE COUNTY

1. American Leprosy Mission International
2. American Federal Bank
3. Baby Superstores, Inc.
4. Bowater Inc.
5. Builder Marts of America Inc.
6. Carolina First Bank
7. Delta Woodside Industries Inc.
8. Ellcon National Inc.
9. First Savings Bank
10. Heckler Manufacturing and Investment Group
11. Henderson Advertising
12. Herbert-Yeargin, Inc.
13. JPS Textile Group Inc.
14. Kemet Electronics Corp.
15. Leslie Advertising
16. Liberty Corp.
17. Mount Vernon Mills Inc.
18. Multimedia Inc.
19. Ryan's Family Steakhouses
20. Span America
21. Steel Heddle Manufacturing Co.
22. Stone Manufacturing Co.
23. Stone International.
24. TNS Mills Inc.
25. Woven Electronics Corp.

> **What ultimately swayed the automaker to choose Greenville? One of the main reasons was physical location.**

The group emphasized the ability to make a profit in Greenville County. The focus of their efforts was turned to creating a sound technical education network along with the flexibility to negotiate packages of incentives to lure large employers. Incentives included negotiable tax and utility rates, plus a strong record of worker reliability due to South Carolina's nonunion tradition, with very few work stoppages. The foresight of this group has paid off handsomely. The majority of the textile mills which provided the backbone of the economy of Greenville County are no longer in business. Many of the old buildings still stand. Ten of the mills built before 1920 now are used in other capacities. American Spinning was built in 1896 and now is used as a warehouse, office space and light manufacturing all under one roof. Most of the mills are used for warehouse space or light manufacturing such as the Brandon Duck Mills, which operated between 1899 and 1977 as a cotton mill. It now houses two small factories which assemble golf clubs and part of the mill is used as a distribution center. The low lease cost (from $1 to $15 per square foot) is an enticement for other businesses to locate in these old buildings.

The old Huguenot Mill, the last water powered mill built in the county, was recently gutted and has been rebuilt as offices for the new 35 million dollar Peace Center entertainment complex in downtown Greenville. The Batesville Mill, the first in the county, was built of wood. It burned and was rebuilt in brick in 1881. It closed its doors in 1912 because the water-powered mill was not competitive. The mill was purchased by a husband and wife in 1983, converted into a restaurant, and was the cornerstone and headquarters of a chain of FATZ Restaurants until it burned again in 1997. So, in considering diversification, one of the first steps was to look for other uses for the facilities which already existed.

Other efforts also met with great success. As businesses began to look south during the 1970s for relocation sites, Greenville began to use its natural advantages to gather some impressive companies into its list of residents. By 1992, the combination of these efforts made Greenville County the wealthiest county in the state of South Carolina. Twenty-five companies have their corporate headquarters in the county, as shown in Table 1.

Forty-nine others have divisional headquarters in the county, as shown in Table 2. This constitutes a sizable investment for the area, yet even this list does not include a 150 million dollar investment by G. E. Gas Turbines in 1992 for expansion of their facility. This was the largest recent investment until 1993.

Along with American companies, foreign investment was sought as well. Companies such as Lucas, Bosch, Michelin, Mita and Hitachi have made major investments in the county. Great effort has been put into reshaping the face of Main Street in Greenville as well. The city is trying to make a place where people want to live and shop. Many specialty stores have opened replacing empty buildings left by such long time mainstays as Woolworths. The Plaza Bergamo was created to encourage people to spend time downtown. The Peace Center Complex provides an array of entertainment choices not usually found in a city the size of Greenville. The Memorial Auditorium, which provided everything from basketball games, to rodeo, concerts, high school graduations and truck pulls has closed its doors and was demolished in 1997 to make way for a new 15,000 seat complex which will be named for its corporate sponsor. It will be called the Bi-Lo Center. Bi-Lo is a grocery store chain and a division of the Dutch Company, Ahold. A new parking garage is being built for this center and two other garages have recently been added to improve the infrastructure of the city. City leaders have traveled to cities such as Portland, Oregon to study how they have handled and managed growth and yet kept the city friendly to its inhabitants.

The largest gamble for Greenville County came in early 1989. The automaker BMW announced that it was considering building a factory in the United States. Greenville County and the state of South Carolina competed against several other sites in the midwest and southeast for nearly two years. On June 23, 1992, the German automaker chose to locate in the Greenville-Spartanburg area. Although the plant is located in Spartanburg County,

Panorama of downtown Greenville, S.C.

Table 2

DIVISIONAL HEADQUARTERS IN GREENVILLE COUNTY

1. Ahold, Bi-Lo.
2. BB&T, BB&T of South Carolina.
3. Bell Atlantic Mobile
4. Canal Insurance Company
5. Coats and Clark, Consumer Sewing Products Division.
6. Cryovac—Div. of W. R. Grace & Co.
7. Dana Corp., Mobile Fluid Products Division.
8. DataStream Systems, Inc.
9. Dodge Reliance Electric
10. Dunlop Slazenger International, Dunlop Slazenger Corp.
11. EuroKera North America, Inc.
12. Fluor Daniel Inc.
13. Fulfillment of America
14. Frank Nott Co.
15. Gates/Arrow Inc.
16. General Nutrition Inc., General Nutrition Products Corp.
17. Gerber Products Co., Gerber Childrenswear, Inc.
18. GMAC
19. Goddard Technology Corp.
20. Greenville Glass
21. Hitachi, Hitachi Electronic Devices (USA)
22. Holzstoff Holding, Fiberweb North America Inc.
23. IBANK Systems, Inc.
24. Insignia Financial Group, Inc.
25. Jacobs-Sirrine Engineering
26. Kaepa, Inc.
27. Kvaerner, John Brown Engineering Corp.
28. Lawrence and Allen
29. LCI Communications
30. Lockheed Martin Aircraft Logistics Center Inc.
31. Manhattan Bagel Co.
32. Mariplast North America, Inc.
33. Michelin Group, Michelin North America.
34. Mita South Carolina, Inc.
35. Moovies, Inc.
36. Munaco Packing and Rubber Company.
37. National Electrical Carbon Corp.
38. O'Neal Engineering
39. Personal Communication Services Dev.
40. Phillips and Goot
41. Pierburg
42. Rust Environment and Infrastructure
43. SC Teleco Federal Credit Union
44. Sodotel
45. South Trust Bank
46. Sterling Diagnostic Imaging, Inc.
47. Umbro, Inc.
48. United Parcel Service
49. Walter Alfmeier GmbH & Co.

the headquarters are in Greenville and both counties will profit greatly. When the announcement was made, the question was: what ultimately swayed the automaker to choose Greenville? One of the main reasons was physical location. The site is only a four hour drive from the deepwater harbor of Charleston, SC. Interstates 26 and 85 are close by for easy transportation of parts to the assembly plant. The Greenville-Spartanburg Airport is being upgraded so that BMW can send fully loaded Boeing 747 cargo planes and have them land within five miles of the factory. Plus, the airport is already designated as a U.S. Customs Port of Entry and the flights from Germany can fly directly to Greenville without having to stop at Customs when entering the country. Other incentives in the form of tax breaks, negotiated utility rates, worker training and state purchased land helped BMW choose the 900 acre site where it will build automobiles.

The large incentive packages might appear self-defeating but BMW's initial investment was scheduled to be between 350 and 400 million dollars. The majority of the companies supplying parts for BMW also looked for sites close enough to satisfy BMW's just-in-time manufacturing needs. The fact that Michelin already made tires for their cars here, Bosch could supply brake and electrical parts from factories already here and J.P. Stevens and others could supply fabrics for automobile carpets and other needs readily from a few miles away also was a factor.

One major BMW supplier, Magna International, which makes body parts for the BMW Roadster and parts for other car manufacturers, located its stamping plant in Greenville County. Magna invested $50 million and will invest $35 million more as BMW expands. Magna needed 100 acres of flat land without any wetlands and large rock formations. This land needed to be close enough to provide delivery to BMW. After studying several sites, Magna chose South Donaldson Industrial Park, formerly an Air Force base, just south of the city of Greenville. The county and state will help prepare the location for their newest employer.

Road improvements, the addition of a rail spur and an updating of water and sewer facilities will all be provided to Magna in this agreement. Also, Magna will receive a reduced 20 year fixed tax rate along with other incentives for each worker hired. These incentive packages

Huguenot Mill (lower left). Built in 1882, on Broad Street, Greenville, S.C., is the last waterpowered mill in Greenville County. It is being refurbished to become part of the Peace Center Complex at right.

may seem unreasonable but they have proven to be necessary in the 1990s when large organizations are deciding where to locate.

The future: location, location and location

From the time of the earliest European settlers, the natural advantages of Greenville County helped bring it to prosperity. The cultural background of the settlers was one of industry and a propensity for changing the physical environment to maximize its industrial potential. Nature provided the swift running rivers and beautiful waterfalls. The cultural background of the settlers caused them to look at these natural resources and see economic potential.

The people worked together to create an environment which led Greenville County to be given the title of "Textile Center of the World" in the 1920s. Then, again taking advantage of transportation opportunities and economic advantages, the area retained its textile center longer than the majority of textile centers.

Today, after 30 years of diversification, economic factors now are normally the deciding factor in the location of a new business or industry. Greenville County with its availability of land, reasonable housing costs, low taxes, willingness to negotiate incentive packages, and positive history of labor relations helped make it a desirable location for business. Proximity to interstate transportation, rail and air transport availability help keep costs low. The county's physical location about half way between the

A combination of physical, environmental and cultural factors greatly influence the location of businesses.

mega-growth centers of Charlotte, North Carolina and Atlanta, Georgia places it in what many experts call the mega-growth center of the next two decades. Now, with BMW as a cornerstone industry for the 1990s and beyond, Greenville County looks to be one of the areas with tremendous growth potential.

Thus, a combination of physical, environmental and cultural factors greatly influence the location of businesses. Transportation costs, wage and benefit packages and technical education availability are all interconnected.

The newest variable involves incentive packages of tax, utility reduction, worker training, site leasing and state and local investment into improving the infrastructure for attracting employers. The equation grows more and more complex with no one factor outweighing another; however, economic costs of plant or office facilities, wages and benefits and transportation seem to be paramount. Greenville County is blessed with everything it needs for success. It will definitely be one the places "to be" in the coming years.

Eugene A. Kennedy is a native of West Virginia who attended Bluefield State College, Bluefield West Virginia, and received an M.A. in geography from Marshall University in Huntington. He is currently a public educator in the Greenville County School system, Greenville S.C. He was awarded a "Golden Apple" by Greenville television station WYFF in 1997, has been a presenter at the South Carolina Science Conference, and a consultant to the South Carolina State Department of Education. He can be reached at GEOGEAK@aol.com

References and further readings

DuPlessis, Jim. 1991. Many Mills Standing 60 Years After Textile Heydays. *Greenville News-Piedmont* July 8. pp. 1c–2c.
Greater Greenville Chamber of Commerce, 1990. *1990 Guide to Greenville.*
Greenville News-Piedmont. 1991–1993. *Fact Book 1991; 1992; 1993.*
Patterson, J. H. 1989. *North America.* Oxford University Press: N.Y. Eighth edition.
Scott, Robert. 1993. Upstate Business. *Greenville-News Piedmont.* 15 August. pp. 2–3.
Shaw, Martha Angelette. 1964. *The Textile Industry in Greenville County.* University of Tennessee Master's Thesis.
Strahler, Arthur. 1989. *Elements of Physical Geography.* John Wiley and Sons: N.Y. Fourth edition.

Flood Hazards and Planning in the Arid West

Can we somehow marry flood control or flood planning with water quality and best management practices? This would require truly integrated watershed management.

By John Cobourn

IN Nevada and other areas in the west, many urbanizing areas have not been inhabited for very long. In these areas, there is a short period of record regarding local flood hazards. Therefore, people commonly lack awareness about flood hazards. One question of particular interest is, "What kind of watershed approach is suitable to address flooding?" In this day-and-age, dams are not necessarily the most popular solution for flooding. What about non-structural techniques? What about best management practices? How can we link flood control or flood management with watershed concerns like water quality and habitat?

In the arid west, there are two main types of floods that are completely different. They are like night and day. There is winter flooding, also called "river valley bottom" flooding. It comes down a river like the Truckee or Carson River from the Sierra Nevada or another mountain range. It produces tremendous flooding in the valleys. Most valley bottoms are Zone A flood zones. The other type of flooding is called alluvial fan flooding or "flash" flooding. These happen in the summer and are called dry

mantle events because they happen when the soil is not saturated. Because desert soils often have poor infiltration capacity, if a cloudburst occurs that lasts even an hour or two, it can cause a flash flood to rush down a normally dry wash. These creeks are normally bone-dry 365 days a year, unless there is a flash flood. The water might come down at 500, 1,000 or 2,000 cfs for 30 minutes or an hour and damage a subdivision on an alluvial fan.

River Valley Floods

Six feet of snow fell in the Sierra Nevada near Lake Tahoe just before Christmas of 1996, and on December 27–28, another storm dumped two to three more feet of snow. On New Year's Eve and New Year's Day a huge rainstorm hit the area. The weather changed from an Alaskan snowstorm to a Hawaiian rainstorm.

As a result, in Nevada and California, major river valley floods occurred on the Truckee, Carson and Walker Rivers in January of 1997. They surprised many people. A rain-on-snow even at high ele-

vations of the Sierra Nevada caused large amounts of runoff to come down the rivers into the local valleys. There was five feet of water in some casinos in downtown Reno. There was also terrible flood damage to the Reno-Tahoe International Airport, to the large industrial section of Sparks, which is next to the river in the floodplain, and to a residential neighborhood in Gardnerville.

The Carson River parallels the Truckee, coming down from the Sierra and going out through Fallon to its terminus, the Stillwater Wildlife Refuge. The Truckee River also fails to reach the sea, as its terminus is the large desert lake, Pyramid Lake, owned by the Paiute Tribe. The watersheds are almost the same size with almost the same water flow. The two rivers are an interesting contrast.

The Truckee River goes through Reno and has many reservoirs on the upper watershed, including Lake Tahoe, which has a six-foot dam on it. The Carson River has very few reservoirs.

The Truckee River has a tremendous urban area right in the middle of the watershed. Reno and Sparks have a population of over 250,000, with much of

From *Land and Water*, May/June 1999, pp. 15-17. © 1999 by Land and Water, Inc. Reprinted by permission.

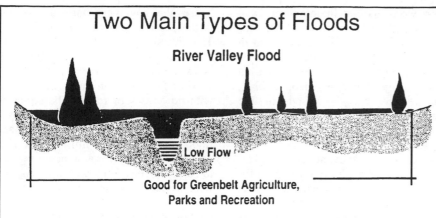

Two Main Types of Floods

River Valley Flood

Low Flow

Good for Greenbelt Agriculture,
Parks and Recreation

In river valley flooding, water spills across the low-lying flood plain, making these lands ideal for open space, agriculture, or recreational uses.

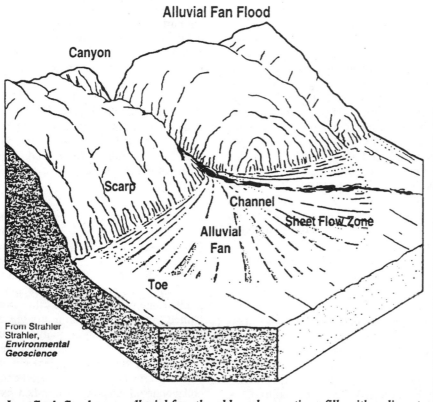

Alluvial Fan Flood

Canyon

Scarp

Channel

Sheet Flow Zone

Alluvial
Fan

Toe

From Strahler
Strahler,
*Environmental
Geoscience*

In a flash flood on an alluvial fan, the old wash sometimes fills with sediment, causing the channel to shift to a new location, disturbing new areas.

ferences on flooding and watershed management that were held in Nevada in 1997 and 1998. Hydrologists and geomorphologists are urging county planners and local ranchers to leave agriculture in the valley floodplains rather than subdivide them.

In the 1997 flood, most Zone A flood zones were inundated in both watersheds. Luckily, there was not a great deal of flooding of homes. One levee did break and flooded about 50 homes in upper Gardnerville.

Water Quality Concerns

The 1997 winter flood scoured and widened the channels of many creeks in the low elevations around Reno. A large amount of sediment washed down many urban tributaries to the Truckee River.

Can we somehow marry flood control or flood planning with water quality and best management practices? This would require truly integrated watershed management.

Steamboat Creek is the largest tributary to the Truckee River in the Reno area. Cooperative Extension and citizen volunteers are working with the Washoe-Storey Conservation District and Nevada Division of Environmental Protection to clean it up, not through structures like dams or levees, but through implementation of best management practices. Nevada is one of the fastest growing states in the nation, so there is a large potential for erosion in this area on new construction sites. In a big event, there is a huge flush of pollution as sediment and nutrients flow downstream.

During a rainstorm, the flow of a creek will increase much more gradually, with a lower flood crest, if there are many wetlands in the watershed than if there are few or no wetlands. The wetlands serve as holding ponds and capture some of the rain before it flows into the river. One golf course in Reno incorporated wetlands as part of the golf course design. The course is located on Steamboat Creek, so some upstream sediment is filtered out at the golf course. When valley floors are used for agriculture, a flooding river can usually flow out of its banks onto its floodplain. When rivers have access to their floodplains, the erosive power of the flood water is dissipated, so that less sedimentation occurs.

their commercial-industrial area situated in the floodplain.

There is not a great deal of urbanization in the floodplain of the Carson River. There is a great deal of agricultural land still in the flood plain. However, the Carson Valley is one of the fastest growing population centers in Nevada. The Carson River flows through the Carson Valley and normally, in late summer, has dwindled to a small stream. In the flood of January 1997 however, it created a temporary lake that

covered Highway 395, the main highway between that area and Reno, and much of the agricultural land on the valley floor.

The population of the Carson Valley is now about 30,000 and is projected to approach 60,000 in the next 20 years or so. There is debate about whether there should be more urbanization in the current floodplain. Should Planners repeat in the Carson Valley what has been done in Reno and Sparks? It is a big question, and it was discussed at two major con-

Alluvial Fan Flash Flooding

Let's consider alluvial fan flooding. It is easy to see how sediment coming down a canyon can create a fan-shaped deposit. Along almost all of western mountain ranges, at the bottom of even small canyons that are dry now, there are alluvial fans. In many parts of the southwest, alluvial fans are being urbanized. Over half of the developable land in Nevada is on alluvial fans, which are subject to flooding. Many people do not recognize that they live on alluvial fans.

One area in Douglas County has had five flash floods between 1990 and 1994. This alluvial fan is one of the fastest growing areas in Douglas County. In the master plan, it is slated for extensive and rapid urban growth. When a thunderstorm occurs, this area may receive intense rain and flash floods.

When subdivisions are built in Nevada, developers are required to put in detention ponds so that the runoff water from the rooftops and pavements in the subdivision can be gathered and possibly create some wildlife habitat. Residents in one subdivision decided after about five years that they didn't like the way their detention pond was working. The pond had a spillway that was high enough so that, when a flash flood occurred, the water it detained backed up into adjacent streets. There was eight inches of water in the street, and the residents didn't like it. The subdivision brought in some bulldozers and lowered the spillway. The spillway still retains some water, but more of the water will flow downstream to where another subdivision will be built someday. It could flood residents of that subdivision in the future.

Many residents and county planners in the west have not yet realized the flash flood dangers of the region's numerous alluvial fans. With their gently sloping topography, they seem to be ideal locations for homesites and subdivisions. Yet the Federal Emergency Management Agency (FEMA) has identified "a critical need to provide guidance to communities, developers and citizens on how to safely accommodate growth while protecting life and property from flood hazards on alluvial fans." (FEMA 165, 1989)

The flash floods that occurred in Douglas County, Nevada in the early 1990s were warning floods. They have not been close in size to a 100-year event. If five similar floods occurred in five years, by definition they couldn't be 100-year events. These floods caused some damage, but many people think, "So, we have had five floods in five years. I'm not going to worry about it. They didn't flood my house." These people don't realize that they haven't seen a true 100-year event yet. The planners and developers keep developing for flood rates of 600 cfs because that is what they have seen in the last five years. They haven't seen the 2,000-cfs flood event yet. Due in part to the recent floods, some of these washes are being restudied by FEMA now.

In 1974, a flood occurred in El Dorado Canyon, located in southern Nevada. A series of warning floods had gone through a national park campground causing damage to picnic tables, and some campers had gotten wet. The Park Service wanted to move the campground because it was in a flood area. Many of the people who had been using the park area for 10 years signed a petition and said they loved that campground. "It is right on the lake and we can bring our boats here. Don't move the campground." Two years later, a flood wiped out the entire campground, demolished every structure, and killed nine people. If people had paid more attention to the small "warning floods", they might have been out of harm's way when the true 100-year event struck the campsite.

One of the things Cooperative Extension is trying to do in Nevada is to educate people about this kind of problem through conferences and publications. In Douglas County, a citizens group called the Buckbrush Wash Flood Safety Coalition has formed. The people there have said, "Let's see if we can work with our county government and local government." These citizens have been using public education and working with their local search and rescue unit to educate the children and their parents about the flash flood areas. They are educating the children not to play in the washes.

This citizens group has also come up with a design for a diversion structure. In Nevada, a diversion structure for flood water is listed as a best management practice. There are several new houses that have been built in the Zone A flood zone near the apex of this fan. The diversion structure, which the Coalition designed with help from NRCS in Nevada, could divert the peak flow from a major flood into an uninhabited basin with no houses or roads. The structure would cost $1.4 million. The county's previous estimate for reducing the flash flood potential was $5 million. The citizens were looking for an innovative way to solve this problem, and they came up with a relatively inexpensive proposal that would also be much better for water quality because it would prevent flood peaks from washing through urban neighborhoods.

There is not much appetite around the nation for dams on rivers. However, with urbanized alluvial fans, because the land surface is convex, non-structural techniques alone will not solve the problem. That is because on most alluvial fans, the entire fan surface below its apex constitutes the floodplain. If you live in an area where people are urbanizing alluvial fans, it would be interesting to see if PL-566 funds or other public grant funds could be used to help create structures to safeguard lives and property.

Conclusion

According to the National Weather Service, floods and flash floods are the number one weather related killer in the United States. To plan appropriately for floods, drought, water quality, urban growth and a sustainable ecosystem, managers of diverse agencies and private concerns need to come together and collaborate on integrated watershed management. When flood problems are viewed as interconnected with pollution, infrastructure, and habitat problems, we will begin the uphill journey to better solutions. If we can learn to cooperate and help each other, we can make substantial progress toward creating safer communities and protecting water quality.

For more information, contact John Cobourn, University of Nevada Cooperative Extension, P.O. Box 8208, Incline Village, NV 89452-8208, (775) 832-4150, fax (775) 832-4139.

Does it matter where you are?

The cliché of the information age is that instantaneous global telecommunications, television and computer networks will soon overthrow the ancient tyrannies of time and space. Companies will need no headquarters, workers will toil as effectively from home, car or beach as they could in the offices that need no longer exist, and events half a world away will be seen, heard and felt with the same immediacy as events across the street—if indeed streets still have any point.

There is something in this. Software for American companies is already written by Indians in Bangalore and transmitted to Silicon Valley by satellite. Foreign-exchange markets have long been running 24 hours a day. At least one California company literally has no headquarters: its officers live where they like, its salesmen are always on the road, and everybody keeps in touch via modems and e-mail.

Yet such developments have made hardly a dent in the way people think and feel about things. Look, for example, at newspapers or news broadcasts anywhere on earth, and you find them overwhelmingly dominated by stories about what is going on in the vicinity of their place of publication. Much has been made of the impact on western public opinion of televised scenes of suffering in such places as Ethiopia, Bosnia and Somalia. Impact, maybe, but a featherweight's worth.

World television graphically displayed first the slaughter of hundreds of thousands of people in Rwanda and then the flight of more than a million Rwandans to Zaire. Not until France belatedly, and for mixed motives, sent in a couple of thousand soldiers did anyone in the West lift so much as a finger to stop the killing; nor, once the refugees had suddenly poured out, did western governments do more than sluggishly bestir themselves to try to contain a catastrophe.

Rwanda, of course, is small (population maybe 8m before the killings began). More important, it is far away. Had it been Flemings killing Walloons in Belgium (population 10m) instead of Hutus slaying Tutsi in Rwanda, European news companies would have vastly increased their coverage, and European governments would have intervened in force. Likewise, the only reason the Clinton administration is even thinking about invading Haiti is that it lies a few hundred miles from American shores. What your neighbours (or your kith and kin) do affects you. The rest is voyeurism.

The conceit that advanced technology can erase the contingencies of place and time ranges widely. Many armchair strategists predicted during the Gulf war that ballistic missiles and smart weapons would make the task of capturing and holding territory irrelevant. They were as wrong as the earlier seers who predicted America could win the Vietnam war from the air.

In business, too, the efforts to break free of space and time have had qualified success at best. American multinationals going global have discovered that—for all their world products, world advertising, and world communications and control—an office in, say, New York cannot except in the most general sense manage the company's Asian operations. Global strengths must be matched by a local feel—and a jet-lagged visit of a few days every so often does not provide one.

Most telling of all, even the newest industries are obeying an old rule of geographical concentration. From the start of the industrial age, the companies in a fast-growing new field have tended to cluster in a small region. Thus, in examples given by Paul Krugman, an American economist, all but one of the top 20 American carpet-makers are located in or near the town of Dalton, Georgia; and, before 1930, the American tire industry consisted almost entirely of the 100 or so firms carrying on that business in Akron, Ohio. Modern technology has not changed the pattern. This is why the world got Silicon Valley in California in the 1960s. It is also why tradable services stay surprisingly concentrated—futures trading (in Chicago), insurance (Hartford, Connecticut), movies (Los Angeles) and currency trading (London).

History's Heavy Hand

This offends not just techno-enthusiasts but also neo-classical economics: for both, the world should tend towards a smooth dispersion of people, skills and economic competence, not towards their concentration. Save for transport costs, it should not matter where a tradable good or service is produced.

The reality is otherwise. Some economists have explained this by pointing to increasing returns to scale (in labour as well as capital markets), geographically uneven patterns of demand and transport costs. The main reason is that history counts: where you are depends very much on where you started from.

The new technologies will overturn some of this, but not much. The most advanced use so far of the Internet, the greatest of the world's computer networks, has not been to found a global village but to strengthen the local business and social ties among people and companies in the heart of Silicon Valley. As computer and communications power grows and its cost falls, people will create different sorts of space and communities from those that exist in nature. But these modern creations will supplement, not displace, the original creation; and they may even reinforce it. Companies that have gone furthest towards linking their global operations electronically report an increase, not a decline in the face-to-face contact needed to keep the firms running well: with old methods of command in ruins, the social glue of personal relations matters more than ever.

The reason lies in the same fact of life that makes it impossible really to understand from statistics alone how exciting, say, China's economic growth is unless you have physically been there to feel it. People are not thinking machines (they absorb at least as much information from sight, smell and emotion as they do from abstract symbols), and the world is not immaterial: "virtual" reality is no reality at all; cyberspace is a pretence at circumventing true space, not a genuine replacement for it. The weight on mankind of time and space, of physical surroundings and history—in short, of geography—is bigger than any earthbound technology is ever likely to lift.

Unit Selections

Key Points to Consider

❖ Describe the spatial form of the place in which you live. Do you live in a rural area, a town, or a city, and why was that particular location chosen?

❖ How does your hometown interact with its surrounding region? With other places in the state? With other states? With other places in the world?

❖ How are places "brought closer together" when transportation systems are improved?

❖ What problems occur when transportation systems are overloaded?

❖ How will public transportation be different in the future? Will there be more or fewer private autos in the next 25 years? Defend your answer.

❖ How good a map reader are you? Why are maps useful in studying a place?

 **Links** **www.dushkin.com/online/**

These sites are annotated on pages 6 and 7.

Geography is the study not only of places in their own right but also of the ways in which places interact. Places are connected by highways, airline routes, telecommunication systems, and even thoughts. These forms of spatial interaction are an important part of the work of geographers.

In "Transportation and Urban Growth: The Shaping of the American Metropolis," Peter Muller considers transportation systems, analyzing their impact on the growth of American cities. Glen Johnson's article outlines the impact of a new high-speed Amtrak route in the Northeastern United States. The next article explores GIS as the technology used to solve a 160-year-old boundary controversy involving Ellis Island. Next, Joseph Kerski reviews the advantages of using USGS maps and other resources in teaching about landscapes. The next article from *American Demographics* uses the power of the choropleth map to tell its story. "Do We Still Need Skyscrapers?" questions the need for high density structures in the new era of extensive communications. The article on Indian gaming presents a series of maps to make its points.

It is essential that geographers be able to describe the detailed spatial patterns of the world. Neither photographs nor words could do the job adequately, because they literally capture too much of the detail of a place. There is no better way to present many of the topics analyzed in geography than with maps. Maps and geography go hand-in-hand. Although maps are used in other disciplines, their association with geography is the most highly developed.

A map is a graphic that presents a generalized and scaled-down view of particular occurrences or themes in an area. If a picture is worth a thousand words, then a map is worth a thousand (or more!) pictures. There is simply no better way to "view" a portion of Earth's surface or an associated pattern than with a map.

Spatial Interaction and Mapping

Transportation and Urban Growth

The shaping of the American metropolis

Peter O. Muller

In his monumental new work on the historical geography of transportation, James Vance states that geographic mobility is crucial to the successful functioning of any population cluster, and that "shifts in the availability of mobility provide, in all likelihood, the most powerful single process at work in transforming and evolving the human half of geography." Any adult urbanite who has watched the American metropolis turn inside-out over the past quarter-century can readily appreciate the significance of that maxim. In truth, the nation's largest single urban concentration today is not represented by the seven-plus million who agglomerate in New York City but rather by the 14 million who have settled in Gotham's vast, curvilinear outer city—a 50-mile-wide suburban band that stretches across Long Island, southwestern Connecticut, the Hudson Valley as far north as West Point, and most of New Jersey north of a line drawn from Trenton to Asbury Park. This latest episode of intrametropolitan deconcentration was fueled by the modern automobile and the interstate expressway. It is, however, merely the most recent of a series of evolutionary stages dating back to

colonial times, wherein breakthroughs in transport technology unleashed forces that produced significant restructuring of the urban spatial form.

The emerging form and structure of the American metropolis has been traced within a framework of four transportation-related eras. Each successive growth stage is dominated by a particular movement technology and transport-network expansion process that shaped a distinctive pattern of intraurban spatial organization. The stages are the Walking/Horsecar Era (pre-1800–1890), the Electric Streetcar Era (1890–1920), the Recreational Automobile Era (1920–1945), and the Freeway Era (1945–present). As with all generalized models of this kind, there is a risk of oversimplification because the building processes of several simultaneously developing cities do not always fall into neat time-space compartments. Chicago's growth over the past 150 years, for example, reveals numerous irregularities, suggesting that the overall metropolitan growth pattern is more complex than a simple, continuous outward thrust. Yet even after developmental ebb and flow, leapfrogging, backfilling, and other departures from the idealized scheme

are considered, there still remains an acceptable correspondence between the model and reality.

Before 1850 the American city was a highly compact settlement in which the dominant means of getting about was on foot, requiring people and activities to tightly agglomerate in close proximity to one another. This usually meant less than a 30-minute walk from the center of town to any given urban point—an accessibility radius later extended to 45 minutes when the pressures of industrial growth intensified after 1830. Within this pedestrian city, recognizable activity concentrations materialized as well as the beginnings of income-based residential congregations. The latter was particularly characteristic of the wealthy, who not only walled themselves off in their large homes near the city center but also took to the privacy of horse-drawn carriages for moving about town. Those of means also sought to escape the city's noise and frequent epidemics resulting from the lack of sanitary conditions. Horse-and-carriage transportation en-

From *Focus*, Summer 1986, pp. 8-17. © 1996 by the American Geographical Society. Reprinted by permission.

abled the wealthy to reside in the nearby countryside for the disease-prone summer months. The arrival of the railroad in the 1830s provided the opportunity for year-round daily commuting, and by 1840 hundreds of affluent businessmen in Boston, New York, and Philadelphia were making round trips from exclusive new trackside suburbs every weekday.

As industrialization and its teeming concentrations of working-class housing increasingly engulfed the mid-nineteenth century city, the deteriorating physical and social environment reinforced the desires of middle-income residents to suburbanize as well. They were unable, however, to afford the cost and time of commuting by steam train, and with the walking city now stretched to its morphological limit, their aspirations intensified the pressures to improve intraurban transport technology. Early attempts involving stagecoach-like omnibuses, cablecar systems, and steam railroads proved impractical, but by 1852 the first meaningful transit breakthrough was finally introduced in Manhattan in the form of the horse-drawn trolley. Light street rails were easy to install, overcame the problems of muddy, unpaved roadways, and enabled horsecars to be hauled along them at speeds slightly (about five mph) faster than those of pedestrians. This modest improvement in mobility permitted the opening of a narrow belt of land at the city's edge for new home construction. Middle-income urbanites flocked to these "horsecar suburbs," which multiplied rapidly after the Civil War. Radial routes were the first to spawn such peripheral development, but the relentless demand for housing necessitated the building of cross-town horsecar lines, thereby filling in the interstices and preserving the generally circular shape of the city.

The less affluent majority of the urban population, however, was confined to the old pedestrian city and its bleak, high-density industrial appendages. With the massive immigration of unskilled laborers, (mostly of European origin after 1870) huge blue-collar communities sprang up

(Library of the Boston Athenaeum)

Horse-drawn trolleys in downtown Boston, circa 1885.

around the factories. Because these newcomers to the city settled in the order in which they arrived—thereby denying them the small luxury of living in the immediate company of their fellow ethnics—social stress and conflict were repeatedly generated. With the immigrant tide continuing to pour into the nearly bursting industrial city throughout the late nineteenth century, pressures redoubled to further improve intraurban transit and open up more of the adjacent countryside. By the late 1880s that urgently needed mobility revolution was at last in the making, and when

it came it swiftly transformed the compact city and its suburban periphery into the modern metropolis.

The key to this urban transport revolution was the invention by Frank Sprague of the electric traction motor, an often overlooked innovation that surely ranks among the most important in American history. The first electrified trolley line opened in Richmond in 1888, was adopted by two dozen other big cities within a year, and by the early 1890s swept across the na-

Electric streetcar lines radiated outward from central cities, giving rise to star-shaped metropolises. Boston, circa 1915.

tion to become the dominant mode of intraurban transit. The rapidity of this innovation's diffusion was enhanced by the immediate recognition of its ability to resolve the urban transportation problem of the day: motors could be attached to existing horsecars, converting them into self-propelled vehicles powered by easily constructed overhead wires. The tripling of average speeds (to over 15 mph) that resulted from this invention brought a large band of open land beyond the city's perimeter into trolley-commuting range.

The most dramatic geographic change of the Electric Streetcar Era was the swift residential development of those urban fringes, which transformed the emerging metropolis into a decidedly star-shaped spatial entity. This pattern was produced by radial streetcar corridors extending several miles beyond the compact city's limits. With so much new space available for homebuilding

within walking distance of the trolley lines, there was no need to extend trackage laterally, and so the interstices remained undeveloped.

Before 1850 the American city was a highly compact settlement in which the dominant means of getting about was on foot, requiring people and activities to tightly agglomerate in close proximity to one another.

The typical streetcar suburb of the turn of this century was a continuous axial corridor whose backbone was the road carrying the trolley line (usually lined with stores and other local commercial facilities), from which gridded residential streets fanned out for several blocks on both sides of the tracks. In general, the quality of housing and prosperity of streetcar subdivisions increased with distance from the edge of the central city. These suburban corridors were populated by the emerging, highly mobile middle class, which was already stratifying itself according to a plethora of minor income and status differences. With frequent upward (and local geographic) mobility the norm, community formation became an elusive goal, a process further retarded by the grid-settlement morphology and the reliance on the distant downtown for employment and most shopping.

Within the city, too, the streetcar sparked a spatial transformation. The ready availability and low fare of the electric trolley now provided every resident with access to the intracity circulatory system, thereby introducing truly "mass" transit to urban America in the final years of the nineteenth century. For nonresidential activities this new ease of movement among the city's various parts quickly triggered the emergence of specialized land-use districts for commerce, manufacturing, and transportation, as well as the continued growth of the multipurpose central business district (CBD) that had formed after mid-century. But the greatest impact of the streetcar was on the central city's social geography, because it made possible the congregation of ethnic groups in their own neighborhoods. No longer were these moderate-income masses forced to reside in the heterogeneous jumble of row-houses and tenements that ringed the factories. The trolley brought them the opportunity to "live with their own kind," allowing the sorting of discrete groups into their own inner-city social territories within convenient and inexpensive traveling distance of the workplace.

By World War I, the electric trolleys had transformed the tracked city into a full-fledged metropolis whose streetcar suburbs, in the larger cases, spread out more than 20 miles from the metropolitan center. It was at this point in time that intrametropolitan transportation achieved its greatest level of efficiency—that the bustling industrial city really "worked." How much closer the American metropolis might have approached optimal workability for all its residents, however, will never be known because the next urban transport revolution was already beginning to assert itself through the increasingly popular automobile. Americans took to cars as wholeheartedly as anything in the nation's long cultural history. Although Lewis Mumford and other scholars vilified the car as the destroyer of the city, more balanced assessments of the role of the automobile recognize its overwhelming acceptance for what it was—the long-awaited attainment of private mass transportation that offered users the freedom to travel whenever and wherever they chose. As cars came to the metropolis in ever greater numbers throughout the interwar decades, their major influence was twofold: to accelerate the deconcentration of population through the development of interstices bypassed during the streetcar era, and to push the suburban frontier farther into the countryside, again producing a compact, regular-shaped urban entity.

While it certainly produced a dramatic impact on the urban fabric by the eve of World War II, the introduction of the automobile into the American metropolis during the 1920s and 1930s came at a leisurely pace. The earliest flurry of auto adoptions had been in rural areas, where farmers badly needed better access to local service centers. In the cities, cars were initially used for weekend outings—hence the term "*Recreational*

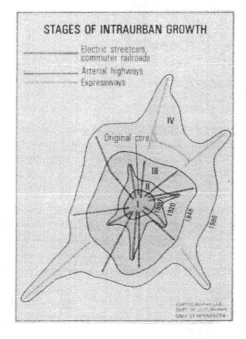

STAGES OF INTRAURBAN GROWTH
— Electric streetcars, commuter railroads
— Arterial highways
-- Expressways

Auto Era"—and some of the earliest paved roadways were landscaped parkways along scenic water routes, such as New York's pioneering Bronx River Parkway and Chicago's Lake Shore Drive. But it was into the suburbs, where growth rates were now for the first time overtaking

The ready availability and low fare of the electric trolley now provided every resident with access to the intracity circulatory system, thereby introducing truly "mass" transit to urban America.

those of the central cities, that cars made a decisive penetration throughout the prosperous 1920s. In fact, the rapid expansion of automobile suburbia by 1930 so adversely affected the

metropolitan public transportation system that, through significant diversions of streetcar and commuter-rail passengers, the large cities began to feel the negative effects of the car years before the auto's actual arrival in the urban center. By facilitating the opening of unbuilt areas lying between suburban rail axes, the automobile effectively lured residential developers away from densely populated traction-line corridors into the suddenly accessible interstices. Thus, the suburban homebuilding industry no longer found it necessary to subsidize privately-owned streetcar companies to provide low-fare access to trolley-line housing tracts. Without this financial underpinning, the modern urban transit crisis quickly began to surface.

The new recreational motorways also helped to intensify the decentralization of the population. Most were radial highways that penetrated deeply into the suburban ring and provided weekend motorists with easy access to this urban countryside. There they obviously were impressed by what they saw, and they soon responded in massive numbers to the sales pitches of suburban subdivision developers. The residential development of automobile suburbia followed a simple formula that was devised in the prewar years and greatly magnified in scale after 1945. The leading motivation was developer profit from the quick turnover of land, which was acquired in large parcels, subdivided, and auctioned off. Understandably, developers much preferred open areas at the metropolitan fringe, where large packages of cheap land could readily be assembled. Silently approving and underwriting this uncontrolled spread of residential suburbia were public policies at all levels of government: financing road construction, obligating lending institutions to invest in new homebuilding, insuring individual mortgages, and providing low-interest loans to FHA and VA clients.

Because automobility removed most of the pre-existing movement

(Boston Public Library)

Afternoon commuters converge at the tunnel leading out of central Boston, 1948.

constraints, suburban social geography now became dominated by locally homogeneous income-group clusters that isolated themselves from dissimilar neighbors. Gone was the highly localized stratification of streetcar suburbia. In its place arose a far more dispersed, increasingly fragmented residential mosaic to which builders were only too eager to cater, helping shape a kaleidoscopic settlement pattern by shrewdly constructing the most expensive houses that could be sold in each locality. The continued partitioning of suburban society was further legitimized by the widespread adoption of zoning (legalized in 1916), which gave municipalities control over lot and building standards that, in turn, assured dwelling prices that would only attract newcomers whose incomes at least

equaled those of the existing local population. Among the middle class, particularly, these exclusionary economic practices were enthusiastically

Americans took to cars as wholeheartedly as anything in the nation's long cultural history.

supported, because such devices extended to them the ability of upper-income groups to maintain their

social distance from people of lower socioeconomic status.

Nonresidential activities were also suburbanizing at an increasing rate during the Recreational Auto Era. Indeed, many large-scale manufacturers had decentralized during the streetcar era, choosing locations in suburban freight-rail corridors. These corridors rapidly spawned surrounding working-class towns that became important satellites of the central city in the emerging metropolitan constellation. During the interwar period, industrial employers accelerated their intraurban deconcentration, as more efficient horizontal fabrication methods replaced older techniques requiring multistoried plants-thereby generating greater space needs that were too expensive to satisfy in the high-density central city. Newly suburbaniz-

Central City-Focused Rail Transit

The widely dispersed distribution of people and activities in today's metropolis makes rail transit that focuses in the central business district (CBD) an obsolete solution to the urban transportation problem. To be successful, any rail line must link places where travel origins and destinations are highly clustered. Even more important is the need to connect places where people really want to go, which in the metropolitan America of the late twentieth century means suburban shopping centers, freeway-oriented office complexes, and the airport. Yet a brief look at the rail systems that have been built in the last 20 years shows that transit planners cannot—or will not—recognize those travel demands, and insist on designing CBD-oriented systems as if we all still lived in the 1920s.

One of the newest urban transit systems is Metrorail in Miami and surrounding Dade County, Florida. It has been a resounding failure since its opening in 1984. The northern leg of this line connects downtown Miami to a number of low- and moderate-income black and Hispanic neighborhoods, yet it carries only about the same number of passengers that used to ride on parallel bus lines. The reason is that the high-skill, service economy of Miami's CBD is about as mismatched as it could possibly be to the modest employment skills and training levels possessed by residents of that Metrorail corridor. To the south, the prospects seemed far brighter because of the possibility of connecting the system to Coral Gables and Dadeland, two leading suburban activity centers. However, both central Coral Gables and the nearby International Airport complex were bypassed in favor of a cheaply available, abandoned railroad corridor alongside U.S. 1. Station locations were poorly planned, particularly at the University of Miami and at Dadeland—where terminal location necessitates a dangerous walk across a six-lane highway from the region's largest shopping mall. Not surprisingly, ridership levels have been shockingly below projections, averaging only about 21,000 trips per day in early 1986. While Dade County's worried officials will soon be called upon to decide the future of the system, the federal government is using the Miami experience as an excuse to withdraw from financially supporting all construction of new urban heavy-rail systems. Unfortunately, we will not be able to discover if a well-planned, high-speed rail system that is congruent with the travel demands of today's polycentric metropolis is capable of solving traffic congestion problems. Hopefully, transportation policy-makers across the nation will heed the lessons of Miami's textbook example of how not to plan a hub-and-spoke public transportation network in an urban era dominated by the multicentered city.

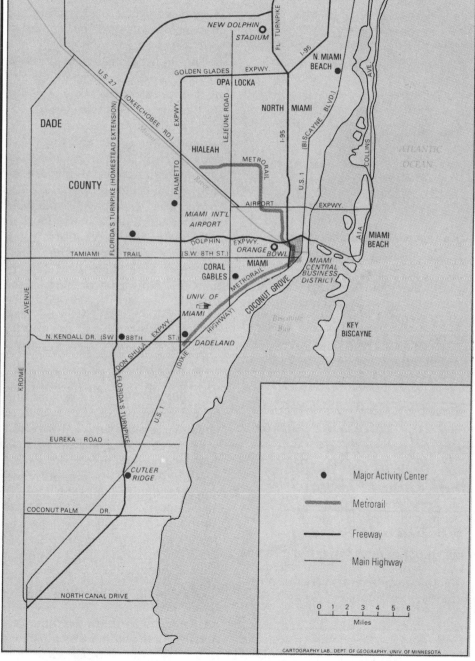

Major Activity Center

Metrorail

Freeway

Main Highway

0 1 2 3 4 5 6
Miles

CARTOGRAPHY LAB. DEPT. OF GEOGRAPHY, UNIV. OF MINNESOTA

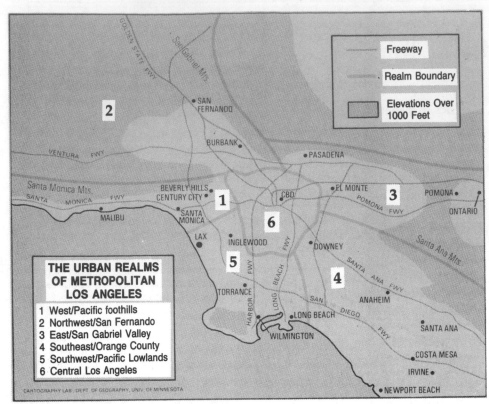

THE URBAN REALMS
OF METROPOLITAN
LOS ANGELES

1 West/Pacific foothills
2 Northwest/San Fernando
3 East/San Gabriel Valley
4 Southeast/Orange County
5 Southwest/Pacific Lowlands
6 Central Los Angeles

CARTOGRAPHY LAB, DEPT. OF GEOGRAPHY, UNIV. OF MINNESOTA

ing manufacturers, however, continued their affiliation with intercity freight-rail corridors, because motor trucks were not yet able to operate with their present-day efficiencies and because the highway network of the outer ring remained inadequate until the 1950s.

The other major nonresidential activity of interwar suburbia was retailing. Clusters of automobile-oriented stores had first appeared in the urban fringes before World War I. By the early 1920s the roadside commercial strip had become a common sight in many southern California suburbs. Retail activities were also featured in dozens of planned automobile suburbs that sprang up after World War I—most notably in Kansas City's Country Club District, where the nation's first complete shopping center was opened in 1922. But these diversified retail centers spread slowly before the suburban highway improvements of the 1950s.

Unlike the two preceding eras, the postwar Freeway Era was not sparked by a revolution in urban transportation. Rather, it represented the coming of age of the now pervasive automobile culture, which coincided with the

emergence of the U.S. from 15 years of economic depression and war. Suddenly the automobile was no longer a luxury or a recreational di-

Retail activities were featured in dozens of planned automobile suburbs that sprang up after World War I—most notably in Kansas City's Country Club District, where the nation's first complete shopping center was opened in 1922.

version: overnight it had become a necessity for commuting, shopping, and socializing, essential to the successful realization of personal op-

portunities for a rapidly expanding majority of the metropolitan population. People snapped up cars as fast as the reviving peacetime automobile industry could roll them off the assembly lines, and a prodigious highway-building effort was launched, spearheaded by high-speed, limited-access expressways. Given impetus by the 1956 Interstate Highway Act, these new freeways would soon reshape every corner of urban America, as the more distant suburbs they engendered represented nothing less than the turning inside-out of the historic metropolitan city.

The snowballing effect of these changes is expressed geographically in the sprawling metropolis of the postwar era. Most striking is the enormous band of growth that was added between 1945 and the 1980s, with freeway sectors pushing the metropolitan frontier deeply into the urban-rural fringe. By the late 1960s, the maturing expressway system began to underwrite a new suburban co-equality with the central city, because it was eliminating the metropolitanwide centrality advantage of the CBD. Now any location on the freeway network could easily be reached by motor vehicle, and intraurban accessibility had become a ubiquitous spatial good. Ironically, large cities had encouraged the construction of radial expressways in the 1950s and 1960s because they appeared to enable the downtown to remain accessible to the swiftly dispersing suburban population. However, as one economic activity after another discovered its new locational flexibility within the freeway metropolis, nonresidential deconcentration sharply accelerated in the 1970s and 1980s. Moreover, as expressways expanded the radius of commuting to encompass the entire dispersed metropolis, residential location constraints relaxed as well. No longer were most urbanites required to live within a short distance of their job: the workplace had now become a locus of opportunity offering access to the best possible residence that an individual could afford anywhere in the urbanized area. Thus, the overall pattern of locally uniform, income-based clusters that had emerged in prewar automobile suburbia was greatly magnified in the Freeway Era, and such new social variables as age and lifestyle produced an ever more balkanized population mosaic.

The revolutionary changes in movement and accessibility introduced during the four decades of the Freeway Era have resulted in nothing less than the complete geographic restructuring of the metropolis. The single-center urban structure of the past has been transformed into a polycentric metropolitan form in which several outlying activity concentrations rival the CBD. These new "suburban downtowns," consisting of vast orchestrations of retailing, office-based business, and light industry, have become common features near the highway interchanges that now encircle every large central city. As these emerging metropolitan-level cores achieve economic and geographic parity with each other, as well as with the CBD of the nearby central city, they provide the totality of urban goods and services to their surrounding populations. Thus each metropolitan sector becomes a self-sufficient functional entity, or *realm*. The application of this model to the Los Angeles region reveals six broad realms. Competition among several new suburban downtowns for dominance in the five outer realms is still occurring. In wealthy Orange County, for example, this rivalry is especially fierce, but Costa Mesa's burgeoning South Coast Metro is winning out as of early 1986.

The new freeways would soon reshape every corner of urban America, as the more distant suburbs they engendered represented nothing less than the turning inside-out of the historic metropolitan city.

The legacy of more than two centuries of intraurban transportation innovations, and the development patterns they helped stamp on the landscape of metropolitan America, is suburbanization—the growth of the edges of the urbanized area at a rate faster than in the already-developed interior. Since the geographic extent of the built-up urban areas has, throughout history, exhibited a remarkably constant radius of about 45 minutes of travel from the center, each breakthrough in higher-speed transport technology extended that radius into a new outer zone of suburban residential opportunity. In the nineteenth century, commuter railroads, horse-drawn trolleys, and electric streetcars each created their own suburbs—and thereby also created the large industrial city, which could not have been formed without incorporating these new suburbs

into the pre-existing compact urban center. But the suburbs that materialized in the early twentieth century began to assert their independence from the central cities, which were ever more perceived as undesirable. As the automobile greatly reinforced the dispersal trend of the metropolitan population, the distinction between central city and suburban ring grew as well. And as freeways eventually eliminated the friction effects of intra-metropolitan distance for most urban functions, nonresidential activities deconcentrated to such an extent that by 1980 the emerging outer suburban city had become co-equal with the central city that spawned it.

As the transition to an information-dominated, postindustrial economy is completed, today's intraurban movement problems may be mitigated by the increasing substitution of communication for the physical movement of people. Thus, the city of the future is likely to be the "wired metropolis." Such a development would portend further deconcentration because activity centers would potentially be able to locate at any site offering access to global computer and satellite networks.

Further Reading

Jackson, Kenneth T. 1985. *Crabgrass Frontier: The Suburbanization of the United States.* New York: Oxford University Press.

Muller, Peter O. 1981. *Contemporary Suburban America.* Englewood Cliffs, N.J.: Prentice-Hall.

Schaeffer, K. H. and Sclar, Elliot. 1975. *Access for All: Transportation and Urban Growth.* Baltimore: Penguin Books.

Amtrak plans high-speed trains as plane alternative

Railroad hopes Acela, slated for East Coast, is ticket to revitalization

GLEN JOHNSON ASSOCIATED PRESS

NEW YORK

Amtrak unveiled a new high-speed train Tuesday that is designed to whisk passengers at 150 mph between Washington, New York and Boston and revitalize the railroad by competing with airlines on such trips.

Named "Acela" (uh-SELL-ah) to hint at both acceleration and excellence, the new trains will travel between Boston and New York in three hours—an improvement of 90 minutes over the current trip—and from New York to Washington in as little as 2½ hours, a savings of a half-hour.

Service is to begin in November or December, and Amtrak officials hope it will be a model for similar trains in the Great Lakes, the Gulf Coast, California and the Pacific Northwest.

"We know we have a product here that will absolutely knock the socks off the competition," Amtrak President George Warrington said at a gala opening attended by more than 1,000 employees. "US Air, Delta, General Motors, Ford, you name it, only Amtrak's Acela will provide a very special journey for customers who will travel downtown to downtown."

In addition to pledging speed, Amtrak promised an unparalleled service. Acela's snub-nosed, silver-and turquoise trains will have business-class seats with audio and power jacks, special check-in areas and concierge service, plus dining cars with meeting tables, upgraded food and beer on tap.

The schedule has not been set, but Amtrak officials said it probably would maintain most current stops, including Baltimore, Philadelphia, New Haven, Conn., and Providence, R.I. The railroad will also retain its slower Northeast direct service.

A one-way trip will cost about $130 to $140 each way between New York and Boston or Washington, an increase from the current Metroliner express fare of $114 but still less than the $199 walk-up fare charged by US Airways and Delta Air Lines. They are the two primary airlines offering shuttle service between Washington, New York and Boston.

Delta spokeswoman Kay Horner said: "There's a lot of traffic in the corridor and Delta offers a different product."

Amtrak projects that Acela will boost its market share in the busy Northeast Corridor from 12 percent to 15 percent annually, or about 14.3 million passengers total. It also expects Acela to generate $180 million in new profits in its first full year of service.

A cash influx is important because Amtrak has not turned a profit since being founded in 1971. The General Accounting Office reported that the railroad lost an average of $47 per passenger in fiscal 1997, renewing criticism in Congress.

In 1997, the House and Senate passed legislation requiring that Amtrak become self-sufficient by 2002. They provided a $2.2 billion cash infusion in 1997 and have been approving steadily declining subsidies since then.

"For years our critics have sat on the sidelines waiting for the high-speed rail program to fade. We will continue to disappoint them" Warrington said. "In fact, we are using the lessons learned on high-speed rail to improve all our services across this country. Acela will be a catalyst for America's 21st century rail renaissance."

GIS Technology Reigns Supreme in Ellis Island Case

Richard G. Castagna, Lawrence L. Thornton and John M. Tyrawski

In 1998, the U.S. Supreme Court ruled on a territorial squabble between the states of New Jersey and New York over Ellis Island that had been brewing for more than 160 years. GIS technology as implemented by the New Jersey Department of Environmental Protection (NJDEP) was instrumental in effecting the outcome. Here's how.

Early agreements between New York and New Jersey set the state boundary line as the middle of the Hudson river down through New York Bay, but the actual boundary around historic Ellis Island was never officially determined. Although the boundary dispute predates the Revolution,

the story officially begins with an 1834 compact between the states, which was approved by the U.S. Congress. This compact set the boundary for Ellis Island as follows: *all non-submerged lands of the island belong to New York, and all submerged lands surrounding the island belong to New Jersey.* At that time, the non-submerged area of the island occupied approximately three acres.

Enlargement of Non-submerged Area

In the intervening years, the boundary question became more muddled. The United States, which owned the island

since 1808, used the island as a fort in the early 19th century and as a powder magazine in the mid-19th century. In 1890, the federal government decided to use Ellis Island as an immigration station. Requiring more space, the federal government began filling the submerged lands around the island. By 1934, the island was enlarged tenfold by successive landfills, from its original 2.75 acres to 27.5 acres. Since 1890, New Jersey has contended that the filling to enlarge and develop the island was done on New Jersey territory.

The NJDEP mapped the natural island as part of the riparian mapping program undertaken by the Bureau of Tidelands in

1934 aerial photo. The upper left side of the island shows active filling in progress. A seawall to contain additional fill material is shown on the right side of the island. (U.S. Supreme Court Evidence)

From *Professional Surveyor,* July/August 1999, pp. 8-14. © 1999 by Professional Surveyor. Reprinted by permission.

1980. Using the riparian map as a preliminary claim, New Jersey officially asked the U.S. Supreme Court in 1993 to adjudicate the boundary dispute. Although both states and several congressional mandates had reviewed the boundary question over the years, the U.S. Supreme Court has sole jurisdiction to resolve boundary issues between states.

State Boundary Case Procedures

The process by which a state boundary case is presented to the Supreme Court is an involved one. The Court first names a Special Master, who reviews all pertinent information submitted by the states to support their claims. The Special Master then submits a report to the full Supreme Court summarizing the evidence and making a recommendation on the settlement. The high court then makes its determination on the final boundary settlement.

In preparation for the trial, the NJDEP Bureau of Tidelands, assisted by the NJDEP GIS Unit, the NJDEP Land Use Regulation Program and the Geodetic section of the New Jersey Department of Transportation, prepared several maps on the DEP's GIS showing the proposed boundary line based

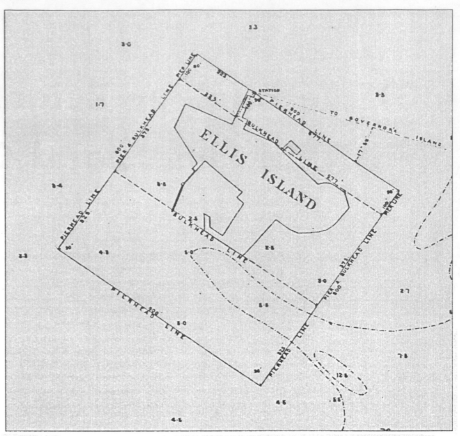

Source: National Archives, June 1890.

(below) "Plan of Ellis' Island 1870," prepared by the Bureau of Ordinance, Navy Department. This map clearly shows the two angles in the Fort Gibson wall used in the 1995 GPS survey. (U.S. Supreme Court Evidence)

(above) This map is titled "Pierhead & Bulkhead Lines for Ellis Island, New Jersey, New York Harbor as recommended by the New York Harbor Line Board." Special Master Paul Verkuil noted, "The significance of this map is that it was approved by the Secretary of War, Elihu Root, and produced over his signature. His signature . . . with the designation Ellis Island, New Jersey, makes this weighty evidence." (U.S. Supreme Court Evidence)

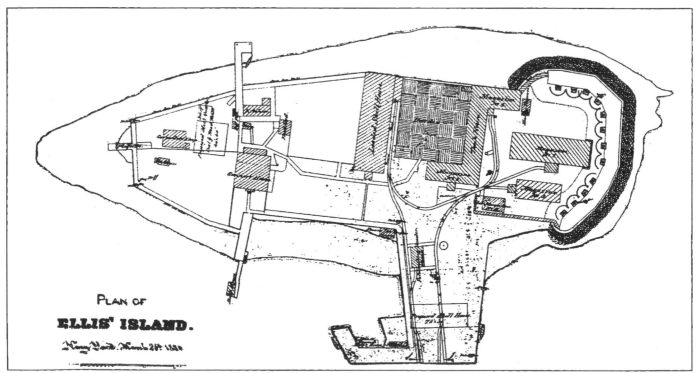

Source; National Archives

Ellis Island in 1995 and in 1857 Showing the Low Water Line

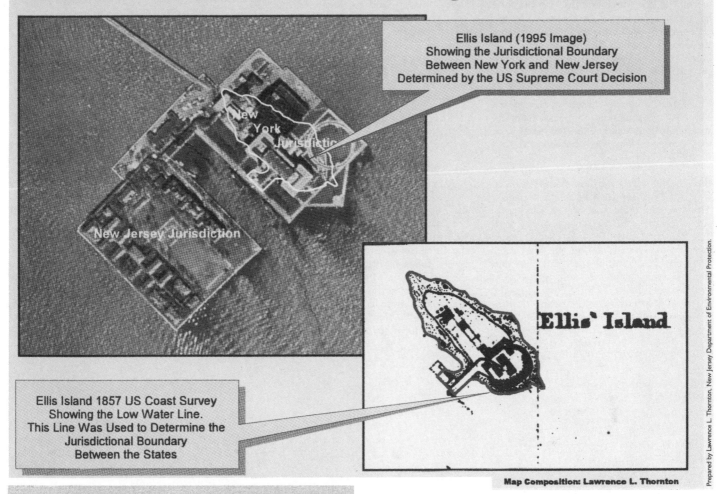

Ellis Island (1995 Image)
Showing the Jurisdictional Boundary
Between New York and New Jersey
Determined by the US Supreme Court Decision

New York Jurisdiction

New Jersey Jurisdiction

Ellis Island

Ellis Island 1857 US Coast Survey
Showing the Low Water Line.
This Line Was Used to Determine the
Jurisdictional Boundary
Between the States

Prepared by Lawrence L. Thornton, New Jersey Department of Environmental Protection.

Map Composition: Lawrence L. Thornton

(above) 1857 U.S. Coast Survey map and the jurisdictional boundary line from the 1857 map superimposed on a 1995 aerial photograph. This exhibit was prepared after the trial and was not used as evidence. (right) Aerial photograph dated about 1993. Note that the open area that straddles the left side of the circular "Wall of Honor" is the exposed wall from Fort Gibson. Two angle points in the excavated sections of the wall were used in a 1995 GPS survey. The wall location was a crucial part of the Supreme Court case. This photo was not used as evidence.

on the 1857 U.S. Coast Survey map. Also mapped were the perimeter of the existing island as determined using GPS, as well as the locations of a portion of the historic fort built on Ellis Island before 1812. The location of the original wall (which has been excavated in part) was helpful in supporting the alignment of the historical maps used to define New Jersey's claim.

1857 Map Chosen to Define Boundary

When the Special Master reviewed all of the evidence, he determined that the 1857 U.S. Coast Survey map should define the boundary between the two states. However, the Special Master disagreed with New Jersey's

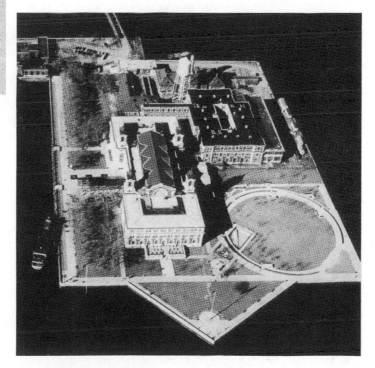

use of the mean high water line on the island as the boundary and directed the state to use the low water line. In May of 1998, after oral arguments were presented by the states, the Supreme Court determined that the 1857 low water line of the natural island should be used to delineate the jurisdictional boundary.

The NJDEP was directed to prepare the Ellis Island boundary based on the Special Master's recommendation. The mapping was completed using GIS. The 1857 U.S. Coast Survey map was scanned and captured as a TIFF image file by the New Jersey Geological Survey. The image file was then brought into ArcView, and the low water line was captured as an edit function. The line depicting low water was represented on the 1857 map by a series of dots. The center of each dot was used to enter each point used to define the low water boundary. Once completed, the points and the line they describe were given geographic referencing by first converting the shape files to coverages—creating tic files for each—and then transforming and projecting the coverages to New Jersey State Plane Feet, NAD83. With this referencing, the line could be plotted on NJDEP's 1995 digital imagery and be integrated into the outer boundary survey, which was completed with the GPS done by NJDOT. The line was then un-generated and sent to NJDOT to prepare a final hard copy map.

New York Officials Approve Map

The map was then presented to New York state officials, who subsequently approved the delineation after several minor changes. The final step in the process is the Special Master's approval of the delineation.

While possibly not the first use of GIS to solve a boundary dispute, this may be a first use of GIS to present and solve a boundary dispute before the U.S. Supreme Court. The success and power of GIS in the Ellis Island case suggests that it will not be the last. After more than 160 years, the jurisdictional fight over Ellis Island is finally over. The high court issued its final decree and approved the boundary line on May 17, 1999. New Jersey was granted sovereign authority over 22.80 acres, and New York was granted authority over the remaining 4.68 acres.

RICHARD G. CASTAGNA is a regional supervisor with the Bureau of Tidelands, NJDEP. He testified as an expert witness before the U.S. Supreme Court in the Ellis Island case on behalf of the State of New Jersey, describing physical changes to the island from the 18th century to the present.

LAWRENCE L. THORNTON is manager of the GIS Unit for NJDEP in Trenton, New Jersey. He delineated a claims line around Ellis Island for the state's riparian claim in 1980 and assisted in the delineation of the low water line, implementing the Supreme Court's decision in 1998.

JOHN M. TYRAWSKI is currently a research scientist with the GIS Unit of NJDEP. For the Ellis Island case, he assisted in the digital creation of the historic 1857 shoreline and in the development of the exhibits used to present the case for the State of New Jersey.

Teaching About Karst Using U.S. Geological Survey Resources

Joseph J. Kerski

All of the resources referred to in this article can be purchased from the USGS by calling 1–800–HELP–MAP, by sending an E-mail to infoservices@usgs.gov, or by using the Global Land Information System (GLIS) at http://edcwww.cr. usgs.gov/webglis/. Each map costs $4, with a volume discount for teachers, plus a $3.50 handling fee per order. The prices of other products, such as professional papers, digital data, and aerial photographs, vary according to the length and the type of media. Most USGS products are also available from the 2,600 authorized USGS map dealers around the world, accessible at http:// mapping. usgs.gov/esic/usimage/dealers.html.

What is karst?

Karst can be broadly defined as all land-forms that are produced primarily by the dissolution of rocks, mainly limestone and dolomite. Landforms in karst terrain include closed surface depressions called sinkholes, disrupted surface drainage, and underground drainage networks that include openings formed from the solution of calcium carbonate in water. These openings range in size from enlarged cracks to large caves. This surface drainage may contain streams that run along the surface for a certain distance and then disappear to follow an underground channel. They may reappear further downstream.

Students can learn to visualize any landscape from analyzing the spacing, elevation, and pattern of contour lines.

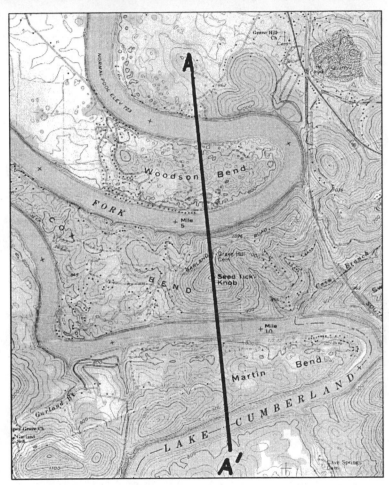

Figure 1. Burnside KY topo map with suggested transect (see text, below).

Why use USGS maps?

Because some of these features are shown on USGS topographic maps, the maps are excellent tools for teaching about the imprint of limestone on the Earth. The predominant scales of USGS topographic maps are 1:24,000, 1:100,000, and 1:250,000, supplemented in some states by a county series at 1:50,000 scale. In Alaska and Hawaii, other scales such as 1:63,360 and 1:25,000 are available. Sinkholes are symbolized by depression contours with tic marks.

Students can learn to visualize any landscape by analyzing the spacing, elevation, and pattern of contour lines.

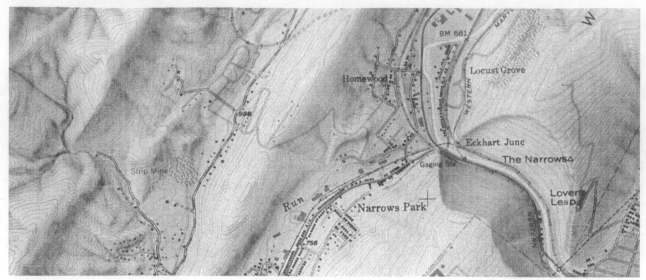

Figure 2. Shaded-relief map of Cumberland, Maryland.

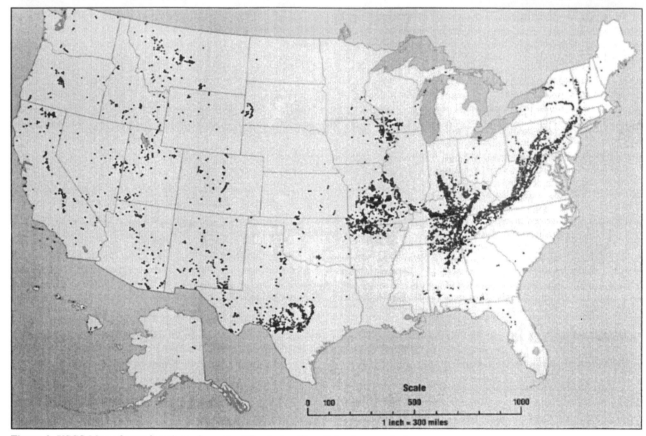

Figure 3. USGS Map of cave locations, from the "Exploring Caves" teachers packet.

Karst landscapes are no exception. Drawing cross-sections of selected transects across the landscape or modeling the landscape with clay, foam, or plaster helps students to understand the three-dimensional shape of the earth's surface and its distinguishing features. For example, a cross-section of transect A to A′ across the topographic map of Burnside, Kentucky, reveals a landscape dissected by river valleys and pitted by holes (figure 1).

The use of a shaded relief map, available for selected areas, most often at the state scale, allows students to better visualize the land surface, as illustrated by this map of Cumberland, Maryland. Students can test a transect that they derived from a topographic map on to a shaded relief map (figure 2).

One resource for discovering the location of karst terrain in the United States is the free USGS teacher's packet named "Exploring Caves." It includes the map (figure 3) which provides a pattern of cave locations, including ones formed by coastal erosion, lava, and limestone.

Free state map indexes can then be used to select topographic maps covering the areas that students will investigate. Topographic maps include famous examples of karst, such as Mammoth Cave, Kentucky (figure 4).

Karst landscapes may include streams that disappear from the surface because the water follows a crack or pit in the

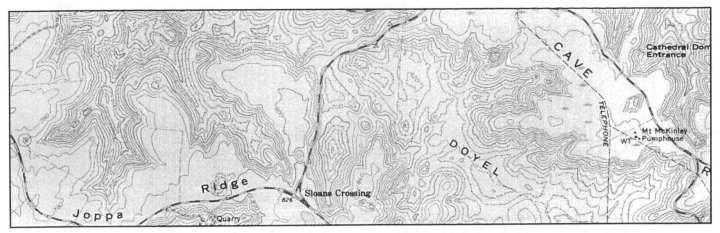

Figure 4. Topographic map of Mammoth Cave, Kentucky.

limestone. These streams may appear on the surface again further downstream (figures 5 and 6).

What can be learned about karst from these maps?

Ways to teach about karst from a map such as this could include posing the following questions: "What percentage of sinks are filled with water?" "Are these sinks perennial or intermittent, and why?" On the basis of their observations, students could estimate the elevation of the water table and identify the places most likely to contain springs. Some springs are shown on USGS topographic maps. Students could also determine the primary direction of water flow in the area. Flow direction observations can be linked with an analysis of drainage basins on the Hydrologic Unit Code maps of each state.

Karst landscapes may include streams that "disappear" from the surface

The effect of limestone on the landscape may be less evident than the effects of coastal erosion, tectonics, or glaciation. Karst can therefore be difficult to detect, even on maps with a scale of 1:24,000, but particularly on maps with a coarser scale and a larger contour interval. When investigating karst, students can view themselves as detectives, uncovering the mysteries of the landscape.

Furthermore, depending on the climate of the area, limestone can act as an erosion-resistant caprock or as an easily-erodible material found on valley floors. Students can investigate maps and data on climate and rainfall for an

Figure 5. Stream disappearing into a sinkhole in karst terrain in Texas. (Photograph by Jon Gilhousen, USGS).

area to determine if the limestone will act as a caprock (predominantly in arid or semiarid regions) or as a weak layer (in regions with a greater amount of rainfall). Students could then be guided in a consideration of the different types of weathering. Karst landscapes are pri-

marily the result of chemical, rather than mechanical, weathering, which explains some of the subtlety of the resultant landforms. Carlsbad Caverns, New Mexico, lies at the top of a Permian-age reef that has been exposed on the surface as a high ridge (figure 8).

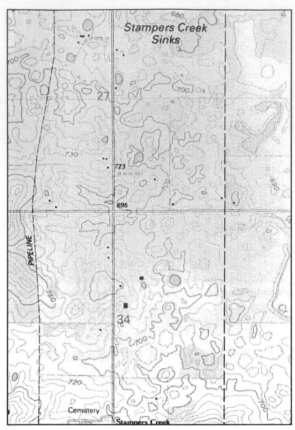

Figure 6. On the Paoli, Indiana topographic map, both Stampers Creek and Wolf Creek disappear below the surface.

soluble limestone beneath. Most land surface depressions containing lakes in this region are formed when the surface deposits slump into sinkholes that have formed in the limestone.

Human impact on karst landscapes can be examined by using prints made from historical editions of topographic maps. These historical editions, available from the USGS Earth Science Information Centers, make population growth and land use change evident. Students can be encouraged to consider how future change might affect cave formations, animal life in caves, and the water quality of the region.

It should be emphasized that the correct interpretation of geomorphic forces operating in a region under study may require the use of additional maps or reports to supplement the topographic map. For instance, in one region, the presence of pits in the landscape may be due to the presence of sand hills or lava covering the surface, rather than from water percolating through limestone.

In that example, a USGS geologic map can aid students in landform study by indicating the ages and types of rock in the region. Geologic maps exist for each state at 1:500,000 or similar scale. In addition, numerous geologic quadrangles exist for selected areas at 1:250,000 or 1:24,000 scale. Most geologic maps also show fault lines, which can be analyzed by the student to determine how they act as controls on landforms. In an arid or semiarid region,

Other available USGS resources

Aerial photographs from the USGS can also aid in teaching about karst. The USGS's PhotoFinder, at http://edc www.cr.usgs.gov/webglis, allows the user to browse data about flight heights, cloud cover, and image types. Aerial photographs are custom-made products that can be ordered from the USGS Earth Science Information Centers. Photographs can be ordered on a variety of media, such as diazo paper or photographic paper. The most common and up-to-date photograph is usually a 1:40,000-scale National Aerial Photography Program (NAPP) product.

Using an aerial photograph, students can consider all surface features, rather than only those features included on topographic maps. For example, they can investigate how karst topography affects vegetation. What kinds of vegetation grow on north-versus south-facing slopes, and in river bottomlands? This orthophotoquad (figure 9) of Carlsbad Caverns, New Mexico, made from aerial photographs, is the same scale as the 1:24,000-scale topographic map.

Satellite images can also aid in karst studies and are available both on paper and in digital form. The section of the

South Florida satellite image map shows a mantled karst region (figure 10). Here, unconsolidated deposits overlie the highly

Figure 7. The presence of karst is evident on the topographic map of Mitchell, Indiana, with its numerous caves, springs, and sinkholes.

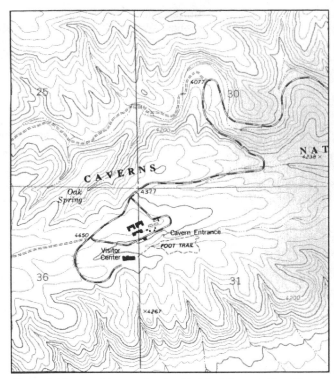

Figure 8. Topographic map showing Carlsbad Caverns, New Mexico.

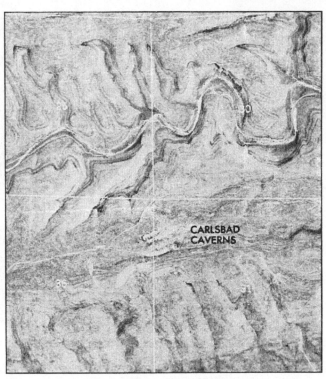

Figure 9. Orthophotoquadrangle showing Carlsbad Caverns, New Mexico.

streams may appear to vanish below the surface, when actually evaporation has caused them to disappear. Climatic maps and data indicating rainfall and temperature can aid in deciphering the most important forces upon the landscape.

Learning about cultural geography and water resources from maps

The study of place names on the landscape incorporates history and human geography into physical geography. The USGS Geographic Names Information System (GNIS) lists the location and type of feature for the name of every stream, town, canyon, sinkhole, and any other feature found on any topographic map for the USA. More than two million names are in the GNIS, available in CD-ROM format or on the Internet at http://mapping.usgs.gov/www/gnis/ Terms such as karst, limestone, cave, or hidden can be input into the system. Due to the Federal Cave Resources Protection Act, the caves and their locations are not listed. However, GNIS displays 988 other features, such as hills, towns, and rivers, that include the word cave and 351 features that include the word sink. The resulting lati-

> ## Students can learn the route and effect of pollutants to the ecosystem.

tude-longitude coordinates for the features can be mapped with stickers, markers, or push-pins on a map of the country, or they can be digitally plotted with a geographic information system.

The use of karst studies in geography and environmental science can be an excellent means of introducing water resources education. Students can learn what a watershed is and the boundaries of the watershed in which they live. Through studying and appreciating the fragile nature of karst ecology, students can learn the route of pollutants and their effects on the ecosystem. The USGS map "Surface Water and Related Land Resources Development in the United States and Puerto Rico" emphasizes river networks, allowing students to derive watershed boundaries. A series of water education posters and publications such as "Ground Water and the Rural Homeowner" are other examples

of water resource materials that could be useful in teaching about rivers and drainage basins.

Publications; paper and digital availability

Circulars, professional papers, bulletins, and open file reports are other USGS resources that could be useful in teaching about karst. For example, Bulletin 1673 is entitled "Selected Caves and Lava-tube Systems in and near Lava Beds National Monument, California." Circular 1139 contains photographs, diagrams, and explanatory text concerning "Ground Water and Surface Water: A Single Resource."

Thematic maps are also useful for karst studies. One of the best examples is Miscellaneous Field Investigations Map 2262 (figure 12), entitled "Sinkholes and Karst-related features of the Shenandoah Valley in the Winchester 30X60-Minute Quadrangle, Virginia and West Virginia."

All USGS topographic maps are available as digital raster graphics (DRGs) which can be viewed on a computer. Digital raster graphics are tagged image format files (tiff) files that can be im-

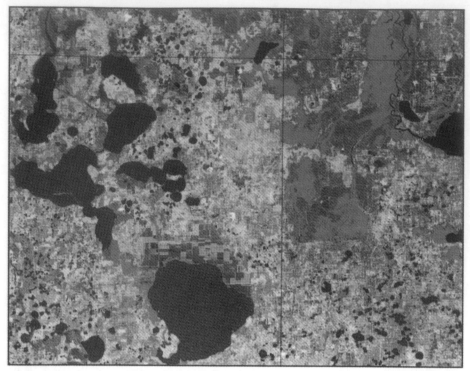

Figure 10. South Florida satellite image map.

ported into a school's geographic information system. To find out more about them, visit the site http://mcm-cweb. er.usgs.gov/drg/. Digital orthophoto- quadrangles (DOQs) are computer versions of aerial photographs that can also be viewed on a computer. Their spatial resolution of 1 meter on the ground allows for a detailed analysis of karst landforms. To find out more about DOQs, visit the site http://mapping.usgs.gov/ www/ndop/. To discover more USGS resources on caves and karst, use the web-based publications search engine at http://www.usgs.gov/pubprod.

Figure 11. Although the unnamed lake in this aerial photograph taken near Sebring, Florida, has a nearly perfect circular shape, indicating its sinkhole origin, many solution-formed lakes do not have this shape (Photograph by E.P. Simonds, USGS).

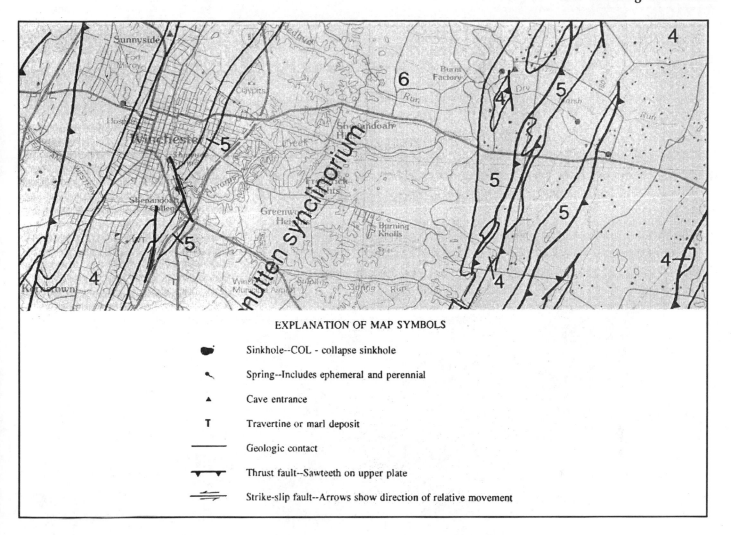

EXPLANATION OF MAP SYMBOLS

Sinkhole--COL - collapse sinkhole

Spring--Includes ephemeral and perennial

Cave entrance

T Travertine or marl deposit

Geologic contact

Thrust fault--Sawteeth on upper plate

Strike-slip fault--Arrows show direction of relative movement

Description of Lithologic Units

6—Interbedded shale, siltstone, and sandstone of the Middle and Upper Ordovician Martinsburg Formation.

5—High calcium limestone of the Middle Ordovician New Market Limestone, limestone and chert of the Middle Ordovician Lincolnshire Limestone, and interbedded limestone, shaly limestone, and calcareous shale of the Middle Ordovician Edinburg and Oranda Formations.

4—Interbedded limestone and dolostone of the Lower and Middle Ordovician Rockdale Run Formation and dolostone and minor limestone of the Middle Ordovician Pinesburg Station Dolomite.

3—Siliceous-laminated limestone of the Lower Ordovician Stonehenge Limestone.

2—Interbedded dolostone, limestone, and dolomitic shale of the Middle and Upper Cambrian Elbrook Dolomite and interbedded limestone, dolostone, and calcareous sandstone of the Upper Cambrian and lowest Ordovician Conococheague Formation.

1—Interbedded dolostone and limestone of the Lower Cambrian Tomstown Dolomite and interbedded limestone, dolostone, siltstone, sandstone, and shale of the Lower and Middle Cambrian Waynesboro Formation.

Figure 12. Section of Shenandoah Valley karst map.

County Buying Power, 1987-97

by Brad Edmondson/Map by Josh Galper

America's biggest cities lost some of their buying power in the last decade, according to Market Statistics estimates of consumer income. Suburbs and small Sunbelt cities picked up the gains.

(changes in inflation-adjusted Effective Buying Income by U.S. county compared with national averages, 1987-92 and 1992-97)

SAN FRANCISCO

Source: Market Statistics, New York, NY (data); ArcView from ESRI, Redlands, CA (mapping software)

Think of the U.S. consumer market as one big economic pie, with America's 3,135 counties as the pieces. The entire pie grew 40 percent in size between 1987 and 1997, according to inflation-adjusted estimates of after-tax income. But the slice given to **(1) Las Vegas** (Clark County, NV) grew 172 percent, while the slice for **(2) Los Angeles County,** CA increased only 17 percent. Las Vegas's share of the national market grew between 1987 and 1992, then grew again between 1992 and 1997. Los Angeles is a two-time loser.

This map is based on Market Statistics' county-by-county estimates of "Effective Buying Income (EBI)," an exclusive measure of money income (wages and salaries; self-employment income; interest; dividends; rent; royalties; retirement payments; and welfare checks) minus personal tax payments (federal, state, and local income taxes; Social Security and other federal retirement payroll deductions; and residential property taxes). In the nation's hot spots, EBI has grown faster than the national average in both 1987–92 (26 percent) and 1992–97 (12 percent). In "turnaround" counties, EBI growth was slower than the national average in 1987–92, but faster than average in 1992–97. In "slowdown" counties, EBI growth was faster than average in 1987–92, but slower in 1992–97. And in "two-time losers," growth in EBI has been slower than the national average in both periods. Although the map shows every county in the U.S., those mentioned in this article are restricted to the 654 where aggregate EBI exceeded $1 million in 1997.

The hottest counties in the U.S. these days are Rocky Mountain boom towns like **(3) Boise** (Ada County, ID) and **(4) Salt Lake**, UT. But the hot list is also dotted with outer-edge suburban counties like **(5) Hamilton**, IN, north of Indianapolis, and smaller eastern metros like **(6) Wilmington**, DE (New Castle County). It also includes the biggest downtown of them all, **(7) Manhattan** (New York County, NY). The other four boroughs of New York City are two-time losers.

While Los Angeles lost share in both periods, buying income in other southern California counties grew faster than

> **SOUTHERN CALIFORNIA SUFFERED A HUGE LOSS OF BUYING INCOME IN THE 1990s.**

the U.S. in 1987–92. But the entire state hit the skids in 1992–97. Only two California counties gained share in both periods, and both were along the Interstate 80 corridor that leads to the nation's center of growth: Nevada. Slowdown counties also include many of the most affluent suburbs. That's why so many upscale retailers in places like **(8) Fairfax**, VA have been struggling.

The sharpest turnaround in the country happened in central **(9) Houston** (Harris County, TX), which lagged slightly behind national growth in 1987–92 (23 percent) but shot ahead in 1992–97 (18 percent). Most of the nation's turnaround counties are in the Ohio Valley and Great Plains, which were the first to recover from the recession of the early 1990s. In **(10) Louisville** (Jefferson County, KY), growth in EBI was 20 percent in 1987–92. Although it slowed to 14 percent in 1992–97, that was still faster than the nation as a whole.

Effective Buying Income is a shorthand way of measuring a market's potential, but the numbers may be misleading unless they are viewed in context. For example, EBI more than tripled in **(11) Esmeralda County**, NV over the last decade, as jobs were created in aerospace and mining. It also tripled in **(12) San Miguel County**, CO, as the rich and famous flocked to the ski slopes of Telluride. But Esmeralda has a total population of about 1,500 people, and San Miguel has just over 5,000 permanent residents. These days, the biggest boom towns are in the boondocks. In the real markets, slow growth is the rule.

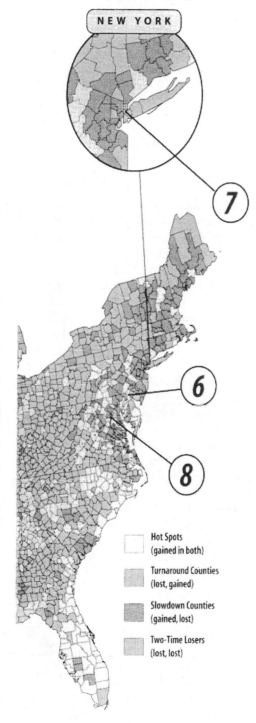

NEW YORK

Hot Spots
(gained in both)

Turnaround Counties
(lost, gained)

Slowdown Counties
(gained, lost)

Two-Time Losers
(lost, lost)

Do We Still Need Skyscrapers?

**The Industrial Revolution made skyscrapers possible.
The Digital Revolution makes them (almost) obsolete**

by William J. Mitchell

O ur distant forebears could create remarkably tall structures by exploiting the compressive strength of stone and brick, but the masonry piles they constructed in this way contained little usable interior space. At 146 meters (480 feet), the Great Pyramid of Cheops is a vivid expression of the ruler's power but inside it is mostly solid rock; the net-to-gross floor area is terrible. On a square base of 230 meters, it encloses the King's Chamber, which is just five meters across. The 52-meter spiraling brick minaret of the Great Mosque of Samarra does not have any interior at all. And the 107-meter stone spires of Chartres Cathedral, though structurally sophisticated, enclose nothing but narrow shafts of empty space and cramped access stairs.

The Industrial Revolution eventually provided ways to open up the interiors of tall towers and put large numbers of people inside. Nineteenth-century architects found that they could achieve greatly improved ratios of open floor area to solid construction by using steel and reinforced concrete framing and thin curtain walls. They could employ mechanical elevators to provide rapid vertical circulation. And they could integrate increasingly sophisticated mechanical systems to heat, ventilate and cool growing amounts of interior space. In the 1870s and 1880s visionary New York and Chicago architects and engineers brought these elements together to produce the modern skyscraper. Among the earliest full-fledged examples were

the Equitable Building (1868–70), the Western Union Building (1872–75) and the Tribune Building (1873–75) in New York City, and Burnham & Root's great Montauk Building (1882) in Chicago.

These newfangled architectural contraptions found a ready market because they satisfied industrial capitalism's growing need to bring armies of office workers together at locations where they could conveniently interact with one another gain access to files and other work materials, and be supervised by their bosses. Furthermore, tall buildings fitted perfectly into the emerging pattern of the commuter city, with its high-density central business district, ring of low-density bedroom suburbs and radial transportation systems for the daily return journey. This centralization drove up property values in the urban core and created a strong economic motivation to jam as much floor area as possible onto every available lot. So as the 20th century unfolded, and cities such as New York and Chicago grew, downtown skylines sprouted higher while the suburbs spread wider.

But there were natural limits to this upward extension of skyscrapers, just as there are constraints on the sizes of living organisms. Floor and wind loads, people, water and supplies must ultimately be transferred to the ground, so the higher you go, the more of the floor area must be occupied by structural supports, elevators and ser-

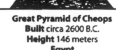

Great Pyramid of Cheops
Built circa 2600 B.C.
Height 146 meters
Egypt

Minaret of Samarra
Built 9th century
Height 52 meters
Iraq

Chartres Cathedral
Built 13th century
Height 107 meters
France

Equitable Building
Built 1870
Height 43 meters
New York

Western Union Building
Built 1875
Height 70 meters
New York

Tribune Building
Built 1875
Height 79 meters
New York

Chrysler Building
Built 1930
Height 319 meters
New York

vice ducts. At some point, it becomes uneconomical to add additional floors; the diminishing increment of usable floor area does not justify the increasing increment of cost.

Urban planning and design considerations constrain height as well. Tall buildings have some unwelcome effects at ground level; they cast long shadows, blot out the sky and sometimes create dangerous and unpleasant blasts of wind. And they generate pedestrian and automobile traffic that strains the capacity of surrounding streets. To control these effects, planning authorities typically impose limits on height and on the ratio of floor area to ground area. More subtly, they may apply formulas relating allowable height and bulk to street dimensions—frequently yielding the stepped-back and tapering forms that so strongly characterize the Manhattan skyline.

The consequence of these various limits is that exceptionally tall buildings—those that really push the envelope—have always been expensive, rare and conspicuous. So organizations can effectively draw attention to themselves and express their power and prestige by finding ways to construct the loftiest skyscrapers in town, in the nation or maybe even in the world. They frequently find this worthwhile, even when it does not make much immediate practical sense.

There has, then, been an ongoing, century-long race for height. The Chrysler Building (319 meters) and the Empire State Building (381 meters) battled it out in New York in the late 1920s, adding radio antennas and even a dirigible mooring mast to gain the last few meters.

The contest heated up again in the 1960s and 1970s, with Lower Manhattan's World Trade Center twin towers (417 meters), Chicago's John Hancock tower (344 meters) and finally Chicago's gigantic Sears Tower (443 meters). More recently, Cesar Pelli's skybridge-linked Petronas Twin Towers (452 meters) in Kuala Lumpur have—for a while at least—taken the title of world's tallest building.

Along the way, there were some spectacular fantasy entrants as well. In 1900 Desiré Despradelle of the Massachusetts Institute of Technology proposed a 457-meter "Beacon of Progress" for the site of the Chicago World's Fair; like Malaysia's Petronas Towers of almost a century later, it was freighted with symbolism of a proud young nation's aspirations. Despradelle's enormous watercolor rendering hung for years in the M.I.T. design studio to inspire the students. Then, in 1956, Frank Lloyd Wright (not much more than five feet in his shoes and cape) topped it with a truly megalomaniac proposal for a 528-story, mile-high tower for the Chicago waterfront.

While this race has been running, though, the burgeoning Digital Revolution has been reducing the need to bring office workers together face-to-face, in expensive downtown locations. Efficient telecommunications have diminished the importance of centrality and correspondingly increased the attractiveness of less expensive suburban sites that are more convenient to the labor force. Digital storage and computer networks have increasingly supported decentralized remote access to data bases rather than reliance on cen-

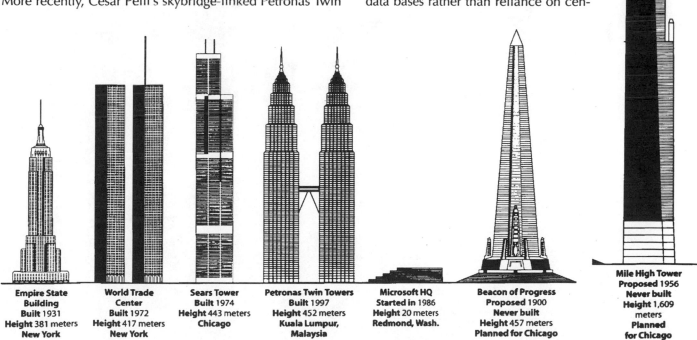

Empire State Building
Built 1931
Height 381 meters
New York

World Trade Center
Built 1972
Height 417 meters
New York

Sears Tower
Built 1974
Height 443 meters
Chicago

Petronas Twin Towers
Built 1997
Height 452 meters
Kuala Lumpur, Malaysia

Microsoft HQ
Started in 1986
Height 20 meters
Redmond, Wash.

Beacon of Progress
Proposed 1900
Never built
Height 457 meters
Planned for Chicago

Mile High Tower
Proposed 1956
Never built
Height 1,609 meters
Planned for Chicago

tralized paper files. And businesses are discovering that their marketing and public-relations purposes may now be better served by slick World Wide Web pages on the Internet and Superbowl advertising spots than by investments in monumental architecture on expensive urban sites.

We now find, more and more, that powerful corporations occupy relatively unobtrusive, low- or medium-rise suburban office campuses rather than flashy downtown towers. In Detroit, Ford and Chrysler spread themselves amid the greenery in this way—though General Motors has bucked the trend by moving into the lakeside Renaissance Center. Nike's campus in Beaverton, Ore., is pretty hard to find, but www.nike.com is not. Microsoft and Netscape battle it out from Redmond, Wash., and Mountain View, Calif., respectively, and—though their logos, the look and feel of their interfaces, and their Web pages are familiar worldwide—few of their millions of customers know or care what the headquarters buildings look like. And—a particularly telling straw in the wind—Sears has moved its Chicago workforce from the great Loop tower that bears its name to a campus in far-suburban Hoffman Estates.

Does this mean that skyscrapers are now dinosaurs? Have they finally had their day? Not quite, as a visit to the fancy bar high atop Hong Kong's prestigious Peninsula Hotel will confirm. Here the washroom urinals are set against the clear plate-glass windows so that powerful men can gaze down on the city while they relieve themselves. Obviously this gesture would not have such satisfying effect on the ground floor. In the 21st century, as in the time of Cheops, there will undoubtedly be taller and taller buildings, built at great effort and often without real economic justification, because the rich and powerful will still sometimes find satisfaction in traditional ways of demonstrating that they're on top of the heap.

WILLIAM J. MITCHELL is dean of the School of Architecture and Planning at the Massachusetts Institute of Technology.

INDIAN GAMING IN THE U.S.

DISTRIBUTION, SIGNIFICANCE AND TRENDS

Dick G. Winchell, John F. Lounsbury and Lawrence M. Sommers

Introduction: The widespread distribution of Indian gaming

Legalized gaming is the most rapidly growing industry in the United States in recent years. Close to $600 billion was wagered in 1996, an increase of over 54 percent from 1993. Although some forms of gaming showed a decrease—charitable bingo, greyhound racing and jai alai there was a significant increase in wagering in casinos, lotteries and on Indian reservations. Indian gaming increased over 125 percent in this three year period compared to a wagering increase in casinos of 48 percent. Although recent entrants to the gaming industry, Indian bingo and more recently tribal casinos were developed rapidly during the 1980s and '90s. In 1993, Indian gaming accounted for $28.96 billion or 7.4 percent of the total money wagered, and grew to $65.18 billion or 11.1 percent by 1996 as shown in Table 1, "Money Wagered on Various Types of Legal Gambling." This growth has been controversial in a number of states because it has been approved only for Indian lands without the potential for development off-reservation, causing complaints from many non-Indian public and private interests. The spread of Indian gaming has led to consideration of legalization of gambling activities for casinos in many states throughout the nation, and has often brought states into conflict with the tribes.

In 1997, 142 Indian tribes or communities operated 281 casinos and bingo

Jamestown Clallam tribe totem Sequim, Wash.
Photo: D. Winchell

halls in 24 states. The widespread distribution of Indian gaming includes major tribal casinos in the east as well as those in the west, where there are more reservations. There are many more tribes preparing to enter into negotiations with the states or actually constructing facilities for casinos based on the success of current tribal casinos. The rapid expansion of Indian gaming dates back to 1988 when the Indian Gaming Regulatory Act (IGRA) was passed by Congress. This article examines the distribution, significance, historical evolution and major issues in recent Indian gaming growth in the U.S. Tribal gaming in the states of Washington, Arizona, Michigan and Connecticut will be used to illustrate the characteristics and selected issues resulting from this trend toward increased Indian gaming.

Background: Expansion of gaming in the United States

The increase nationally in gaming activities is seen in the gross revenues in 1996—what's left over after payouts and prizes—which was $41.8 billion, an increase of 20 percent in 3 years. In 1996, gross gaming revenues were $47.6 billion, about the same as the total amount of money spent on all the following major forms of entertainment combined—recorded music ($12.5 billion), video games ($7.1 billion), theme parks ($7.2 billion), movies ($5.9 billion), spectator sports ($6.3 billion), and non-sport live entertainment ($9.2 billion). Once con-

From *Focus*, Winter/Spring 1997, pp. 1-10. © 1997 by the American Geographical Society. Reprinted by permission.

Table 1

Money Wagered on Various Types of Legal Gambling

Activity	1996 (in billions)	Percent of Total
Casinos	$438.68	74.8
Indian Reservations	$65.18	11.1
Lotteries	$42.93	7.3
Horse Racing	$14.10	2.4
Card Rooms	$9.86	1.7
Charitable Games	$5.68	1.0
Charitable Bingo	$4.04	0.7
Bookmaking	$2.61	0.5
Greyhound Racing	$2.31	0.4
Jai Alai	$0.24	0.1
Totals	**$586.52**	**100.0**

Source: International Gaming and Wagering Business Magazine, "1996 Gross Annual Wager," August, 1997, p. 8.

sidered an immoral and reckless activity, gaming is now increasingly seen as an acceptable form of recreation, and a potential source of tax revenue for state governments. Further, as the type of hotels and theme parks built recently in Las Vegas indicate, attempts are being made to provide entertainment for the whole family.

At this time, casino gaming far surpasses other forms of legalized gaming, accounting for about three-fourths of the total dollars wagered. Gaming in 1996 employed nearly 350,000 workers in the ten largest gaming states, including an estimated 42,500 employed in Indian gaming. Lotteries now operate in 37 states and the District of Columbia. Collectively, they generate millions of dollars in revenue which is generally earmarked for education in state budgets. Many states hungry for revenue and afraid of higher taxes favor lotteries as an easier alternative. In some states, such as Michigan, competition by casinos has decreased the lottery intake.

The amount of money wagered varies greatly from one state to another as shown in Table 2 "Gaming by State—1996." This variation is the result of differences in gaming regulations, the nature of gaming facilities and the population of the state and the region. In 1996, there were 31 states in which gross wagering exceeded $1 billion. In order of amounts wagered, the top five were Nevada, New Jersey, Mississippi, Louisiana and Illinois. Nevada has had well-established full-scale casinos since 1933, and New Jersey since 1978. Las Vegas, Reno, Lake Tahoe and Laughlin in Nevada, and Atlantic City in New Jersey have elaborate casino/resort hotels that attract people from all parts of the country and abroad, while other "boom towns" have grown rapidly in Nevada as

discussed in a 1991 *Focus* article by two of the present authors. The Mississippi Gaming Control Act was enacted in 1990, and nine counties have subsequently approved dockside casino gaming. Thirty casinos operated in 1997 with a yearly total gross revenue of nearly $2 billion. These casinos are drawing people from all over the state and adjacent states as well. Illinois and California, with popular race tracks and lotteries in existence for many years, have large populations from which to obtain customers.

The states in which gross wagering is less than $100 million are Hawaii, Tennessee and Utah, which had no legal gaming in 1996, and North Carolina, Vermont and Wyoming. The latter states have small populations or do not have elaborate gaming complexes at this time.

Indian gaming historical background

There are over 500 American Indian Tribes which are officially recognized by the United States Government, with extremes in size from those with no tribal land base to the Navajo Nation with approximately 25,000 square miles; and extremes in population from tribes with less than ten tribal members to over 150,000 Navajo tribal members. Generally, Indian reservations were established in rural areas, and in recent decades the expansion of urban areas, especially in the west, has resulted in many reservations now being surrounded by or adjoining cities including San Diego and Palm Springs, California; Phoenix and Tucson, Arizona; and Seattle, Washington. For the most part, reservation economies suffer from persistent poverty, and lacking independent tax

bases, tribal governments have remained dependent upon federal resources.

To alleviate this poverty, to try to build the local tribal economies and tribal governments, and to respond to the sovereignty of tribes to control their own communities, several tribes sought to introduce gambling as tribal enterprises in the 1960s and 1970s. By the 1980s several Supreme Court rulings provided tribes with the rights to establish bingo operations, where legal for other organizations by state law. Efforts by tribes to expand these operations resulted in the passage of the Indian Gaming Regulatory Act (IGRA) in 1988, which permits tribes to operate a wide range of gaming operations including casinos with some restrictions.

Many American Indian tribes had gambling activities as integral functions of community interaction prior to European contact. Stick games, foot races, and ball games are just a few of the traditional Indian gambling activities which often provided social opportunities for interaction and exchange. During the period of contact with Europeans, many traditional tribal members were quick to adopt European forms of gambling, especially dice games. Europeans also introduced the non-Indian moral attitude that gambling is evil, especially when done by Indians. Laws were passed to control gambling, and this argument is still used to try to restrict or limit Indian gaming "for their own good."

Not only gambling but even those traditional ceremonials which involved giveaways, such as the potlatches of the American Northwest tribes, were made illegal in the reservation period of the late 1800s up to the 1960s. During the 1960s and 1970s some tribes began to openly carry out traditional gaming activities, as well as bingo operations for profit. By the late 1970s a number of tribal governments operated bingo facilities in states where such activities were legal for other organizations, and some developments, such as the Seminole Bingo Hall in Florida and the Cabazon Band facility near San Diego, established high stakes bingo for non-Indian markets. When the states tried to stop these two operations, the courts ruled in favor of the tribes.

The Supreme Court left standing a lower court in the Seminole case which established that, according to the *Congressional Quarterly Almanac*, "Florida ...and by implication other states . . . could not regulate bingo on Indian reservations if the game was legal elsewhere in the state." This was reinforced by a decision on the Cabazon Band when the legality of tribal high stakes

Table 2
Gaming by State–1996

Gross wagering in the USA (amount players spent on gaming), and gross revenues (the amounts wagered minus the winnings returned to the players) in 1996[1] (in millions).

State	Wagering	Revenue
Ala.	264.0	63.7
Alaska	264.9	58.1
Ariz.	760.4	219.6
Ark.	267.9	55.8
Calif.	14,416.2	2,393.2
Colo.	8,003.1	654.1
Conn.	1,134.7	406.6
Del.	2,169.3	284.2
D.C.	234.0	117.9
Fla.	3,986.9	1,510.7
Ga.	1,713.8	860.5
Idaho	145.9	49.1
Ill.	26,099.2	2,312.3
Ind.	8,050.6	788.8
Iowa	11,835.3	773.8
Kan.	367.6	128.0
Ky.	1,689.3	501.3
La.	28,667.5	2,070.9
Maine	272.4	94.6
Md.	2,318.6	739.4
Mass.	3,783.5	1,121.6
Mich.	2,254.9	865.8
Minn.	1,807.8	417.4
Miss.	35,075.1	1,894.4
Mo.	11,748.4	831.3
Mont.	2,328.8	249.3
Neb.	493.3	137.2
Nev.	222,253.9	7,452.8
N.H.	477.7	135.9
N.J.	89,412.0	4,860.8
N.M.	220.1	64.4
N.Y.	6,535.2	2,431.3
N.C.	34.7	8.4
N.D.	551.3	67.2
Ohio	3,677.4	1,347.3
Okla.	278.1	61.2
Ore.	3,871.8	540.1
Pa.	2,959.9	1,141.1
R.I.	1,221.7	193.6
S.C.	1,446.5	419.3
S.D.	2,778.4	241.0
Texas	4,739.4	1,821.5
Vt.	90.0	34.5
Va.	1,151.3	489.0
Wash.	1,554.0	480.7
W.Va	1,131.0	185.4
Wis.	664.2	247.1
Wyo.	29.2	7.6
Total	**514,234.5**	**41,829.9**

[1]Does not include gaming on oceangoing cruise ships or on Indian reservations. Note: Hawaii, Tennessee and Utah have no legal gambling.

Source: International Gaming & Wagering Business. "1996 Gross Annual Wager." August, 1997, p. 20, 39.

bingo was established in *California v. Cabazon Band of Mission Indians.* As summarized in the *Congressional Quarterly Almanac:*

> Thus, after the 1987 Cabazon ruling, "Games on the reservation still had to abide by federal law and by state criminal laws. If a particular form of gambling, such as *roulette*, was prohibited altogether under state law it was also illegal on Indian reservations. But 45 states permitted bingo, while regulating where the games could be played and what prizes could be offered. On a reservation, those restrictions did not apply. The state could ban bingo outright, but then other groups that used the games for fund raising would be shut out as well.

After the Florida decision in 1982, many tribes started to develop bingo operations, and by 1986 over 100 tribes operated bingo facilities.

IGRA: Indian Gaming Regulatory Act

The Indian Gaming Regulatory Act of 1988 (PL. 1000–497) was an attempt to reach a compromise between state efforts to restrict tribal gambling, particularly Class III gaming, and tribal efforts to seek gaming operations as a means of badly needed economic development. The IGRA first established definitions for

different types of gambling, and, provided for Class I gaming, "traditional ceremonial gaming or social games for prizes of limited value" which come under the sole control of the tribes; Class II gaming including bingo, lotto, and certain card games (but not blackjack, chemin de fer or baccarat), which are subject to oversight by a five-member National Indian Gaming Commission (NIGC); and Class III gaming which, again quoting the *CQA*, includes "casino gambling, slot machines, horse and dog racing, and jai alai and which are prohibited unless they are legal in the state and the state and tribe enter into a compact for their operation."

Problems with the Act have centered upon the slow time frame for establishing the NIGC, for its delays in the adoption of regulations for tribal gaming, and for the complex requirement for compacts between tribes and states to authorize tribal gaming. Although the Act provides for tribes to establish their own gaming authority if states do not negotiate in good faith, states argue they cannot be forced to sign an agreement with the tribes. This provision of the Act both reduces tribal sovereign powers and has led states to delay signing compacts or to require extensive control over tribal gaming. U.S. Senator Pete Domenici stated in Congressional hearings that:

> ...we didn't have a commission (National Indian Gaming Commission) in

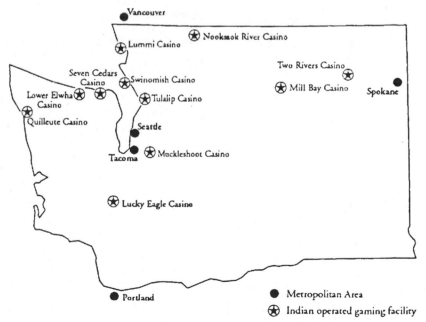

Washington Indian Gaming Casinos

- ● Metropolitan Area
- ✪ Indian operated gaming facility

Map by: Michigan State Cartography Lab

a timely manner and we do not even have regulations and regulatory distinctions between Class II and Class III. So to the extent that some of us are frustrated about the statute (IGRA) we drew . . . (we are still in the state of) determining what is criminal and what isn't (in Class III gaming under IGRA).

Some of the problems may be resolved with the empowerment of the NIGC, which has only been fully operationalized for a few years. In the NIGC structure, the state's role might again be diminished, particularly if the NIGC's role includes enforcement capabilities. Despite not having the Commission or its regulations in place, many tribes have pursued compacts for Class III gaming, opened additional Class II gaming facilities, and have sought to create their own regulations which will assure the success of their programs. Particularly for those tribes which entered the Class III gaming early, their programs have been extremely successful.

The 1988 IGRA, as it now reads, is itself receiving criticism from various sources. Some feel that it is too vague and, as a consequence, 49 state governors have petitioned Congress to clarify the act seeking tighter regulations and clearer negotiating guidelines. The tribes, on the other hand, want less intrusion. It is a political battle pitting sovereign state rights against sovereign tribal rights. It also is an economic battle over entertainment dollars—Indian-operated casinos versus state lotteries, race tracks, privately owned casinos, charitable games and bingo. Billions of dollars are at stake.

Another criticism is that the Bureau of Indian Affairs does not have the expertise and labor to provide the necessary oversight, and that as a consequence, Indian-operated casinos could become havens for money laundering and organized crime. Although states have often fallen back on the argument that organized crime and unrestricted gaming by Indians is a dangerous problem, both the FBI and Justice Department say there is no evidence of substantial organized crime in Indian gaming at this time.

The Indian Affairs Committee of Congress has worked to establish a minimum set of standards for regulating Indian gaming, and the National Indian Gaming Commission (NIGC) has been formed to serve as the regulatory agency over tribal gaming, working in partnership with other regulatory bodies, mostly tribal gaming commissions as noted in *Indian Country Today*. Numerous bills to change the IGRA are introduced each year, but so far none have resolved the complex issues which gen-

Photo by: D. Winchell

Totem poles symbolize traditional designs of the Jamestown Clallam Tribe, and welcome visitors at the entrance of the tribe's Seven Cedars Casino near Sequim, WA.

erally pit state interests against tribal interests. The states won a partial victory in a 1996 Supreme Court Decision, *Seminole Tribe of Florida v. Florida,* which upheld a lower court ruling that the tribe could not sue the state to force them to negotiate to permit specific gaming activities on their reservation, as noted in *Facts on File.*

The significance of Class III Indian gaming developments

The operation of Class III Gaming Facilities has provided a whole new arena for tribal economic development and self-sufficiency. Gambling operations have brought employment opportunities to tribal members, revenues to tribal gov-

Table 3
Washington State Indian Gaming Operations

Tribe or Indian Community	Casino	Location
Chehalis Reservation	Lucky Eagle Casino	Olympia Area
Colville Confederated Tribes	Mill Bay Casino (Completed 1995)	Lake Chelan
Jamestown Clallam Res.	Seven Cedars Casino	Sequim
Lower Elwha Tribe	Bingo/Casino	Port Angeles Area
Lummi Tribal Council	Lummi Casino	Bellingham Area
Muckleshoot Reservation	Muckleshoot Casino	Greater Tacoma Area
Nooksack Indian Tribe	Nooksack River Casino	Greater Seattle Area
Quileute Reservation	Quileute Casino	La Push
Spokane Tribe	Two Rivers Casino	Greater Spokane Area
Swinomish Tribe	Bingo, Casino	LaConner Area
Tulalip Tribe	Casino	Everette Area

Table 4
Arizona Indian Gaming Operations

Tribe or Indian Community	Casino	Location
Ak Chin	Harrah's Ak Chin Casino	Greater Phoenix Area
Cocopah	Cocopah Bingo/Casino	Greater Yuma Area
Mohave-Apache	Fort McDowell Gaming Center	Greater Phoenix Area
Gila River	Gila River Casino	Greater Phoenix Area
Pascus Yaqui	Casino of The Sun	Greater Tucson Area
Quechan	Quechan Bingo	Greater Yuma Area
San Carlos Apache	Apache Gold Casino	Globe-San Carlos
Tohono O'odham	Desert Diamond Casino	Greater Tucson Area
Tonto-Apache	Mazatzal Casino	Payson
White Mountain Apache	Hon Dah Casino	Pinetop
Yavapai-Prescott	Yavapai Gaming Center	Prescott

ernments through taxes or as profits to the tribes, and have produced capital for investment and development not just for casinos but for other community programs, activities and facilities.

Tribal casino development has often taken advantage of tribal locations near urban centers, with the largest tribal casino at Ledyard, Connecticut, a short distance from New York City, as shown in Figure 3, "Foxwoods Casino/Hotel, Connecticut." Other early tribal casinos include Lummi and Tulalip in the state of Washington, which draw from the Seattle and Vancouver, British Columbia urban areas; and tribal facilities near Minneapolis, San Diego, Palm Springs, Phoenix and Upper Peninsula, Michigan.

A special issue of *Indian Country Today* reported the degree of success by tribes in Class III gaming. The Cabazon Band of California grossed $50 million in 1992, with tax revenues to the tribe of over $530,000. The article stated that "Gaming revenues have enabled the Cabazons to become self-sufficient, with a true and functioning tribal government and court system."

The Oneida Tribal gaming operation in Wisconsin has had a $650 million annual impact on the surrounding area, generating over $43 million in revenues and employing over 1,100 people, approximately 75 percent of whom are Native Americans. The Mashantucket Pequot Tribe in Connecticut provided over 10,000 jobs through gaming, while the Mille Lacs Band of the Chippewa in Minnesota eliminated unemployment.

The distribution of Class III gaming facilities correlates closely with the location of reservations themselves, but the largest and most successful have been linked to urban populations or to recreation/resort amenities in rural areas. Some facilities, however, have been able to create their own destination attraction based on the existence of a casino alone. The following case studies in Connecticut, Washington, Arizona and Michigan illustrate the significance of Indian reservation gaming developments and the associated issues, changes, and trends in these four representative states.

Mashantucket Pequots: the landmark Connecticut case

Perhaps no other Indian-operated gaming facility has received as much publicity and notoriety in recent years as the Foxwoods Casino on the Mashantucket Pequot reservation near Ledyard, Connecticut. It wasn't until 1983 that the small tribe (179 enrolled members) obtained federal recognition. In 1989, the spotlight focused on the tribe as it entered into a bitter two-year battle with the state of Connecticut to open a full-scale casino based on the claim that charitable "Las Vegas" nights were legal in the state. After months of highly publicized confrontations, the U.S. Supreme Court refused to hear the state's challenge to the tribal plans. This case served as a legal landmark, and opened the door to casino developments on Indian lands nationwide.

In February, 1992, the Foxwoods Casino opened and expanded rapidly. It now is the largest casino in the western hemisphere, larger than the MGM Grand in Las Vegas. Located about half way between Boston and New York, the casino had a total handle of nearly $8 billion and a gross revenue of nearly $600 million in slot machine income alone. The facilities include two hotels, a virtual reality 1,500-seat theater, and a sporting events complex. The tribe is expanding its existing 1,794 acre development to over 9,000 acres with another 1,100 room hotel, two golf courses, and a theme park. They are investing in a variety of other local businesses such as shipbuilding, spas and gravel quarries. As is often the case in areas undergoing rapid land use changes, neighboring communities oppose the expansion, fearing the loss of the region's rural character. No doubt the conflict will be resolved over a period of time.

The tribe, now numbering over 300 members, has become a major political

Table 5
Michigan Gaming Operations

Tribe or Indian Community	Casino	Location
Lower Peninsula		
Isabella Indian Reservation	Soaring Eagle Casino	Near Mt. Pleasant
Grand Traverse Band of Ottawa & Chippewa Indians	Leelanau Sands Casino	Suttons Bay, North of Traverse City
Upper Peninsula		
Sault Ste. Marie Tribe of Chippewa Indians	Kewadin Shores Casino	St. Ignace
Sault Ste. Marie Tribe of Chippewa Indians	Vegas Kewadin Plus Casino	Sault Ste. Marie, Christmas, Hessel & Manistique
Bay Mills Indian Community	King's Club Casino	Near Brimley
Potawatomi Tribe	Chip-In Casino	Near Harris
Keeweenaw Bay Indian Community	Ojibwa Casino	Near Baraga
Lac Vieux Desert Tribe	Lac Vieux Desert	Watersmeet

and economic force in the state and region. The tribe employs more than 10,000 workers and pumps more than $100 million a year from slot machine revenue into state coffers. It has pledged $10 million to the Smithsonian Institution for a Museum of the American Indian— the largest single donation ever received by the Smithsonian; and has poured millions into lobbying and campaign contributions.

A second Indian casino, the Mohegan Sun operated by the Mohegan Tribe, opened in 1996 just 10 miles from the Foxwoods. This facility is much smaller than the Foxwoods, but nonetheless is a growing competitor.

The State of Washington and the Lummi Indian Nation Casino

Legalized gaming by American Indian tribes within the state of Washington was initiated even before the adoption of the IGRA in 1988. By summer, 1994, six tribes operated casinos as shown in Table 3, "Washington State Indian Gaming Operations" and Figure 4, "Washington Indian Gaming Casinos," while three additional tribes had casinos under development, two of which have been completed and are now open. Fifteen tribes had bingo operations. Total employment in 1996 from tribal gaming has been significant for Washington tribes, with over 6,000 jobs created on reservations since 1990 according to studies by author Winchell. Gaming employment in 1994 represented approximately 10 percent of total reservation employment in the state of Washington, and completion of new casinos increased this employment rate threefold by 1996. In addition, tribal governments have received revenues from gaming which has allowed expansion of tribal services and programs. Many tribes have developed special programs

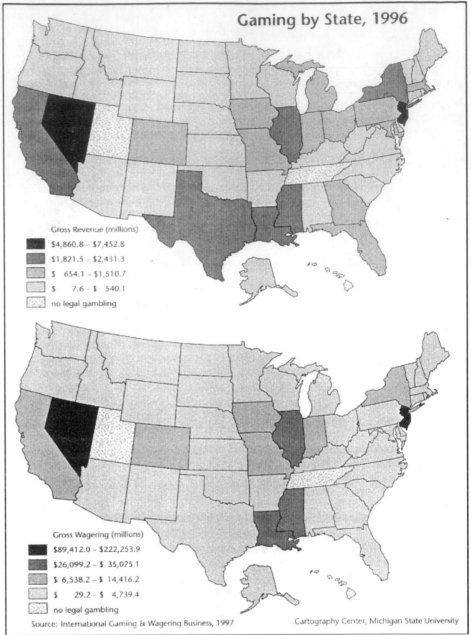

Map by: Michigan State Cartography Lab

Gaming is good for economic development only if you can import the gamblers.

for gaming-related job training, but have also increased support for general education and established programs to directly address potential negative problems such as gaming addiction and substance

abuse for tribal members. Tribal leaders report that negative impacts are not as great as anticipated from casino gaming, and most tribes have sought non-Indian clients as their target. To better illustrate the impacts of gaming in Washington, the development of the Lummi Casino is described below.

The Lummi tribe in northwest Washington between Seattle, Washington and Vancouver, British Columbia, took advantage of its proximity to two major urban centers and the growing autonomy of tribes to initiate bingo and other gaming activities in the early 1980s. The Lummi tribe began gaming activities in 1983, when it converted a community center into a blackjack gaming room, and used a bank loan to pave the road

onto the reservation so patrons could get to the facility. A federal judge closed the tribe's "casino" within two months for violation of state laws, but when the IGRA was passed in 1988, it allowed previously operating facilities to be "grandfathered into operation" without being required to complete a compact with the state.

The Lummi tribe reopened its casino in 1992 with a facility which held 42 blackjack tables, and which employed 413 people, of whom about 70 percent were tribal members. The Lummi Casino grossed over $3 million in revenues in its first year of operation, and the success of this casino was the stimulus for other tribes within the state to pursue Class III gaming.

Arizona Indian Gaming Casinos

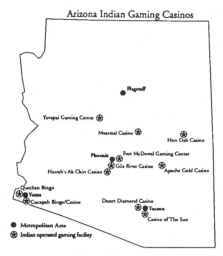

Map by: Michigan State Cartography Lab

The Lummi Casino operated under a separate management board which the tribe established, and the tribe took an active role to create positive employment opportunities for tribal members, and established a dealer training program and a gaming/resort degree program on the reservation in conjunction with the Northwest Indian Community College. It established rules that prevented tribal elected leaders and casino employees from gambling at the facility, and used some of its profits to establish its own security force. The casino also provided funds for two tribal policemen and a police car to offset the impact of increased traffic through the reservation.

The tribal government, in addition to creating jobs and training programs for tribal members, received special program support and tribal revenues of approximately $1 million a year. This dramatically contributed to the effectiveness of its programs and services.

According to interviews with the casino manager and tribal officials, the social impacts were more positive than negative. There seems to have actually been a decrease in social problems as a result of more people being employed. There was a strong recognition of potential problems, however, and the tribe established a special counseling and Gambler's Anonymous program run by American Indian counselors. The casino is clearly oriented to tourists, and has been especially successful in capturing a strong market from Canadians in the Vancouver area, less than an hour's drive away. This orientation toward non-Indians has helped lessen the negative social impact.

The Lummi Casino is also an example of the volatility in the Indian gaming industry. A new casino/hotel operated by the Nooksack River Tribe in conjunction with Harrah's only 30 miles away had

some impact on the Lummi Casino, but more significantly Canada liberalized its gaming laws, and Canadians no longer need to travel across the border for gambling. By mid-year 1997, plans were being carried out to close the Lummi Casino. The Lummi Casino did produce short-term employment and revenues, and many tribal employees have been able to find employment in other nearby casinos. The tribe was able to pay off its capital expenses over the short but successful operation of the casino, and make considerable gains for the tribe.

The Arizona case

Tourism has been a major part of Arizona's economy for several decades. The climate, scenery, resort hotels, golf courses and national-level sporting events attract millions of visitors each year. In addition, legalized gaming has been in existence in the state for many years in the form of horse and greyhound racing, lotteries, and bingo halls. Indian reservations make up 28 percent of the state's land area, and some are located in close proximity to the populous urban areas of Phoenix, Tucson, Yuma and Flagstaff. Considering this mix of circumstances, it is not surprising that Indian-operated gaming would develop shortly after the passage of the 1988 Indian Gaming Regulatory Act.

As of January 1995, sixteen of Arizona's twenty-one tribes had signed gaming compacts with the state. Thirteen of these tribes or communities operated casinos in 1997, which generate tribal revenues of millions of dollars. The spatial distribution of the thirteen shows a distinct relationship to the metropolitan areas of Phoenix, Tucson, and Yuma or to major tourist areas as shown in Table 4, "Arizona Indian Gaming Operations" and Figure 5, "Arizona Indian Gaming Casinos."

The establishment and rise of Indian gaming has not developed without opposition. The horse and greyhound racing industries have not welcomed the competition, and understandably so. Inasmuch as these industries have made large contributions to most legislators' campaign funds over many years, their influence is felt in the state legislature. Also, other groups have voiced concerns regarding the impact of increased traffic congestion and crime on neighboring communities. After months of confrontations that included lawsuits, use of federal mediators, the temporary banning of charitable "casino" or "Las Vegas" nights, standoffs with federal agents and the granting of special tax benefits to the racetracks, many issues were resolved to some degree. However, other areas of dispute between the Indian tribes and the state remain. Foremost was the manner of determining what types of gaming are legal. Also, there was confusion over the jurisdiction regarding legal matters related to casinos. Whether state or tribal courts have the ultimate

Photo by: R. D. Winchell

The Seven Cedars Casino includes a gaming area, restaurant, a native arts gallery, arts and crafts sales and a large showroom for dining, dancing and performances.

authority was not clearly defined. In November, 1994, the tribes that operate casinos established the Arizona Indian Gaming Association to provide a united front to challenge the state gaming agency concerning these disputes and other issues that might arise.

It is likely that one of the most recent developments may have the most far-reaching effect on the Phoenix metropolitan area. The Salt River Pima-Maricopa Indians recently voted overwhelmingly in favor of going into the casino business. The reservation abuts Scottsdale, Mesa and Tempe, and within a five-mile radius there are several of the area's largest resort hotels. The establishment of a large high quality casino on Indian lands will have a major impact on Scottsdale and Tempe, for better or worse. A collaborative effort between the communities is essential to maximize the economic benefits and to head off whatever adverse impacts that will arise.

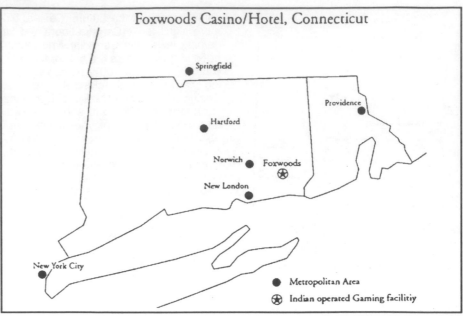

Foxwoods Casino/Hotel, Connecticut

● Metropolitan Area
✪ Indian operated Gaming facilitiy

Map by: Michigan State Cartography Lab

The Michigan case

Indian commercial gaming in Michigan, as in most other states, started with bingo and moved gradually into casino type operations. The initial casino efforts were small and created much controversy over their legality. The first Indian-owned casino was a minuscule

Photo by: R. D, Winchell
Seven Cedars Casino electronic sign advertising "The Platters," Sequim, WA.

operation in a remodeled garage near Baraga in the Upper Peninsula in 1984. Indian casino gaming mushroomed in number and size with the signing of a compact between the tribes and the state in August of 1993. By 1995, six Indian operated casino complexes were in operation in the Upper Peninsula as shown in Table 5 and Figure 6, "Michigan Indian Gaming Operations." Growth has taken place at a rapid rate and is characterized by major casino floor space expansion, the building and enlargement of adjoining motels, hotels, and restaurants and the possibility of marinas and golf courses.

The nature of gaming activities has evolved rapidly from bingo halls to casinos and from blackjack and slot machines to all table and video games. In the compact with the state, the tribes agreed to give the state eight percent and the local government two percent of the video poker and slot machine profits. This was done in order to be able to keep those highly profitable machines in their casinos.

The locations of Indian gaming in Michigan present a marketing problem, in that only the Soaring Eagle Casino near Mt. Pleasant, operated by the Saginaw Chippewa Indian Tribe, is within easy daily driving distance of the major population concentrations of the southern one-third of Michigan, especially the Grand Rapids, Saginaw Bay, Flint, Lansing and Detroit urban areas. The other seven casinos are in thinly populated but highly popular summer and winter tourist areas of Michigan. The Leelanau Sands Casino, on Suttons Bay, about 20 miles north of the important Traverse City tourist area, is a good ex-

ample of a location possessing excellent summer water-related activities for tourists as well as the winter attractions of ski areas, such as the nearby Sugarloaf Ski Resort. This casino is operated by the Grand Traverse Band of the Ottawa and Chippewa Indians and includes over 20,000 square feet of gambling space in two buildings, including all types of casino gaming, two restaurants and a 51-unit motel. Profits from this casino gaming complex have been used for improving Indian housing, building a health center and a tribal center, as well as plowing funds into casino expansion and reducing debt obligations. Some Michigan tribes are also distributing some gaming profits to official tribal members on a quarterly basis.

Much controversy has existed over whether casino gaming should be developed in the Detroit area. State referendums on casinos in Detroit were defeated until the Casino Windsor opened in 1994, just across the Detroit River in Windsor, Ontario, Canada. This facility draws thousands of dollars and thousands of Michiganians daily, mostly from Detroit. It is estimated that 80 percent of the Casino Windsor customers are from Detroit. As a result, a referendum on casino gambling passed easily in 1996.

The 1996 referendum permits three casinos to be developed in Detroit. Much discussion took place in the Michigan legislature on how to adequately control the developments and the amount of required financial returns to the state. Agreements were reached in July, 1997. The mayor of Detroit screened applications from major gaming enterprises, casino locations are being decided,

Photo by: L. Sommers

Soaring Eagle Casino and Hotel near Mount Pleasant, MI. This huge establishment, operated by the Isabella Indian Reservation, is one of the largest between the U.S. east coast and Las Vegas, NV.

including consideration of the mayor's recommendation of a 100-acre site in northwest Detroit, and construction will begin in 1998. Casino gambling alone will not solve Detroit's economic problems, but it will provide considerable employment and reduce the flow of gaming money to Ontario.

Gaming on Indian reservation lands is having a major economic and social impact on Michigan Indians and the areas in which Indian operated casinos exist. About half of the workers in the casinos are Indians. Before casino gaming, more than half of the Michigan Indians were unemployed or on welfare. This figure has been drastically reduced and some Indians and Indian tribes are now much more economically self-sufficient. The economic impact on the areas where casinos have been built is considerable in terms of employment and the increase in income of nearby businesses supplying goods and services to the casinos and their patrons.

Income from Indian gaming has been used to enlarge casinos and associated buildings. The Roaring Eagle Casino near Mt. Pleasant is a good example. A major casino, one of the largest between the east coast and Las Vegas, has been built, and a hotel with over 500 beds was completed in 1997. Considerable funds also have been funneled to improve community, economic, and social conditions. Existing housing for Indians was renovated and new homes are being built. Other funds are being used to build health facilities, social service and substance abuse centers, day-care centers, elderly housing, and to improve and fund education.

Some Indian strategies to broaden the long-term impacts of gaming are demonstrated by investments in industry and activities other than gaming that will benefit Indians as well as others. The overall result is that there has been marked improvement in the economic and social well-being of Michigan Indians in areas where casino gaming is located. Indians are also playing

Michigan Indian Gaming Casinos

Ojibwa Casino
La Vieux Desert Casino
Kings Club Casino
Vegas Kewadin Casino
Kewadin Shores Casino
Chip-In Casino
Leelanau Sands Casino
Soaring Eagle Casino
Muskegon
Saginaw
Grand Rapids
Flint
Lansing
Detroit

● Metropolitan Area
✺ Indian operated gaming facility

Map by: Michigan State Cartography Lab

a more important role in the policies and decisions of local and state government. Problems such as drug and alcohol abuse exist, as in the rest of society, but the tribes are assuming more responsibility and making inputs which augur well for their communities as well as for their contributions to the social and economic health of Michigan.

Political, legal, economic and social questions

The advent of gaming on Indian lands has resulted in political and legal issues—federal, state and local—that will take years to resolve. At the federal level, several members of Congress wish to change the 1988 IGRA, although at this time there is no clear consensus as to what these changes should be. Other members of Congress would repeal the Act, while still others are satisfied with it as it now stands. The tribes involved in gaming obviously resist more federal or state intrusion into their sovereignty. In any case, any bill introduced in Congress regarding changes or amendments to the 1988 Act is bound to be argued for a long period of time.

At the state level, it becomes a battle of state sovereign rights versus tribal sovereign rights. States resist Indian gaming because it diverts gamblers from state-sponsored games and other forms of recreation, and adds to the state's cost of roads, police and community services. Perhaps the major issue is who has the ultimate say as to which games can be offered and the magnitude of those operations. There is a great diversity among the states regarding this issue and the manner of negotiations with tribal units. Unfortunately, it is likely that many of these issues will be taken to the courts for decision.

At the local level, counties, townships and municipalities near Indian gaming facilities are concerned with rural-to-urban changes in lifestyle, and confusion over jurisdiction regarding enforcement of crimes associated with the increase of people and traffic such as speeding, drunken driving, fraud and car burglary. Again, many of these issues across the country will likely end up in long drawn-out court cases.

In addition to political issues, economic and social impacts of Indian gaming are of concern in many areas. Economically, other forms of recreation, particularly charitable bingo and card games, horse and greyhound racing, nightclubs, and bars do not welcome the competition. Whether or not a given Indian gaming operation is a boon to the

Indian Reservation Gaming Establishments
1994

Map by: Michigan State Cartography Lab

local area depends on whether it is a *basic* or *non-basic* economic activity. Basic economic activities generate local or regional growth by exporting goods or services beyond the borders of the locality or area. Conversely, nonbasic eco-

In the majority of cases, the positive economic and social gains have greatly outweighed the negative factors.

nomic activities serve the local area markets only: they bring in or import money from the outside. In this context, gaming is good for economic development only if you can import the gamblers. There are gaming operations that are non-basic in nature which have not

greatly benefited the local region, such as in Colorado, but many have proven to be basic to their region and have produced tremendous economic growth. The Foxwoods in Connecticut is a prime example.

From the social standpoint, the economic gains have drastically helped the living standards of many tribes. Increased income, employment, new schools, water and sewer facilities, roads, social centers and the like have exceeded the wildest dreams of many tribes in various parts of the country. Conversely, there are cases of increased crime, unwise spending and investments, and breakdowns of tribal cultures. Overall, in the majority of cases, the positive economic and social gains have greatly outweighed the negative factors, at least in the short run.

Another major issue is whether Indian gaming developments will be able to compete in the gaming industry over the long run. As Indian reservation casinos increase in number, size and complexity of gaming, and the number of non-Indian casinos increase, the competition for customers will also intensify. It remains to be seen whether the nationwide increase in casino gaming can continue to

be successful in its various forms and levels. It is likely that some developments will fail as has been the case in Las Vegas. As that city has evolved to megacasinos and theme parks, some of the smaller and less appealing casinos have failed to remain competitive. The Indian gaming industry undoubtedly will also go through this kind of evolution. In the meantime it has provided the single-greatest boon to reservation economies and tribal government resources since Indian

Dick G. Winchell *is Professor of Planning at Eastern Washington University. His interests include rural development and ethnicity in America. Email: DWINCHELL@ewu.edu*

John F. Lounsbury *is Professor Emeritus of Geography at Arizona State University. He and Larry Sommers wrote "Border Boom Towns of Nevada," FOCUS 41(4), in 1991.*

Lawrence M. Sommers *is Professor in the Department of Geography, Michigan State University. His interests include applied geography and economic development. He and John Lounsbury wrote "Border Boom Towns of Nevada," FOCUS 41(4), in 1991.*

reservations were first established in the United States.

References and Further Readings

Devereaux, Francis and Asmara Habte. 1997. The Planning Impacts of Tribal ming in Washington. The Western Planner.

Eadington, W. (ed.) 1990. Indian Gaming and the Law. Reno: Institute for the Study of Gambling and Commercial Gaming, University of Nevada, Reno.

Facts on File. Supreme Court: Indian Gaming Act Overturned. April 4, 1996. v. 56, n 2882, p. 220.

Indian Country Today. 1992. Gaming: Winner's Circle Special Edition. November 5, 1992. 38 pp.

1996. Bloom Is Off American Indian Gaming Rose. 1994, 1997. July 14–21, p. A5.

International Gaming and Wagering Business. 1994, 1997. New York.

Meyer-Arendt, K. and R. Hartmann (eds.). 1998. Casino Gambling in America: Origins, Trends and Impacts. Cognizant C.C.: New York.

Minnesota Indian Gaming Association. 1992. Economic Benefits of Tribal Gaming in Minnesota. Minneapolis, Minnesota: Minnesota Indian Gaming Association Report, March, 1992 (also reprinted in Senate Select Committee, 1992, part 2, pp. 273–298).

The New York Times Magazine. 1994. Gambling Nation. The New York Times Magazine. July 17, 1994, pp. 34–61.

Senate Select Committee on Indian Affairs. 1992. Implementation of Indian Gaming Regulatory Act. Hearing before the Select Committee on Indian Affairs United States Senate. March 18, 1992. Parts 1, 2, 3 and 4. Washington, D.C. U.S. Government Printing Office.

Sommers, Lawrence M. and John F. Lounsbury. 1991. Border Boom Towns of Nevada. FOCUS 41 (4), Winter, 1991, pp. 12–18.

White, R. 1990. The Rebirth of Native America. New York: Henry Holt and Co.

Winchell, Dick G. and Francis Devereaux. 1993. The Impact of Gaming on the Colville Confederated Tribes. Nespelem, WA.: The Planning Department, Colville Confederated Tribes.

Legal Cases

California v. Cabazon Band of Mission Indians,] 480 U.S. 208 (1987).

Confederated Tribes v. State of Washington, 938 F. 2nd 146(9th Cir. 1990).

Seminole Tribe of Florida v. Florida. March 27,1996.

Legislation

Public Law 100–497 Indian Gaming Regulatory Act (IGRA), 25 U.S.C. sec. 2701, et. seq. and 18 U.S.C. sec. 1166–1168, enacted October, 1988.

Unit 5

Key Points to Consider

❖ How are you personally affected by the population explosion?

❖ Give examples of how economic development adversely affects the environment. How can such adverse effects be prevented?

❖ How do you feel about the occurrence of starvation in developing world regions?

❖ What might it be like to be a refugee?

❖ In what forms is colonialism present today?

❖ How is Earth a system?

❖ For how long are world systems sustainable?

❖ What is your scenario of the world in the year 2010?

 **Links** ## www.dushkin.com/online/

These sites are annotated on pages 6 and 7.

The final unit of this anthology includes discussions of several important problems facing humankind. Geographers are keenly aware of regional and global difficulties. It is hoped that their work with researchers from other academic disciplines and representatives of business and government will help bring about solutions to these serious problems.

Probably no single phenomenon has received as much attention in recent years as the so-called population explosion. World population continues to increase at unacceptably high rates. The problem is most severe in the less developed countries, where in some cases, populations are doubling in less than 20 years.

The human population of the world passed the 6 billion mark in 1999. It is anticipated that population increase will continue well into the twenty-first century, despite a slowing in the rate of population growth globally since the 1960s. The first four articles in this section deal with related issues of population growth and declining petroleum reserves. Then, freshwater is considered as a scarce resource in "A Rare and Precious Resource." The next article deals with aspects of the Hispanic population. "Russia's Population Sink" reports on the dilemma of death rates exceeding birth rates in this postcommunist country. Gerhard Heilig outlines changes in land use in China as that country shifts its economic focus. The last article, "Helping the World's Poorest," makes a plea for developed world assistance to the poorest countries.

Before the Next Doubling

Nearly 6 billion people now inhabit the Earth—almost twice as many as in 1960. At some point over the course of the next century, the world's population could double again. But we don't have anything like a century to prevent that next doubling; we probably have less than a decade.

by Jennifer D. Mitchell

In 1971, when Bangladesh won independence from Pakistan, the two countries embarked on a kind of unintentional demographic experiment. The separation had produced two very similar populations: both contained some 66 million people and both were growing at about 3 percent a year. Both were overwhelmingly poor, rural, and Muslim. Both populations had similar views on the "ideal" family size (around four children); in both cases, that ideal was roughly two children smaller than the actual average family. And in keeping with the Islamic tendency to encourage large families, both generally disapproved of family planning.

But there was one critical difference. The Pakistani government, distracted by leadership crises and committed to conventional ideals of economic growth, wavered over the importance of family planning. The Bangladeshi government did not: as early as 1976, population growth had been declared the country's number one problem, and a national network was established to educate people about family planning and supply them with contraceptives. As a result, the proportion of couples using contraceptives rose from around 6 percent in 1976 to about 50 percent today, and fertility rates have dropped from well over six children per woman to just over three. Today, some 120 million people people live in Bangladesh, while 140 million live in Pakistan—a difference of 20 million.

Bangladesh still faces enormous population pressures—by 2050, its population will probably have increased by nearly 100 million. But even so, that 20 million person "savings" is a colossal achievement, especially given local conditions. Bangladeshi officials had no hope of producing the classic "demographic transition," in which improvements in education, health care, and general living standards tend to push down the birth rate. Bangladesh was—and is—one of the poorest and most densely populated countries on earth. About the size of England and Wales, Bangladesh has twice as many people. Its per capita GDP is barely over $200. It has one doctor for every 12,500 people and nearly three-quarters of its adult population are illiterate. The national diet would be considered inadequate in any industrial country, and even at current levels of population growth, Bangladesh may be forced to rely increasingly on food imports.

All of these burdens would be substantially heavier than they already are, had it not been for the family planning program. To appreciate the Bangladeshi achievement, it's only necessary to look at Pakistan: those "additional" 20 million Pakistanis require at least 2.5 million more houses, about 4 million more tons of grain each year, millions more jobs, and significantly greater investments in health care—or a significantly greater burden of disease. Of the two nations, Pakistan has the more robust economy—its per capita GDP is twice that of Bangladesh. But the Pakistani economy is still primarily agricultural, and the size of the average farm is shrinking, in part because of the expanding population. Already, one fourth of the country's farms are under 1 hectare, the standard minimum size for economic viability, and Pakistan is looking increasingly towards the international grain markets to feed its people. In 1997, despite its third consecutive year of near-record harvests, Pakistan attempted to double its wheat imports but was not able to do so because it had exhausted its line of credit.

And Pakistan's extra burden will be compounded in the next generation. Pakistani women still bear an average of well over five children, so at the current birth rate, the 10 million or so extra couples would produce at least 50 million children. And these in turn could bear nearly 125 million children of their own. At its current fertility rate, Pakistan's population will double in just 24 years—that's more than twice as fast as Bangladesh's population is growing. H. E. Syeda Abida Hussain, Pakistan's Minister of Population Welfare, explains the problem bluntly: "If we achieve success in lowering our population growth substantially, Pakistan has a future. But if, God forbid, we should not—no future."

The Three Dimensions of the Population Explosion

Some version of Mrs. Abida's statement might apply to the world as a whole. About 5.9 billion people currently inhabit the Earth. By the middle of the next century, according to U.N. projections, the population will probably reach 9.4 billion—and all of the net increase is likely to occur in the developing world. (The total population of the industrial countries is expected to decline slightly over the next 50 years.) Nearly 60 percent of the increase will occur in Asia, which will grow from 3.4 billion people in 1995 to more than 5.4 billion in 2050. China's population will swell from 1.2 billion to 1.5 billion, while India's is projected to soar from 930 million to 1.53 billion. In the Middle East and North Africa, the population will probably more than double, and in sub-Saharan Africa, it will triple. By 2050, Nigeria alone is expected to have 339 million people—more than the entire continent of Africa had 35 years ago.

Despite the different demographic projections, no country will be immune to the effects of population growth. Of course, the countries with the highest growth rates are likely to feel the greatest immediate burdens—on their educational and public health systems, for instance, and on their forests, soils, and water as the struggle to grow more food intensifies. Already some 100 countries must rely on grain imports to some degree, and 1.3 billion of the world's people are living on the equivalent of $1 a day or less.

But the effects will ripple out from these "front-line" countries to encompass the world as a whole. Take the water predicament in the Middle East as an example. According to Tony Allan, a water expert at the University of London, the Middle East "ran out of water" in 1972, when its population stood at 122 million. At that point, Allan argues, the region had begun to draw more water out of its aquifers and rivers than the rains were replenishing. Yet today, the region's population is twice what it was in 1972 and still growing. To some degree, water management now determines political destiny. In Egypt, for example, President Hosni Mubarak has announced a $2 billion diversion project designed to pump water from the Nile River into an area that is now desert. The project—Mubarak calls it a "necessity imposed by population"—is designed to resettle some 3 million people outside the Nile flood plain, which is home to more than 90 percent of the country's population.

Elsewhere in the region, water demands are exacerbating international tensions; Jordan, Israel, and Syria, for instance, engage in uneasy competition for the waters of the Jordan River basin. Jordan's King Hussein once said that water was the only issue that could lead him to declare war on Israel. Of course, the United States and the western European countries are deeply involved in the region's antagonisms and have invested heavily in its fragile states. The western nations have no realistic hope of escaping involvement in future conflicts.

Yet the future need not be so grim. The experiences of countries like Bangladesh suggest that it is possible to build population policies that are a match for the threat. The first step is to understand the causes of population growth. John Bongaarts, vice president of the Population Council, a non-profit research group in New York City, has identified three basic factors. (See figure on the next page.)

Unmet demand for family planning. In the developing world, at least 120 million married women—and a large but undefined number of unmarried women—want more control over their pregnancies, but cannot get family planning services. This unmet demand will cause about one-third of the projected population growth in developing countries over the next 50 years, or an increase of about 1.2 billion people.

Desire for the large families. Another 20 percent of the projected growth over the next 50 years, or an increase of about 660 million people, will be caused by couples who may have access to family planning services, but who choose to have more than two children. (Roughly two children per family is the "replacement rate," at which a population could be expected to stabilize over the long term.)

Population momentum. By far the largest component of population growth is the least commonly understood. Nearly one-half of the increase projected for the next 50 years will occur simply because the next reproductive generation—the group of people currently entering puberty or younger—is so much larger than the current reproductive generation. Over the next 25 years, some 3 billion people—a number equal to the entire world population in 1960—will enter their reproductive years, but only about 1.8 billion will leave that phase of life. Assuming that the couples in this reproductive bulge begin to have children at a fairly early age, which is the global norm, the global population would still expand by 1.7 billion, even if all of those couples had only two children—the longterm replacement rate.

Meeting the Demand

Over the past three decades, the global percentage of couples using some form of family planning has increased dramatically—from less than 10 to more than 50 percent. But due to the growing population, the absolute number of women not using family planning is greater today than it was 30 years ago. Many of these women fall into that first category above—they want the services but for one reason or another, they cannot get them.

Sometimes the obstacle is a matter of policy: many governments ban or restrict valuable methods of contraception. In Japan, for instance, regulations discourage the use of birth control pills in favor of condoms, as a public health measure against sexually transmitted diseases. A study conducted in 1989 found that some 60 countries required a husband's permission before a woman can be sterilized; several required a husband's consent for all forms of birth control.

Elsewhere, the problems may be more logistical than legal. Many developing countries lack clinics and pharmacies in rural areas. In some rural areas of sub-Saharan Africa, it takes an average of two hours to reach the nearest contraceptive provider. And often contraceptives are too expensive for most people. Sometimes the products or services are of such poor quality that they are not simply ineffective, but dangerous. A woman who has been injured by a badly made or poorly inserted IUD may well be put off by contraception entirely.

In many countries, the best methods are simply unavailable. Sterilization is often the only available nontraditional option, or the only one that has gained wide acceptance. Globally, the procedure accounts for about 40 percent of contraceptive use and in some countries the fraction is much higher: in the Dominican Republic and India, for example, it stands at 69 percent. But women don't generally resort to sterilization until well into their childbearing years, and in some countries, the procedure isn't permitted until a woman reaches a certain age or bears a certain number of children. Sterilization is therefore no substitute for effective temporary methods like condoms, the pill, or IUDs.

There are often obstacles in the home as well. Women may be prevented from seeking family planning services by disapproving husbands or in-laws. In Pakistan, for example, 43 percent of husbands object to family planning. Frequently,

Population of Developing Countries, 1950–95, with Projected Growth to 2050

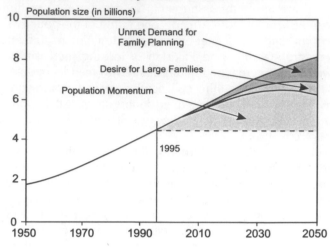

Source: U.N., *World Population Prospects: The 1996 Revision* (New York: October 1998); and John Bongaarts, "Population Policy Options in the Developing World," *Science,* 11 February 1994.

such objections reflect a general social disapproval inculcated by religious or other deeply-rooted cultural values. And in many places, there is a crippling burden of ignorance: women simply may not know what family planning services are available or how to obtain them.

Yet there are many proven opportunities for progress, even in conditions that would appear to offer little room for it. In Bangladesh, for instance, contraception was never explicitly illegal, but many households follow the Muslim custom of *purdah,* which largely secludes women in their communities.

Since it's very difficult for such women to get to family planning clinics, the government brought family planning to them: some 30,000 female field workers go door-to-door to explain contraceptive methods and distribute supplies. Several other countries have adopted Bangladesh's

approach. Ghana, for instance, has a similar system, in which field workers fan out from community centers. And even Pakistan now deploys 12,000 village-based workers, in an attempt to reform its family planning program, which still reaches only a quarter of the population.

Reducing the price of contraceptives can also trigger a substantial increase in use. In poor countries, contraceptives can be an extremely price-sensitive commodity even when they are very cheap. Bangladesh found this out the hard way in 1990, when officials increased contraceptive prices an average of 60 percent. (Under the increases, for example, the cheapest condoms cost about 1.25 U.S. cents per dozen). Despite regular annual sales increases up to that point, the market slumped immediately: in 1991, condom sales fell by 29 percent and sales of the pill by 12 percent. The next year, prices were rolled back; sales rebounded and have grown steadily since then.

Additional research and development can help broaden the range of contraceptive options. Not all methods work for all couples, and the lack of a suitable method may block a substantial amount of demand. Some women, for instance, have side effects to the pill; others may not be able to use IUDs because of reproductive tract infections. The wider the range of available methods, the better the chance that a couple will use one of them.

Planning the Small Family

Simply providing family planning services to people who already want them won't be enough to arrest the population juggernaut. In many countries, large families are still the ideal. In Senegal, Cameroon, and Niger, for example, the average woman still wants six or seven children. A few countries have tried to legislate such desires away. In India, for example, the Ministry of Health and Family Welfare is interested in promoting a policy that would bar people who have more than two children from political careers, or deny them promotion if they work within the civil service bureaucracy. And China's well-known policy allows only one child per family.

But coercion is not only morally questionable—it's likely to be ineffective because of the backlash it invites. A better starting point for policy would be to try to understand why couples want large families in the first place. In many developing countries, having lots of children still seems perfectly rational: children are a source of security in old age and may be a vital part of the family economy. Even when they're very young, children's labor can make them an asset rather than a drain on family income. And in countries with high child mortality rates, many births may be viewed as necessary to compensate for the possible deaths (of course, the cumulative statistical effect of such a reaction is to *over*-compensate).

Religious or other cultural values may contribute to the big family ideal. In Pakistan, for instance, where 97 percent of the population is Muslim, a recent survey of married women found that almost 60 percent of them believed that the number of children they have is "up to God." Preference for sons is another widespread factor in the big

family psychology: many large families have come about from a perceived need to bear at least one son. In India, for instance, many Hindus believe that they need a son to perform their last rites, or their souls will not be released from the cycle of births and rebirths. Lack of a son can mean abandonment in this life too. Many husbands desert wives who do not bear sons. Or if a husband dies, a son is often the key to a woman's security: 60 percent of Indian women over 60 are widows, and widows tend to rely on their sons for support. In some castes, a widow has no other option since social mores forbid her from returning to her birth village or joining a daughter's family. Understandably, the fear of abandonment prompts many Indian women to continue having children until they have a son. It is estimated that if son preference were eliminated in India, the fertility rate would decline by 8 percent from its current level of 3.5 children per woman.

Yet even deeply rooted beliefs are subject to reinterpretation. In Iran, another Muslim society, fertility rates have dropped from seven children per family to just over four in less than three decades. The trend is due in some measure to a change of heart among the government's religious authorities, who had become increasingly concerned about the likely effects of a population that was growing at more than 3 percent per year. In 1994, at the International Conference on Population and Development (ICPD) held in Cairo, the Iranian delegation released a "National Report on Population" which argued that according to the "quotations from prophet Mohammad . . . and verses of [the] holy Quran, what is standing at the top priority for the Muslims' community is the social welfare of Muslims." Family planning, therefore, "not only is not prohibited but is emphasized by religion."

Promotional campaigns can also change people's assumptions and behavior, if the campaigns fit into the local social context. Perhaps the most successful effort of this kind is in Thailand, where Mechai Viravidaiya, the founder of the Thai Population and Community Development Association, started a program that uses witty songs, demonstrations, and ads to encourage the use of contraceptives. The program has helped foster widespread awareness of family planning throughout Thai society. Teachers use population-related examples in their math classes; cab drivers even pass out condoms. Such efforts have paid off: in less than three decades, contraceptive use among married couples has risen from 8 to 75 percent and population growth has slowed from over 3 percent to about 1 percent—the same rate as in the United States.

Better media coverage may be another option. In Bangladesh, a recent study found that while local journalists recognize the importance of family planning, they do not understand population issues well enough to cover them effectively and objectively. The study, a collaboration between the University Research Corporation of Bangladesh and Johns Hopkins University in the United States, recommended five ways to improve coverage: develop easy-to-use information for journalists (press releases, wall charts, research summaries), offer training and workshops, present awards for population journalism, create a forum for communication between journalists and family planning professionals, and establish a population resource center or data bank.

Often, however, the demand for large families is so tightly linked to social conditions that the conditions themselves must be viewed as part of the problem. Of course, those conditions vary greatly from one society to the next, but there are some common points of leverage:

Reducing child mortality helps give parents more confidence in the future of the children they already have. Among the most effective ways of reducing mortality are child immunization programs, and the promotion of "birth spacing"—lengthening the time between births. (Children born less than a year and a half apart are twice as likely to die as those born two or more years apart.)

Improving the economic situation of women provides them with alternatives to child-bearing. In some countries, officials could reconsider policies or customs that limit women's job opportunities or other economic rights, such as the right to inherit property. Encouraging "micro-leaders" such as Bangladesh's Grameen Bank can also be an effective tactic. In Bangladesh, the Bank has made loans to well over a million villagers—mostly impoverished women—to help them start or expand small businesses.

Improving education tends to delay the average age of marriage and to further the two goals just mentioned. Compulsory school attendance for children undercuts the economic incentive for larger families by reducing the opportunities for child labor. And in just about every society, higher levels of education correlate strongly with small families.

Momentum: The Biggest Threat of All

The most important factor in population growth is the hardest to counter—and to understand. Population momentum can be easy to overlook because it isn't directly captured by the statistics that attract the most attention. The global growth rate, after all, is dropping: in the mid-1960s, it amounted to about a 2.2 percent annual increase; today the figure is 1.4 percent. The fertility rate is dropping too: in 1950, women bore an average of five children each; now they bear roughly three. But despite these continued declines, the absolute number of births won't taper off any time soon. According to U.S. Census Bureau estimates, some 130 million births will still occur annually for the next 25 years, because of the sheer number of women coming into their child-bearing years.

The effects of momentum can be seen readily in a country like Bangladesh, where more than 42 percent of the population is under 15 years old—a typical proportion for many poor countries. Some 82 percent of the population growth projected for Bangladesh over the next half century will be caused by momentum. In other words, even if from now on, every Bangladeshi couple were to have only two children, the country's population would

still grow by 80 million by 2050 simply because the next reproductive generation is so enormous.

The key to reducing momentum is to delay as many births as possible. To understand why delay works, its helpful to think of momentum as a kind of human accounting problem in which a large number of births in the near term won't be balanced by a corresponding number of deaths over the same period of time. One side of the population ledger will contain those 130 million annual births (not all of which are due to momentum, of course), while the other side will contain only about 50 million annual deaths. So to put the matter in a morbid light, the longer a substantial number of those births can be delayed, the longer the death side of the balance sheet will be when the births eventually occur. In developing countries, according to the Population Council's Bongaarts, an average 2.5-year delay in the age when a woman bears her first child would reduce population growth by over 10 percent.

One way to delay childbearing is to postpone the age of marriage. In Bangladesh, for instance, the median age of first marriage among women rose from 14.4 in 1951 to 18 in 1989, and the age at first birth followed suit. Simply raising the legal age of marriage may be a useful tactic in countries that permit marriage among the very young. Educational improvements, as already mentioned, tend to do the same thing. A survey of 23 developing countries found that the median age of marriage for women with secondary education exceeded that of women with no formal education by four years.

Another fundamental strategy for encouraging later childbirth is to help women break out of the "sterilization syndrome" by providing and promoting high-quality, temporary contraceptives. Sterilization might appear to be the ideal form of contraception because it's permanent. But precisely because it is permanent, women considering sterilization tend to have their children early, and then resort to it. A family planning program that relies heavily on sterilization may therefore be working at cross purposes with itself: when offered as a primary form of contraception, sterilization tends to promote early childbirth.

What Happened to the Cairo Pledges?

At the 1994 Cairo Conference, some 180 nations agreed on a 20-year reproductive health package to slow population growth. The agreement called for a progressive rise in annual funding over the life of the package; according to U.N. estimates, the annual price tag would come to about $17 billion by 2000 and $21.7 billion by 2015. Developing countries agreed to pay for two thirds of the program, while the developed countries were to pay for the rest. On a global scale, the package was fairly modest: the annual funding amounts to less than two weeks' worth of global military expenditures.

Today, developing country spending is largely on track with the Cairo agreement, but the developed countries are not keeping their part of the bargain. According to a recent study by the U.N. Population Fund (UNFPA), all forms of developed country assistance (direct foreign aid, loans from multilateral agencies, foundation grants, and so on) amounted to only $2 billion in 1995. That was a 24 percent increase over the previous year, but preliminary estimates indicate that support declined some 18 percent in 1996 and last year's funding levels were probably even lower than that.

The United States, the largest international donor to population programs, is not only failing to meet its Cairo commitments, but is toying with a policy that would undermine international family planning efforts as a whole. Many members of the U.S. Congress are seeking reimposition of the "Mexico City Policy" first enunciated by President Ronald Reagan at the 1984 U.N. population conference in Mexico City, and repealed by the Clinton administration in 1993. Essentially, a resurrected Mexico City Policy would extend the current U.S. ban on funding abortion services to a ban on funding any organization that:

- funds abortions directly, or
- has a partnership arrangement with an organization that funds abortions, or
- provides legal services that may facilitate abortions, or
- engages in any advocacy for the provision of abortions, or
- participates in any policy discussions about abortion, either in a domestic or international forum.

The ban would be triggered even if the relevant activities were paid for entirely with non-U.S. funds. Because of its draconian limits even on speech, the policy has been dubbed the "Global Gag Rule" by its critics, who fear that it could stifle, not just abortion services, but many family planning operations involved only incidentally with abortion. Although Mexico City proponents have not managed to enlist enough support to reinstate the policy, they have succeeded in reducing U.S. family planning aid from $547 million in 1995 to $385 million in 1997. They have also imposed an unprecedented set of restrictions that meter out the money at the rate of 8 percent of the annual budget per month—a tactic that *Washington Post* reporter Judy Mann calls "administrative strangulation."

If the current underfunding of the Cairo program persists, according to the UNFPA study, 96 million fewer couples will use modern contraceptives in 2000 than if commitments had been met. One-third to one-half of these couples will resort to less effective traditional birth control methods; the rest will not use any contraceptives at all. The result will be an additional 122 million unintended pregnancies. Over half of those pregnancies will end in births, and about 40 percent will end in abortions. (The funding shortfall is expected to produce 16 million more abortions in 2000 alone.) The unwanted pregnancies will kill about 65,000 women by 2000, and injure another 844,000.

Population funding is always vulnerable to the illusion that the falling growth rate means the problem is going away. Worldwide, the annual population increase had dropped from a high of 87 million in 1988 to 80 million today. But dismissing the problem with that statistic is like comforting someone stuck on a railway crossing with the news that an oncoming train has slowed from 87 to 80 kilometers an hour, while its weight has increased. It will now take 12.5 years instead of 11.5 years to add the next billion people to the world. But that billion will surely arrive—and so will at least one more billion. Will still more billions follow? That, in large measure, depends on what policymakers do now. Funding alone will not ensure that population stabilizes, but lack of funding will ensure that it does not.

The Next Doubling

In the wake of the Cairo conference, most population programs are broadening their focus to include improvements in education, women's health, and women's social status among their many goals. These goals are worthy in their own right and they will ultimately be necessary for bringing population under control. But global population growth has gathered so much momentum that it could simply overwhelm a development agenda. Many countries now have little choice but to tackle their population problem in as direct a fashion as possible—even if that means temporarily ignoring other social problems. Population growth is now a global social emergency. Even as officials in both developed and developing countries open up their program agendas, it is critical that they not neglect their single most effective tool for dealing with that emergency: direct expenditures on family planning.

The funding that is likely to be the most useful will be constant, rather than sporadic. A fluctuating level of commitment, like sporadic condom use, can

end up missing its objective entirely. And wherever it's feasible, funding should be designed to develop self-sufficiency—as, for instance, with UNFPA's $1 million grant to Cuba, to build a factory for making birth control pills. The factory, which has the capacity to turn out 500 million tablets annually, might eventually even provide the country with a new export product. Self-sufficiency is likely to grow increasingly important as the fertility rate continues to decline. As Tom Merrick, senior population advisor at the World Bank explains, "while the need for contraceptives will not go away when the total fertility rate reaches two—the donors will."

Even in narrow, conventional economic terms, family planning offers one of the best development investments available. A study in Bangladesh showed that for each birth prevented, the government spends $62 and saves $615 on social services expenditures—nearly a tenfold return. The study estimated that the Bangladesh program prevents 890,000 births a year, for a net annual savings of $547 million. And that figure does not include savings resulting from lessened pressure on natural resources.

Over the past 40 years, the world's population has doubled. At some point in the latter half of the next century, today's population of 5.9 billion could double again. But because of the size of the next reproductive generation, we probably have only a relatively few years to stop that next doubling. To prevent all of the damage—ecological, economic, and social—that the next doubling is likely to cause, we must begin planning the global family with the same kind of urgency that we bring to matters of trade, say, or military security. Whether we realize it or not, our attempts to stabilize population—or our failure to act—will likely have consequences that far outweigh the implications of the military or commercial crisis of the moment. Slowing population growth is one of the greatest gifts we can offer future generations.

Jennifer D. Mitchell is a staff researcher at the Worldwatch Institute.

The Population Surprise

The old assumptions about world population trends need to be rethought. One thing is clear: in the next century the world is in for some rapid downsizing

by Max Singer

FIFTY years from now the world's population will be declining, with no end in sight. Unless people's values change greatly, several centuries from now there could be fewer people living in the entire world than live in the United States today. The big surprise of the past twenty years is that in not one country did fertility stop falling when it reached the replacement rate—2.1 children per woman. In Italy, for example, the rate has fallen to 1.2. In Western Europe as a whole and in Japan it is down to 1.5. The evidence now indicates that within fifty years or so world population will peak at about eight billion before starting a fairly rapid decline.

Because in the past two centuries world population has increased from one billion to nearly six billion, many people still fear that it will keep "exploding" until there are too many people for the earth to support. But that is like fearing that your baby will grow to 1,000 pounds because its weight doubles three times in its first seven years. World population was growing by two percent a year in the 1960s; the rate is now down to one percent a year, and if the patterns of the past century don't change radically, it will head into negative numbers. This view is coming to be widely accepted among population experts, even as the public continues to

Max Singer was a founder of the Hudson Institute. He is a co-author, with Aaron Wildavsky, of *The Real World Order* (1996).

focus on the threat of uncontrolled population growth.

As long ago as September of 1974 *Scientific American* published a special issue on population that described what demographers had begun calling the "demographic transition" from traditional high rates of birth and death to the low ones of modern society. The experts believed that birth and death rates would be more or less equal in the future, as they had been in the past, keeping total population stable after a level of 10–12 billion people was reached during the transition.

Developments over the past twenty years show that the experts were right in thinking that population won't keep going up forever. They were wrong in thinking that after it stops going up, it will stay level. The experts' assumption that population would stabilize because birth rates would stop falling once they matched the new low death rates has not been borne out by experience. Evidence from more than fifty countries demonstrates what should be unsurprising: in a modern society the death rate doesn't determine the birth rate. If in the long run birth rates worldwide do not conveniently match death rates, then population must either rise or fall, depending on whether birth or death rates are higher. Which can we expect?

The rapid increase in population during the past two centuries has been the result of lower death rates, which have produced an increase in worldwide life expectancy from about thirty to about sixty-two. (Since the maximum—if we

do not change fundamental human physiology—is about eighty-five, the world has already gone three fifths as far as it can in increasing life expectancy.) For a while the result was a young population with more mothers in each generation, and fewer deaths than births. But even during this population explosion the average number of children born to each woman—the fertility rate—has been falling in modernizing societies. The prediction that world population will soon begin to decline is based on almost universal human behavior. In the United States fertility has been falling for 200 years (except for the blip of the Baby Boom), but partly because of immigration it has stayed only slightly below replacement level for twenty-five years.

Obviously, if for many generations the birth rate averages fewer than 2.1 children per woman, population must eventually stop growing. Recently the United Nations Population Division estimated that 44 percent of the world's people live in countries where the fertility rate has already fallen below the replacement rate, and fertility is falling fast almost everywhere else. In Sweden and Italy fertility has been below replacement level for so long that the population has become old enough to have more deaths than births. Declines in fertility will eventually increase the average age in the world, and will cause a decline in world population forty to fifty years from now.

Because in a modern society the death rate and the fertility rate are largely independent of each other, world population need not be stable. World population can be stable only if fertility rates around the world average out to 2.1 children per woman. But why should they average 2.1, rather than 2.4, or 1.8, or some other number? If there is nothing to keep each country exactly at 2.1, then there is nothing to ensure that the overall average will be exactly 2.1.

The point is that the number of children born depends on families' choices about how many children they want to raise. And when a family is deciding whether to have another child, it is usually thinking about things other than the national or the world population. Who would know or care if world population were to drop from, say, 5.85 billion to 5.81 billion? Population change is too slow and remote for people to feel in their lives—even if the total population were to double or halve in only a century (as a mere 0.7 percent increase or decrease each year would do). Whether world population is increasing or decreasing doesn't necessarily affect the decisions that determine whether it will increase or decrease in the future. As the systems people would say, there is no feedback loop.

WHAT does affect fertility is modernity. In almost every country where people have moved from traditional ways of life to modern ones, they are choosing to have too few children to replace themselves. This is true in Western and in Eastern countries, in Catholic and in secular societies. And it is true in the richest parts of the richest countries. The only exceptions seem to be some small religious communities. We can't be sure what will happen in Muslim countries, because few of them have become modern yet, but so far it looks as if their fertility rates will respond to modernity as others' have.

Nobody can say whether world population will ever dwindle to very low numbers; that depends on what values people hold in the future. After the approaching peak, as long as people continue to prefer saving effort and money by having fewer children, population will continue to decline. (This does not imply that the decision to have fewer children is selfish; it may, for example, be motivated by a desire to do more for each child.) Some people may have values significantly different from those of the rest of the world, and therefore different fer-

tility rates. If such people live in a particular country or population group, their values can produce marked changes in the size of that country or group, even as world population changes only slowly. For example, the U.S. population, because of immigration and a fertility rate that is only slightly below replacement level, is likely to grow from 4.5 percent of the world today to 10 percent of a smaller world over the next two or three centuries. Much bigger changes in share are possible for smaller groups if they can maintain their difference from the average for a long period of time. (To illustrate: Korea's population could grow from one percent of the world to 10 percent in a single lifetime if it were to increase by two percent a year while the rest of the world population declined by one percent a year.)

World population won't stop declining until human values change. But human values may well change—values, not biological imperatives, are the unfathomable variable in population predictions. It is quite possible that in a century or two or three, when just about the whole world is at least as modern as Western Europe is today, people will start to value children more highly than they do now in modern societies. If they do, and fertility rates start to climb, fertility is no more likely to stop climbing at an average rate of 2.1 children per woman than it was to stop falling at 2.1 on the way down.

In only the past twenty years or so world fertility has dropped by 1.5 births per woman. Such a degree of change, were it to occur again, would be enough to turn a long-term increase in world population of one percent a year into a long-term decrease of one percent a year. Presumably fertility could someday increase just as quickly as it has declined in recent decades, although such a rapid change will be less likely once the world has completed the transition to modernity. If fertility rises only to 2.8, just 33 percent over the replacement rate, world population will eventually grow by one percent a year again—doubling in seventy years and multiplying by twenty in only three centuries.

The decline in fertility that began in some countries, including the United States, in the past century is taking a long time to reduce world population because when it started, fertility was very much higher than replacement level. In addition, because a preference for fewer children is associated with

modern societies, in which high living standards make time valuable and children financially unproductive and expensive to care for and educate, the trend toward lower fertility couldn't spread throughout the world until economic development had spread. But once the whole world has become modern, with fertility everywhere in the neighborhood of replacement level, new social values might spread worldwide in a few decades. Fashions in families might keep changing, so that world fertility bounced above and below replacement rate. If each bounce took only a few decades or generations, world population would stay within a reasonably narrow range—although probably with a long-term trend in one direction or the other.

The values that influence decisions about having children seem, however, to change slowly and to be very widespread. If the average fertility rate were to take a long time to move from well below to well above replacement rate and back again, trends in world population could go a long way before they reversed themselves. The result would be big swings in world population—perhaps down to one or two billion and then up to 20 or 40 billion.

Whether population swings are short and narrow or long and wide, the average level of world population after several cycles will probably have either an upward or a downward trend overall. Just as averaging across the globe need not result in exactly 2.1 children per woman, averaging across the centuries need not result in zero growth rather than a slowly increasing or slowly decreasing world population. But the long-term trend is less important than the effects of the peaks and troughs. The troughs could be so low that human beings become scarcer than they were in ancient times. The peaks might cause harm from some kinds of shortages.

One implication is that not even very large losses from disease or war can affect the world population in the long run nearly as much as changes in human values do. What we have learned from the dramatic changes of the past few centuries is that regardless of the size of the world population at any time, people's personal decisions about how many children they want can make the world population go anywhere—to zero or to 100 billion or more.

The End of Cheap Oil

Global production of conventional oil will begin to decline sooner than most people think, probably within 10 years

by Colin J. Campbell and Jean H. Laherrère

In 1973 and 1979 a pair of sudden price increases rudely awakened the industrial world to its dependence on cheap crude oil. Prices first tripled in response to an Arab embargo and then nearly doubled again when Iran dethroned its Shah, sending the major economies sputtering into recession. Many analysts warned that these crises proved that the world would soon run out of oil. Yet they were wrong.

Their dire predictions were emotional and political reactions; even at the time, oil experts knew that they had no scientific basis. Just a few years earlier oil explorers had discovered enormous new oil provinces on the north slope of Alaska and below the North Sea off the coast of Europe. By 1973 the world had consumed, according to many experts' best estimates, only about one eighth of its endowment of readily accessible crude oil (so-called conventional oil). The five Middle Eastern members of the Organization of Petroleum Exporting Countries (OPEC) were able to hike prices not because oil was growing scarce but because they had managed to corner 36 percent of the market. Later, when demand sagged, and the flow of fresh Alaskan and North Sea oil weakened OPEC's economic stranglehold, prices collapsed.

The next oil crunch will not be so temporary. Our analysis of the discovery and production of oil fields around the world suggests that within the next decade, the supply of conventional oil will be unable to keep up with demand. This conclusion contradicts the picture one gets from oil industry reports, which boasted of 1,020 billion barrels of oil (Gbo) in "proved" reserves at the start of 1998. Dividing that figure by the current production rate of about 23.6 Gbo a year might suggest that crude oil could remain plentiful and cheap for 43 more years—probably longer, because official charts show reserves growing.

Unfortunately, this appraisal makes three critical errors. First, it relies on distorted estimates of reserves. A second mistake is to pretend that production will remain constant. Third and most important, conventional wis-

dom erroneously assumes that the last bucket of oil can be pumped from the ground just as quickly as the barrels of oil gushing from wells today. In fact, the rate at which any well—or any country—can produce oil always rises to a maximum and then, when about half the oil is gone, begins falling gradually back to zero.

From an economic perspective, when the world runs completely out of oil is thus not directly relevant: what matters is when production begins to taper off. Beyond that point, prices will rise unless demand declines commensurately. Using several different techniques to estimate the current reserves of conventional oil and the amount still left to be discovered, we conclude that the decline will begin before 2010.

Digging for the True Numbers

We have spent most of our careers exploring for oil, studying reserve figures and estimating the amount of oil left to discover, first while employed at major oil companies and later as independent consultants. Over the years, we have come to appreciate that the relevant statistics are far more complicated than they first appear.

Consider, for example, three vital numbers needed to project future oil production. The first is the tally of how much oil has been extracted to date, a figure known as cumulative production. The second is an estimate of reserves, the amount that companies can pump out of known oil fields before having to abandon them. Finally, one must have an educated guess at the quantity of conventional oil that remains to be discovered and exploited. Together they add up to ultimate recovery, the total number of barrels that will have been extracted when production ceases many decades from now.

The obvious way to gather these numbers is to look them up in any of several publications. That approach works well enough for

cumulative production statistics because companies meter the oil as it flows from their wells. The record of production is not perfect (for example, the two billion barrels of Kuwaiti oil wastefully burned by Iraq in 1991 is usually not included in official statistics), but errors are relatively easy to spot and rectify. Most experts agree that the industry had removed just over 800 Gbo from the earth at the end of 1997.

Getting good estimates of reserves is much harder, however. Almost all the publicly available statistics are taken from surveys conducted by the *Oil and Gas Journal* and *World Oil*. Each year these two trade journals query oil firms and governments around the world. They then publish whatever production and reserve numbers they receive but are not able to verify them.

The results, which are often accepted uncritically, contain systematic errors. For one, many of the reported figures are unrealistic. Estimating reserves is an inexact science to begin with, so petroleum engineers assign a probability to their assessments. For example, if, as geologists estimate, there is a 90 percent chance that the Oseberg field in Norway contains 700 million barrels of recoverable oil but only a 10 percent chance that it will yield 2,500 million more barrels, then the lower figure should be cited as the so-called P90 estimate (P90 for "probability 90 percent") and the higher as the P10 reserves.

In practice, companies and countries are often deliberately vague about the likelihood of the reserves they report, preferring instead to publicize whichever figure, within a P10 to P90 range, best suits them. Exaggerated estimates can, for instance, raise the price of an oil company's stock.

The members of OPEC have faced an even greater temptation to inflate their reports because the higher their reserves, the more oil they are allowed to export. National companies, which have exclusive oil rights in the main OPEC countries, need not (and do not) release detailed statistics on each field that could be used to verify the country's total reserves. There is thus good reason to suspect that when, during the late

FLOW OF OIL starts to fall from any large region when about half the crude is gone. Adding the output of fields of various sizes and ages (*bottom curves at right*) usually yields a bell-shaped production curve for the region as a whole. M. King Hubbert (*left*), a geologist with Shell Oil, exploited this fact in 1956 to predict correctly that oil from the lower 48 American states would peak around 1969.

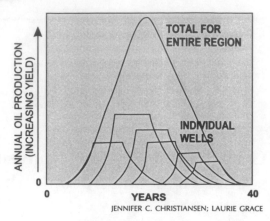

1980s, six of the 11 OPEC nations increased their reserve figures by colossal amounts, ranging from 42 to 197 percent, they did so only to boost their export quotas.

Previous OPEC estimates, inherited from private companies before governments took them over, had probably been conservative, P90 numbers. So some upward revision was warranted. But no major new discoveries or technological breakthroughs justified the addition of a staggering 287 Gbo. That increase is more than all the oil ever discovered in the U.S.—plus 40 percent. Non-OPEC countries, of course, are not above fudging their numbers either: 59 nations stated in 1997 that their reserves were unchanged from 1996. Because reserves naturally drop as old fields are drained and jump when new fields are discovered, perfectly stable numbers year after year are implausible.

Unproved Reserves

Another source of systematic error in the commonly accepted statistics is that the definition of reserves varies widely from region to region. In the U.S., the Securities and Exchange Commission allows companies to call reserves "proved" only if the oil lies near a producing well and there is "reasonable certainty" that it can be recovered profitably at current oil prices, using existing technology. So a proved reserve estimate in the U.S. is roughly equal to a P90 estimate.

Regulators in most other countries do not enforce particular oil-reserve definitions. For many years, the former Soviet countries have routinely released wildly optimistic figures—essentially P10 reserves. Yet analysts have often misinterpreted these as estimates of "proved" reserves. *World Oil* reckoned reserves in the former Soviet Union amounted to 190 Gbo in 1996, whereas the *Oil and Gas Journal* put the number at 57 Gbo. This large discrepancy shows just how elastic these numbers can be.

Using only P90 estimates is not the answer, because adding what is 90 percent likely for each field, as is done in the U.S., does not in fact yield what is 90 percent likely for a country or the entire planet. On the contrary, summing many P90 reserve estimates always understates the amount of proved oil in a region. The only correct way to total up reserve numbers is to add the mean, or average, estimates of oil in each field. In practice, the median estimate, often called "proved and probable," or P50 reserves, is more widely used and is good enough. The P50 value is the number of barrels of oil that are as likely as not to come out of a well during its lifetime, assuming prices remain within a limited range. Errors in P50 estimates tend to cancel one another out.

We were able to work around many of the problems plaguing estimates of conventional reserves by using a large body of statistics maintained by Petroconsultants in Geneva. This information, assembled over 40 years from myriad sources, covers some 18,000 oil fields worldwide. It, too, con-

tains some dubious reports, but we did our best to correct these sporadic errors.

According to our calculations, the world had at the end of 1996 approximately 850 Gbo of conventional oil in P50 reserves—substantially less than the 1,019 Gbo reported in the *Oil and Gas Journal* and the 1,160 Gbo estimated by *World Oil*. The difference is actually greater than it appears because our value represents the amount most likely to come out of known oil fields, whereas the larger number is supposedly a cautious estimate of proved reserves.

For the purposes of calculating when oil production will crest, even more critical than the size of the world's reserves is the size of ultimate recovery—all the cheap oil there is to be had. In order to estimate that, we need to know whether, and how fast, reserves are moving up or down. It is here that the official statistics become dangerously misleading.

Diminishing Returns

According to most accounts, world oil reserves have marched steadily upward over the past 20 years. Extending that apparent trend into the future, one could easily conclude, as the U.S. Energy Information Administration has, that oil production will continue to rise unhindered for decades to come, increasing almost two thirds by 2020.

Such growth is an illusion. About 80 percent of the oil produced today flows

EARTH'S CONVENTIONAL CRUDE OIL is almost half gone. Reserves (defined here as the amount as likely as not to come out of known fields) and future discoveries together will provide little more than what has already been burned.

UNDISCOVERED: 150 BILLION BARRELS

RESERVES: 850 BILLION BARRELS

GLOBAL PRODUCTION OF OIL, both conventional and unconventional, recovered after falling in 1973 and 1979. But a more permanent decline is less than 10 years away, according to the authors' model, based in part on multiple Hubbert curves (*thin lines*). U.S. and Canadian oil topped out in 1972; production in the former Soviet Union has fallen 45 percent since 1987. A crest in the oil produced outside the Persian Gulf region now appears imminent.

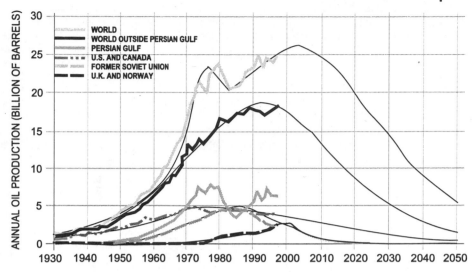

LAURIE GRACE; SOURCE: JEAN H. LAHERRÈRE

from fields that were found before 1973, and the great majority of them are declining. In the 1990s oil companies have discovered an average of seven Gbo a year; last year they drained more than three times as much. Yet official figures indicated that proved reserves did not fall by 16 Gbo, as one would expect—rather they expanded by 11 Gbo. One reason is that several dozen governments opted not to report declines in their reserves, perhaps to enhance their political cachet and their ability to obtain loans. A more important cause of the expansion lies in revisions: oil companies replaced earlier estimates of the reserves left in many fields with higher numbers. For most purposes, such amendments are harmless, but they seriously distort forecasts extrapolated from published reports.

To judge accurately how much oil explorers will uncover in the future, one has to backdate every revision to the year in which the field was first discovered—not to the year in which a company or country corrected an earlier estimate. Doing so reveals that global discovery peaked in the early 1960s and has been falling steadily ever since. By extending the trend to zero, we can make a good guess at how much oil the industry will ultimately find.

We have used other methods to estimate the ultimate recovery of conventional oil for each country [*see box*], EARTH's CONVENTIONAL CRUDE OIL., and we calculate that the oil industry will be able to recover only about another 1,000 billion barrels of conventional oil. This number, though great, is little more than the 800 billion barrels that have already been extracted.

It is important to realize that spending more money on oil exploration will not change this situation. After the price of crude hit all-time highs in the early 1980s,

explorers developed new technology for finding and recovering oil, and they scoured the world for new fields. They found few: the discovery rate continued its decline uninterrupted. There is only so much crude oil in the world, and the industry has found about 90 percent of it.

Predicting the Inevitable

Predicting when oil production will stop rising is relatively straightforward once one has a good estimate of how much oil there is left to produce. We simply apply a refinement of a technique first published in 1956 by M. King Hubbert. Hubbert observed that in any large region, unrestrained extraction of a finite resource rises along a bell-shaped curve that peaks when about half the resource is gone. To demonstrate his theory, Hubbert fitted a bell curve to production statistics and projected that crude oil production in the lower 48 U.S. states would rise for 13 more years, then crest in 1969, give or take a year. He was right: production peaked in 1970 and has continued to follow Hubbert curves with only minor deviations. The flow of oil from several other regions, such as the former Soviet Union and the collection of all oil producers out-

side the Middle East, also follows Hubbert curves quite faithfully.

The global picture is more complicated, because the Middle East members of OPEC deliberately reined back their oil exports in the 1970s, while other nations continued producing at full capacity. Our analysis reveals that a number of the largest producers, including Norway and the U.K., will reach their peaks around the turn of the millennium unless they sharply curtail production. By 2002 or so the world will rely on Middle East nations, particularly five near the Persian Gulf (Iran, Iraq, Kuwait, Saudi Arabia and the United Arab Emirates), to fill in the gap between dwindling supply and growing demand. But once approximately 900 Gbo have been consumed, production must soon begin to fall. Barring a global recession, it seems most likely that world production of conventional oil will peak during the first decade of the 21st century.

Perhaps surprisingly, that prediction does not shift much even if our estimates are a few hundred billion barrels high or low. Craig Bond Hatfield of the University of Toledo, for example, has conducted his own analysis based on a 1991 estimate by the U.S. Geological Survey of 1,550 Gbo remaining—55 percent higher than our figure. Yet he similarly concludes that the world will hit maximum oil production within the next 15 years. John D. Edwards of the University of Colorado publish-

PRODUCED: 800 BILLION BARRELS

How Much Oil Is Left to Find?

We combined several techniques to conclude that about 1,000 billion barrels of conventional oil remain to be produced. First, we extrapolated published production figures for older oil fields that have begun to decline. The Thistle field off the coast of Britain, for example, will yield about 420 million barrels (*a*). Second, we plotted the amount of oil discovered so far in some regions against the cumulative number of exploratory wells drilled there. Because larger fields tend to be found first—they are simply too large to miss—the curve rises rapidly and then flattens, eventually

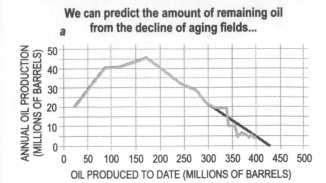

We can predict the amount of remaining oil from the decline of aging fields...

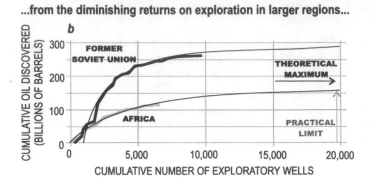

...from the diminishing returns on exploration in larger regions...

ed last August one of the most optimistic recent estimates of oil remaining: 2,036 Gbo. (Edwards concedes that the industry has only a 5 percent chance of attaining that very high goal.) Even so, his calculations suggest that conventional oil will top out in 2020.

Smoothing the Peak

Factors other than major economic changes could speed or delay the point at which oil production begins to decline. Three in particular have often led economists and academic geologists to dismiss concerns about future oil production with naive optimism.

First, some argue, huge deposits of oil may lie undetected in far-off corners of the globe. In fact, that is very unlikely. Exploration has pushed the frontiers back so far that only extremely deep water and polar regions remain to be fully tested, and even their prospects are now reasonably well understood. Theoretical advances in geochemistry and geophysics have made it possible to map productive and prospective fields with impressive accuracy. As a result, large tracts can be condemned as barren. Much of the deepwater realm, for example, has been shown to be absolutely nonprospective for geologic reasons.

What about the much touted Caspian Sea deposits? Our models project that oil production from that region will grow until around 2010. We agree with analysts at the USGS World Oil Assessment program and elsewhere who rank the total resources there as roughly equivalent to those of the North Sea—that is, perhaps 50 Gbo but certainly not several hundreds of billions as sometimes reported in the media.

A second common rejoinder is that new technologies have stead- ily increased the fraction of oil that can be recovered from fields in a basin—the so-called recovery factor. In the 1960s oil companies assumed as a rule of thumb that only 30 percent of the oil in a field was typically recoverable; now they bank on an average of 40 or 50 percent. That progress will continue and will extend global reserves for many years to come, the argument runs.

Of course, advanced technologies will buy a bit more time before production starts to fall [see "Oil Production in the 21st Century," by Roger N. Anderson*]. But most of the apparent improvement in recovery factors is an artifact of reporting. As oil fields grow old, their owners often deploy newer technology to slow their decline. The falloff also allows engineers to gauge the size of the field more accurately and to correct previous underestimation—in particular P90 estimates that by definition were 90 percent likely to be exceeded.

Another reason not to pin too much hope on better recovery is that oil companies routinely count on technological progress when they compute their reserve estimates. In truth, advanced technologies can offer little help in draining the largest basins of oil, those onshore in the Middle East where the oil needs no assistance to gush from the ground.

Last, economists like to point out that the world contains enormous caches of unconventional oil that can substitute for crude oil as soon as the price rises high enough to make them profitable. There is no question that the resources are ample: the Orinoco oil belt in Venezuela has been assessed to contain a staggering 1.2 trillion barrels of the sludge known as heavy oil. Tar sands and shale deposits in

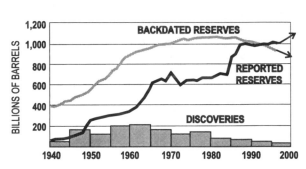

LAURIE GRACE; SOURCE: PETROCONSULTANTS, *OIL AND GAS JOURNAL* AND U.S. GEOLOGICAL SURVEY

GROWTH IN OIL RESERVES since 1980 is an illusion caused by belated corrections to oil-field estimates. Back-dating the revisions to the year in which the fields were discovered reveals that reserves have been falling because of a steady decline in newfound oil (*bottom bars*).

Canada and the former Soviet Union may contain the equivalent of more than 300 billion barrels of oil [see "Mining for Oil," by Richard L. George *]. Theoretically, these unconventional oil reserves could quench the world's thirst for liquid fuels as conventional oil passes its prime. But the industry will be hard-pressed for the time and money needed to ramp up production of unconventional oil quickly enough.

Such substitutes for crude oil might also exact a high environmental price. Tar sands

reaching a theoretical maximum: for Africa, 192 Gbo. But the time and cost of exploration impose a more practical limit of perhaps 165 Gbo (*b*). Third, we analyzed the distribution of oil-field sizes in the Gulf of Mexico and other provinces. Ranked according to size and then graphed on a logarithmic scale, the fields tend to fall along a parabola that grows predictably over time. (*c*). (Interestingly, galaxies, urban populations and other

natural agglomerations also seem to fall along such parabolas.) Finally, we checked our estimates by matching our projections for oil production in large areas, such as the world outside the Persian Gulf region, to the rise and fall of oil discovery in those places decades earlier (*d*).

—C.J.C. and J.H.L.

...by extrapolating the size of new fields into the future...

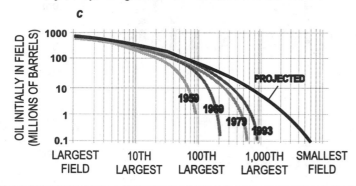

...and by matching production to earlier discovery trends.

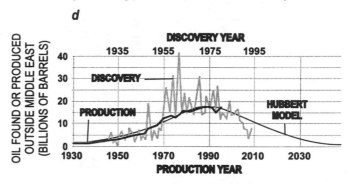

LAURIE GRACE; SOURCE: JEAN H. LAHERRÈRE

typically emerge from strip mines. Extracting oil from these sands and shales creates air pollution. The Orinoco sludge contains heavy metals and sulfur that must be removed. So governments may restrict these industries from growing as fast as they could. In view of these potential obstacles, our skeptical estimate is that only 700 Gbo will be produced from unconventional reserves over the next 60 years.

On the Down Side

Meanwhile global demand for oil is currently rising at more than 2 percent a year. Since 1985, energy use is up about 30 percent in Latin America, 40 percent in Africa and 50 percent in Asia. The Energy Information Administration forecasts that worldwide demand for oil will increase 60 percent (to about 40 Gbo a year) by 2020.

The switch from growth to decline in oil production will thus almost certainly create economic and political tension. Unless alternatives to crude oil quickly prove themselves, the market share of the OPEC states in the Middle East will rise rapidly. Within two years, these nations' share of the global oil business will pass 30 percent, nearing the level reached during the oil-price shocks of the 1970s. By 2010 their share will quite probably hit 50 percent.

The world could thus see radical increases in oil prices. That alone might be sufficient to curb demand, flattening production for perhaps 10 years. (Demand fell more than 10 percent after the 1979 shock and took 17 years to recover.) But by 2010 or so, many Middle Eastern nations will themselves be past the midpoint. World production will then have to fall.

With sufficient preparation, however, the transition to the post-oil economy need not be traumatic. If advanced methods of producing liquid fuels from natural gas can be made profitable and scaled up quickly, gas could become the next source of transportation fuel [see "Liquid Fuels from

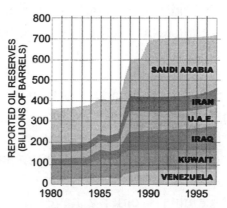

LAURIE GRACE; SOURCE: *OIL AND GAS JOURNAL*

SUSPICIOUS JUMP in reserves reported by six OPEC members added 300 billion barrels of oil to official reserve tallies yet followed no major discovery of new fields.

Natural Gas," by Safaa A. Fouda *]. Safer nuclear power, cheaper renewable energy, and oil conservation programs could all help postpone the inevitable decline of conventional oil.

Countries should begin planning and investing now. In November a panel of energy experts appointed by President Bill Clinton strongly urged the administration to increase funding for energy research by $1 billion over the next five years. That is a small step in the right direction, one that must be followed by giant leaps from the private sector.

The world is not running out of oil—at least not yet. What our society does face, and soon, is the end of the abundant and cheap oil on which all industrial nations depend.

The Authors

COLIN J. CAMPBELL and JEAN H. LAHERRÈRE have each worked in the oil industry for more than 40 years. After completing his Ph.D. in geology at the University of Oxford, Campbell worked for Texaco as an exploration geologist and then at Amoco as chief geologist for Ecuador. His decade-long study of global oil-production trends has led to two books and numerous papers. Laherrère's early work on seismic refraction surveys contributed to the discovery of Africa's largest oil field. At Total, a French oil company, he supervised exploration techniques worldwide. Both Campbell and Laherrère are currently associated with Petroconsultants in Geneva.

Further Reading

UPDATED HUBBERT CURVES ANALYZE WORLD OIL SUPPLY. L. F. Ivanhoe in *World Oil*, Vol. 217, No. 11, pages 91–94; November 1996.

THE COMING OIL CRISIS. Colin J. Campbell. Multi-Science Publishing and Petroconsultants, Brentwood, England, 1997.

OIL BACK ON THE GLOBAL AGENDA. Craig Bond Hatfield in *Nature*, Vol. 387, page 121; May 8, 1997.

Editor's note: All of the articles mentioned in this article can be found in Scientific American, March 1998.

Gray Dawn:
The Global Aging Crisis

Peter G. Peterson

DAUNTING DEMOGRAPHICS

THE LIST of major global hazards in the next century has grown long and familiar. It includes the proliferation of nuclear, biological, and chemical weapons, other types of high-tech terrorism, deadly super-viruses, extreme climate change, the financial, economic, and political aftershocks of globalization, and the violent ethnic explosions waiting to be detonated in today's unsteady new democracies. Yet there is a less-understood challenge—the graying of the developed world's population—that may actually do more to reshape our collective future than any of the above.

Over the next several decades, countries in the developed world will experience an unprecedented growth in the number of their elderly and an unprecedented decline in the number of their youth. The timing and magnitude of this demographic transformation have already been determined. Next century's elderly have already been born and can be counted—and their cost to retirement benefit systems can be projected.

Unlike with global warming, there can be little debate over whether or when global aging will manifest itself. And unlike with other challenges, even the struggle to preserve and strengthen unsteady new democracies, the costs of global aging will be far beyond the means of even the world's wealthiest nations—unless retirement benefit systems are radically reformed. Failure to do so, to prepare early and boldly enough, will spark economic crises that will dwarf the recent meltdowns in Asia and Russia.

Reprinted by permission of *Foreign Affairs*, January/February 1999, pp. 42-55. © 1999 by the Council on Foreign Relations, Inc.

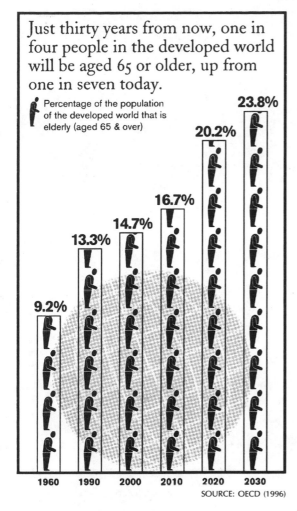

Just thirty years from now, one in four people in the developed world will be aged 65 or older, up from one in seven today.

Percentage of the population of the developed world that is elderly (aged 65 & over)

9.2% — 1960
13.3% — 1990
14.7% — 2000
16.7% — 2010
20.2% — 2020
23.8% — 2030

SOURCE: OECD (1996)

How we confront global aging will have vast economic consequences costing quadrillions of dollars over the next century. Indeed, it will greatly influence how we manage, and can afford to manage, the other major challenges that will face us in the future.

For this and other reasons, global aging will become not just the transcendent economic issue of the 21st century, but the transcendent political issue as well. It will dominate and daunt the public-policy agendas of developed countries and force the renegotiation of their social contracts. It will also reshape foreign policy strategies and the geopolitical order.

The United States has a massive challenge ahead of it. The broad outlines can already be seen in the emerging debate over Social Security and Medicare reform. But ominous as the fiscal stakes are in the United States, they loom even larger in Japan and Europe, where populations are aging even faster, birthrates are lower, the influx of young immigrants from developing countries is smaller, public pension benefits are more generous, and private pension systems are weaker.

Aging has become a truly global challenge, and must therefore be given high priority on the global policy agenda. A gray dawn fast approaches. It is time to take an unflinching look at the shape of things to come.

The Floridization of the developed world. Been to Florida lately? You may not have realized it, but the vast concentration of seniors there—nearly 19 percent of the population—represents humanity's future. Today's Florida is a demographic benchmark that every developed nation will soon pass. Italy will hit the mark as early as 2003, followed by Japan in 2005 and Germany in 2006. France and Britain will pass present-day Florida around 2016; the United States and Canada in 2021 and 2023.

Societies much older than any we have ever known. Global life expectancy has grown more in the last fifty years than over the previous five thousand. Until the Industrial Revolution, people aged 65 and over never amounted to more than 2 or 3 percent of the population. In today's developed world, they amount to 14 percent. By the year 2030, they will reach 25 percent and be closing in on 30 in some countries.

An unprecedented economic burden on working-age people. Early in the next century, working-age populations in most developed countries will shrink. Between 2000 and 2010, Japan, for example, will suffer a 25 percent drop in the number of workers under age 30. Today the ratio of working taxpayers to nonworking pensioners in the developed world is around 3:1. By 2030, absent reform, this ratio will fall to 1.5:1, and in some countries, such as Germany and Italy, it will drop all the way down to 1:1 or even lower. While the longevity revolution represents a miraculous triumph of modern medicine and the extra years of life will surely be treasured by the elderly and their families, pension plans and other retirement benefit programs were not designed to provide these billions of extra years of payouts.

The aging of the aged: the number of "old old" will grow much faster than the number of "young old." The United Nations projects that by 2050, the number of people aged 65 to 84 worldwide will grow from 400 million to 1.3 billion (a threefold increase), while the number of people aged 85 and over will grow from 26 million to 175 million (a sixfold increase)—and the number aged 100 and over from 135,000 to 2.2 million (a sixteenfold increase). The "old old" consume far more health care than the "young old"—about two to three times as much. For nursing-home care, the ratio is roughly 20:1. Yet little of this cost is figured in the official projections of future public expenditures.

Falling birthrates will intensify the global aging trend. As life spans increase, fewer babies are being born. As recently as the late 1960s, the worldwide total fertility rate (that is, the average number of lifetime births per woman) stood at about 5.0, well within the historical range. Then came a behavioral revolution, driven by growing affluence, urbanization, feminism, rising female participation in the workforce, new birth control technologies, and legalized abortion. The result: an unprecedented and unexpected decline in the global fertility rate to about 2.7—a drop fast approaching the replacement rate of 2.1 (the rate required merely to maintain a constant population). In the developed world alone, the average fertility rate has plummeted to 1.6. Since 1995, Japan has had fewer births annually than in any year since 1899. In Germany; where the rate has fallen to 1.3, fewer babies are born each year than in Nepal, which has a population only one-quarter as large.

A shrinking population in an aging developed world. Unless their fertility rates rebound, the total populations of western Europe and Japan will shrink to about one-half of their current size before the end of the next century. In 1950, 7 of the 12 most populous nations were in the developed world: the United States, Russia, Japan, Germany, France, Italy, and the United Kingdom. The United Nations projects that by 2050, only the United States will remain on the list. Nigeria, Pakistan, Ethiopia, Congo, Mexico, and the Philippines will replace the others. But since developing countries are also experiencing a drop in fertility, many are now actually aging faster than the typical developed country. In France, for example, it took over a century for the elderly to grow from 7 to 14 percent of the population. South Korea, Taiwan, Singapore, and China are projected to traverse that distance in only 25 years.

From worker shortage to rising immigration pressure. Perhaps the most predictable consequence of the gap in fertility and population growth rates between developed and developing countries will be the rising demand for immigrant workers in older and wealthier societies facing labor shortages. Immigrants are typically young and tend to bring with them the family practices of their native culture—including higher fertility rates. In many European countries, non-European foreigners already make up roughly 10 percent of the population. This includes 10 million to 13 million Muslims, nearly all of whom are working-age or younger. In Germany, foreigners will make up 30 percent of the total population by 2030, and over half the population of major cities like Munich and Frankfurt. Global aging and attendant labor shortages will therefore ensure that immigration remains a major issue in developed countries for decades to come. Culture wars could erupt over the balkanization of language and religion; electorates could divide along ethnic lines; and émigré leaders could sway foreign policy.

GRAYING MEANS PAYING

OFFICIAL PROJECTIONS suggest that within 30 years, developed countries will have to spend at least an extra 9 to 16 percent of GDP simply to meet their old-age benefit promises. The unfunded liabilities for pensions (that is, benefits already earned by today's workers for which nothing has been saved) are already almost $35 trillion. Add in health care, and the total jumps to at least twice as much. At minimum, the global aging issue thus represents, to paraphrase the old quiz show, a $64 trillion question hanging over the developed world's future.

To pay for promised benefits through increased taxation is unfeasible. Doing so would raise the total tax burden by an unthinkable 25 to 40 percent of every worker's taxable wages—in countries where payroll tax rates sometimes already exceed 40 percent. To finance the costs of these benefits by borrowing would be just as disastrous. Governments would run unprecedented deficits that would quickly consume the savings of the developed world.

And the $64 trillion estimate is probably low. It likely underestimates future growth in longevity and health care costs and ignores the negative effects on the economy of more borrowing, higher interest rates, more taxes, less savings, and lower rates of productivity and wage growth.

There are only a handful of exceptions to these nightmarish forecasts. In Australia, total public retirement costs as a share of GDP are expected to rise only slightly, and they may even decline in Britain and Ireland. This fiscal good fortune is not due to any special demographic trend, but to timely policy reforms—including tight limits on public health spending, modest pension benefit formulas, and new personally owned savings programs that allow future public benefits to shrink as a share of average wages. This approach may yet be emulated elsewhere.

Failure to respond to the aging challenge will destabilize the global economy, straining financial and political institutions around the world. Consider Japan, which today runs a large current account surplus making up well over half the capital exports of all the surplus nations combined. Then imagine a scenario in which Japan leaves its retirement programs and fiscal policies on autopilot. Thirty years from now, under this scenario, Japan will be importing massive amounts of capital to prevent its domestic economy from collapsing under the weight of benefit outlays. This will require a huge reversal in global capital flows. To get some idea of the potential volatility, note that over the next decade, Japan's annual pension deficit is projected to grow to roughly 3 times the size of its recent and massive capital exports to the United States; by 2030, the annual deficit is expected to be 15 times as large. Such reversals will cause wildly fluctuating interest and exchange rates, which may in turn short-circuit financial institutions and trigger a serious market crash.

As they age, some nations will do little to change course, while others may succeed in boosting their national savings rate, at least temporarily, through a combination of fiscal restraint and household thrift. Yet this too could result in a volatile disequilibrium in supply and demand for global capital. Such imbalance could wreak havoc with international institutions such as the European Union.

In recent years, the EU has focused on monetary union, launched a single currency (the euro), promoted cross-border labor mobility, and struggled to harmonize fiscal, monetary, and trade policies. European leaders expect to have their hands full smoothing out differences between members of the Economic and Monetary Union (EMU)—from the timing of their business cycles to the diversity of their credit institutions and political cultures. For this reason, they established official public debt and deficit criteria (three percent of GDP for EMU membership) in order to discourage maverick nations from placing undue economic burdens on fellow members. But the EU has yet to face up to the biggest challenge to its future viability: the likelihood of varying national responses to the fiscal pressures of demographic aging. Indeed, the EU does not even include unfunded pension liabilities in the official EMU debt and deficit criteria—which is like measuring icebergs without looking beneath the water line.

When these liabilities come due and move from "off the books" to "on the books," the EU will, under current constraints, be required to penalize EMU members that exceed the three percent deficit cap. As a recent IMF report concludes, "over time it will become increasingly difficult for most countries to meet the deficit ceiling without comprehensive social security reform." The EU could, of course, retain members by raising the deficit limit. But once the floodgates are opened, national differences in fiscal policy may mean that EMU members rack up deficits at different rates. The European Central Bank, the euro, and a half-century of progress toward European unity could be lost as a result.

The total projected cost of the age wave is so staggering that we might reasonably

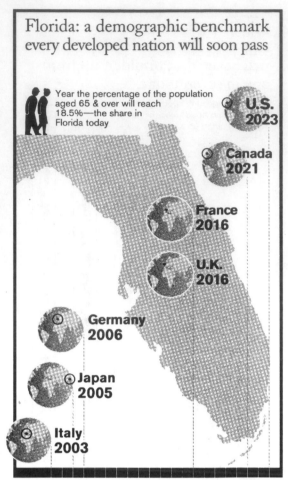

Florida: a demographic benchmark every developed nation will soon pass

Year the percentage of the population aged 65 & over will reach 18.5%—the share in Florida today

U.S. 2023
Canada 2021
France 2016
U.K. 2016
Germany 2006
Japan 2005
Italy 2003

SOURCE: OECD (1996)
SOURCE: OECD (1996); author's calculations

conclude it could never be paid. After all, these numbers are projections, not predictions. They tell us what is likely to happen if current policy remains unchanged, not whether it is likely or even possible for this condition to hold. In all probability, economies would implode and governments would collapse before the projections ever materialize. But this is exactly why we must focus on these projections, for they call attention to the paramount question: Will we change course sooner, when we still have time to control our destiny and reach a more sustainable path? Or later, after unsustainable economic damage and political and social trauma cause a wrenching upheaval?

A GRAYING NEW WORLD ORDER

WHILE THE fiscal and economic consequences of global aging deserve serious discussion, other important consequences must also be examined. At the top of the list is the impact of the age wave on foreign policy and international security.

Will the developed world be able to maintain its security commitments? One need not be a Nobel laureate in economics to understand that a country's GDP growth is the

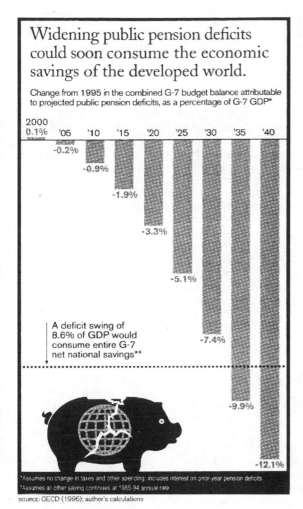

Widening public pension deficits could soon consume the economic savings of the developed world.

Change from 1995 in the combined G-7 budget balance attributable to projected public pension deficits, as a percentage of G-7 GDP*

2000 0.1%
'05 -0.2%
'10 -0.9%
'15 -1.9%
'20 -3.3%
'25 -5.1%
'30 -7.4%
'35 -9.9%
'40 -12.1%

A deficit swing of 8.6% of GDP would consume entire G-7 net national savings**

*Assumes no change in taxes and other spending; includes interest on prior-year pension deficits
*Assumes all other saving continues at 1985-94 annual rate

SOURCE: OECD (1996); author's calculations

smaller armies. And how many parents will allow their only child to go off to war?

With fewer soldiers, total capability can be maintained only by large increases in technology and weaponry. But boosting military productivity creates a Catch-22. For how will governments get the budget resources to pay for high-tech weaponry if the senior-weighted electorate demands more money for high-tech medicine? Even if military capital is successfully substituted for military labor, the deployment options may be dangerously limited. Developed nations facing a threat may feel they have only two extreme (but relatively inexpensive) choices: a low-level response (antiterrorist strikes and cruise-missile diplomacy) or a high-level response (an all-out attack with strategic weapons).

Will Young/Old become the next North/South fault line? Historically, the richest industrial powers have been growing, capital-exporting, philanthropic giants that project their power and mores around the world. The richest industrial powers of the future may be none of these things. Instead, they may be demographically imploding, capital-importing, fiscally starving neutrals who twist and turn to avoid expensive international entanglements. A quarter-century from now, will the divide between today's "rich" and "poor" nations be better described as a divide between growth and decline, surplus and deficit, expansion and retreat, future and past? By the mid-2020s, will the contrast between North and South be better described as a contrast between Young and Old?

If today's largest low-income societies, especially China, set up fully funded retirement systems to prepare for their own future aging, they may well produce ever larger capital surpluses. As a result, today's great powers could someday depend on these surpluses to keep themselves financially afloat. But how should we expect these new suppliers of capital to use their newly acquired leverage? Will they turn the tables in international diplomacy? Will the Chinese, for example, someday demand that the United States shore up its Medicare system the way Americans once demanded that China reform its human rights policies as a condition for foreign assistance?

As Samuel Huntington recently put it, "the juxtaposition of a rapidly growing people of one culture and a slowly growing or stagnant people of another culture generates pressure for economic and/or political adjustments in both societies." Countries where populations are still exploding rank high on

any list of potential trouble spots, whereas the countries most likely to lose population—and to see a weakening of their commitment to expensive defense and global security programs—are the staunchest friends of liberal democracy.

In many parts of the developing world, the total fertility rate remains very high (7.3 in the Gaza Strip versus 2.7 in Israel), most people are very young (49 percent under age 15 in Uganda), and the population is growing very rapidly (doubling every 26 years in Iran). These areas also tend to be the poorest, most rapidly urbanizing, most institutionally unstable—and most likely to fall under the sway of rogue leadership. They are the same societies that spawned most of the military strongmen and terrorists who have bedeviled the United States and Europe in recent decades. The Pentagon's long-term planners predict that outbreaks of regional anarchy will occur more frequently early in the next century. To pinpoint when and where, they track what they call "youth bulges" in the world's poorest urban centers.

Is demography destiny, after all? Is the rapidly aging developed world fated to decline? Must it cede leadership to younger and faster-growing societies? For the answer to be no, the developed world must redefine that role around a new mission. And what better way to do so than to show the younger, yet more tradition-bound, societies—which will soon age in their turn—how a world dominated by the old can still accommodate the young.

WHOSE WATCH IS IT, ANYWAY?

FROM PRIVATE discussions with leaders of major economies, I can attest that they are well briefed on the stunning demographic trends that lie ahead. But so far they have responded with paralysis rather than action. Hardly any country is doing what it should to prepare. Margaret Thatcher confesses that she repeatedly tried to raise the aging issue at G-7 summit meetings. Yet her fellow leaders stalled. "Of course aging is a profound challenge," they replied, "but it doesn't hit until early in the next century—after my watch."

Americans often fault their leaders for not acknowledging long-term problems and for not facing up to silent and slow-motion challenges. But denial is not a peculiarly American syndrome. In 1995, Silvio Berlusconi's *Forza Italia* government was buffeted by a number of political storms, all of which it weathered—except for pension reform, which shattered the coalition. That same year, the Dutch parliament was forced to repeal a recent cut in retirement benefits after a strong Pension Party, backed by the elderly, emerged from nowhere to punish the reformers. In 1996, the French government's modest proposal to trim pensions triggered

product of workforce and productivity growth. If workforces shrink rapidly, GDP may drop as well, since labor productivity may not rise fast enough to compensate for the loss of workers. At least some developed countries are therefore likely to experience a long-term decline in total production of goods and services—that is, in real GDP.

Economists correctly focus on the developed world's GDP per capita, which can rise even as its workforce and total GDP shrink. But anything with a fixed cost becomes a national challenge when that cost has to be spread over a smaller population and funded out of shrinking revenues. National defense is the classic example. The West already faces grave threats from rogue states armed with biological and chemical arsenals, terrorists capable of hacking into vulnerable computer systems, and proliferating nuclear weapons. None of these external dangers will shrink to accommodate our declining workforce or GDP.

Leading developed countries will no doubt need to spend as much or more on defense and international investments as they do today. But the age wave will put immense pressure on governments to cut back. Falling birthrates, together with a rising demand for young workers, will also inevitably mean strikes and even riots. A year later the Socialists overturned the ruling government at the polls.

Each country's response, or nonresponse, is colored by its political and cultural institutions. In Europe, where the welfare state is more expansive, voters can hardly imagine that the promises made by previous generations of politicians can no longer be kept. They therefore support leaders, unions, and party coalitions that make generous unfunded pensions the very cornerstone of social democracy. In the United States, the problem has less to do with welfare-state dependence than the uniquely American notion that every citizen has personally earned and is therefore entitled to whatever benefits government happens to have promised.

How governments ultimately prepare for global aging will also depend on how global aging itself reshapes politics. Already some of the largest and most strident interest groups in the United States are those that claim to speak for senior citizens, such as the American Association of Retired Persons, with its 33 million members, 1,700 paid employees, ten times that many trained volunteers, and an annual budget of $5.5 billion.

Senior power is rising in Europe, where it manifests itself less through independent senior organizations than in labor unions and (often union-affiliated) political parties that formally adopt pro-retiree platforms. Could age-based political parties be the wave of the future? In Russia, although the Communist resurgence is usually ascribed to nationalism and nostalgia, a demographic bias is at work as well. The Communists have repositioned themselves as the party of retirees, who are aggrieved by how runaway inflation has slashed the real value of their pensions. In the 1995 Duma elections, over half of those aged 55 and older voted Communist, versus only ten percent of those under age 40.

Commenting on how the old seem to trump the young at every turn, Lee Kuan Yew once proposed that each taxpaying worker be given two votes to balance the lobbying clout of each retired elder. No nation, not even Singapore, is likely to enact Lee's suggestion. But the question must be asked: With ever more electoral power flowing into the hands of elders, what can motivate political leaders to act on behalf of the long-term future of the young?

A handful of basic strategies, all of them difficult, might enable countries to overcome the economic and political challenges of an aging society: extending work lives and postponing retirement; enlarging the workforce through immigration and increased labor force participation; encouraging higher fertility and investing more in the education and productivity of future workers; strengthening intergenerational bonds of responsibility within families; and targeting government-paid benefits to those most in need while encouraging and even requiring workers to save for their own retirements. All of these strategies unfortunately touch raw nerves—by amending existing social contracts, by violating cultural expectations, or by offending entrenched ideologies.

TOWARD A SUMMIT ON GLOBAL AGING

ALL COUNTRIES would be well served by collective deliberation over the choices that lie ahead. For that reason I propose a Summit on Global Aging. Few venues are as well covered by the media as a global summit. Leaders have been willing to convene summits to discuss global warming. Why not global aging, which will hit us sooner and with greater certainty? By calling attention to what is at stake, a global aging summit could shift the public discussion into fast forward. That alone would be a major contribution. The summit process would also help provide an international framework for voter education, collective burden-sharing, and global leadership. Once national constituencies begin to grasp the magnitude of the global aging challenge, they will be more inclined to take reform seriously. Once governments get into the habit of cooperating on what in fact is a global challenge, individual leaders will not need to incur the economic and political risks of acting alone.

This summit should launch a new multilateral initiative to lend the global aging agenda a visible institutional presence: an Agency on Global Aging. Such an agency would examine how developed countries should reform their retirement systems and how developing countries should properly set them up in the first place. Perhaps the most basic question is how to weigh the interests and well-being of one generation against the next. Then there is the issue of defining the safety-net standard of social adequacy. Is there a minimum level of retirement income that should be the right of every citizen? To what extent should retirement security be left to people's own resources? When should government pick up the pieces, and how can it do so without discouraging responsible behavior? Should government compel people in advance to make better life choices, say, by enacting a mandatory savings program?

Another critical task is to integrate research about the age wave's timing, magnitude, and location. Fiscal projections should be based on assumptions that are both globally consistent and—when it comes to longevity, fertility, and health care costs—more realistic than those now in use. Still to be determined: Which countries will be hit earliest and hardest? What might happen to interest rates, exchange rates, and cross-border capital flows under various political and fiscal scenarios?

But this is not all the proposed agency could do. It could continue to build global awareness, publish a high-visibility annual report that would update these calculations, and ensure that the various regular multilateral summits (from the G-7 to ASEAN and APEC) keep global aging high on their discussion agendas. It could give coherent voice to the need for timely policy reform around the world, hold up as models whatever major steps have been taken to reduce unfunded liabilities, help design funded benefit programs, and promote generational equity. On these and many other issues, nations have much to learn from each other, just as those who favor mandatory funded pension plans are already benefiting from the examples of Chile, Britain, Austria, and Singapore.

Global aging could trigger a crisis that engulfs the world economy. This crisis may even threaten democracy itself. By making tough choices now world leaders would demonstrate that they genuinely care about the future, that they understand this unique opportunity for young and old nations to work together, and that they comprehend the price of freedom. The gray dawn approaches. We must establish new ways of thinking and new institutions to help us prepare for a much older world.

PETER G. PETERSON is the author of *Gray Dawn: How the Coming Age Wave Will Transform America—and the World*. He is Chairman of The Blackstone Group, a private investment bank, Chairman of The Institute for International Economics, Deputy Chairman of The Federal Reserve Bank of New York, Co-founder and President of The Concord Coalition, and Chairman of The Council on Foreign Relations.

A rare and precious resource

Fresh water is a scarce commodity. Since it's impossible to increase supply, demand and waste must be reduced. But how?

Houria Tazi Sadeq*

■Water is a bond between human beings and nature. It is ever-present in our daily lives and in our imaginations. Since the beginning of time, it has shaped extraordinary social institutions, and access to it has provoked many conflicts.

But most of the world's people, who have never gone short of water, take its availability for granted. Industrialists, farmers and ordinary consumers blithely go on wasting it. These days, though, supplies are diminishing while demand is soaring. Everyone knows that the time has come for attitudes to change.

Few people are aware of the true extent of fresh water scarcity. Many are fooled by the huge expanses of blue that feature on maps

Sharper vision

Desalinization, state of the art irrigation systems, techniques to harvest fog—technological solutions like these are widely hailed as the answer to water scarcity. But in searching for the "miracle" solution, hydrologists and policy-makers often lose sight of the question: how can we use and safeguard this vital resource? UNESCO's International Hydrological Programme (IHP) takes an interdisciplinary approach to this question. On the one hand, IHP brings together scientists from 150 countries to develop global and regional assessments of water supplies and, for example, inventories of groundwater contamination. At the same time, the programme focuses on the cultural and socio-economic factors involved in effective policy-making. For example, groundwater supplies in Gaza (Palestinian Authority) are coming under serious strain, partly because of new business investment in the area. IHP has a two-pronged approach. First, train and help local hydrologists accurately assess the supplies. Second, work with government officials to set up a licensing system for pumping groundwater.

By joining forces with the World Water Council, an international think-tank on hydrological issues, IHP is now hosting one of the most ambitious projects in the field: World Water Vision. Hundreds of thousands of hydrologists, policy-makers, farmers, business leaders and ordinary citizens will take part in public consultations to develop regional scenarios as to how key issues like contamination will evolve in the next 25 years.

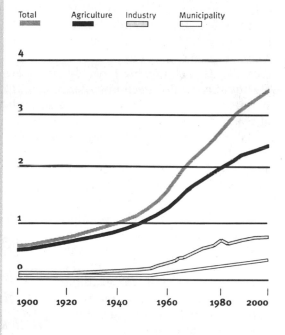

Global water withdrawals (1900-2000) in thousands of km³ per year

*Moroccan jurist, president of the Maghreb–Machrek Water Union, vice-president of the International Water Secretariat

Reprinted with permission from *The Unesco Courier*, February 1999, pp. 18-21.

of the world. They do not know that 97.5 per cent of the planet's water is salty—and that most of the world's fresh water—the remaining 2.5 per cent—is unusable: 70 per cent of it is frozen in the icecaps of Antarctica and Greenland and almost all the rest exists in the form of soil humidity or in water tables which are too deep to be tapped. In all, barely one per cent of fresh water—0.007 per cent of all the water in the world, is easily accessible.

Over the past century, population growth and human activity have caused this precious resource to dwindle. Between 1900 and 1995, world demand for water increased more than sixfold—compared with a threefold increase in world population. The ratio between the stock of fresh water and world population seems to show that in overall terms there is enough water to go round. But in the most vulnerable regions, an estimated 460 million people (8 per cent of the world's population) are short of water, and another quarter of the planet's inhabitants are heading for the same fate. Experts say that if nothing is done, two-thirds of humanity will suffer from a moderate to severe lack of water by the year 2025.

Inequalities in the availability of water—sometimes even within a single country—are reflected in huge differences in consumption levels. A person living in rural Madagascar uses 10 litres a day, the minimum for survival, while a French person uses 150 litres and an American as many as 425.

Scarcity is just one part of the problem. Water quality is also declining alarmingly. In some areas, contamination levels are so high that water can no longer be used even for industrial purposes. There are many reasons for this—untreated sewage, chemical waste, fuel leakages, dumped garbage, contamination of soil by chemicals used by farmers. The worldwide extent of such pollution is hard to assess because data are lacking for several countries. But some figures give an idea of the problem. It is thought for example that 90 per cent of waste water in developing countries is released without any kind of treatment.

Things are especially bad in cities, where water demand is exploding. For the first time in human history, there will soon be more people living in cities than in the countryside and so water consumption will continue to increase. Soaring urbanization will sharpen the rivalry between the different kinds of water users.

Curbing the explosion in demand

Today, farming uses 69 per cent of the water consumed in the world, industry 23 per cent and households 8 per cent. In developing countries, agriculture uses as much as 80 per cent. The needs of city-dwellers, industry and tourists are expected to increase rapidly, at least as much as the need to produce more farm products to feed the planet. The problem of increasing water supply has long been seen as a technical one, calling for technical solutions such as building more dams and desalination plants. Wild ideas like towing chunks of icebergs from the poles have even been mooted.

But today, technical solutions are reaching their limits. Economic and socio-ecological arguments are levelled against building new dams, for example: dams are costing more and more because the best sites have already been used, and they take millions of people out of their environment and upset ecosystems. As a result, twice as many dams were built on average between 1951 and 1977 than during the past decade, according to the US environmental research body Worldwatch Institute.

Hydrologists and engineers have less and less room for manoeuvre, but a new consensus with new actors is taking shape. Since supply can no longer be expanded—or only at prohibitive cost for many countries—the explosion in demand must be curbed along with wasteful practices. An estimated 60 per cent of the water used in irrigation is lost through inefficient systems, for example.

Economists have plunged into the debate on water and made quite a few waves. To obtain "rational use" of water, i.e. avoiding waste and maintaining quality, they say consumers must be made to pay for it. Out of the question, reply those in favour of free water, which some cultures regard as "a gift from heaven." And what about the poor, ask the champions of human rights and the right to water? Other important and prickly questions being asked by decisionmakers are how to

The water from the fountain glides, flows and dreams as, almost dumb, it licks the mossy stone.

Antonio Machado (1875–1939), Spain

calculate the "real price" of water and who should organize its sale.

The state as mediator

The principle of free water is being challenged. For many people, water has become a commodity to be bought and sold. But management of this shared resource cannot be left exclusively to market forces. Many elements of civil society—NGOs, researchers, community groups—are campaigning for the cultural and social aspects of water management to be taken into account.

Even the World Bank, the main advocate of water privatization, is cautious on this point. It recognizes the value of the partnerships between the public and private sectors which have sprung up in recent years. Only the state seems to be in a position to ensure that practices are fair and to mediate between the parties involved—consumer groups, private firms and public bodies. At any rate, water regulation and management systems need to be based on other than purely financial criteria. If they aren't, hundreds of millions of people will have no access to it.

➕ •••

- A person can survive for about a month without food, but only about a week without water.

- About 70 per cent of human skin consists of water.

- Women and children in most developing regions travel an average of 10 to 15 kilometres each day to get water.

- Some 34,000 people die a day from water-related diseases like diarrhoea and parasitic worms. This is the equivalent to casualties from 100 jumbo jets crashing every day!

- A person needs five litres of water a day for drinking and cooking and another 25 litres for personal hygiene.

- The average Canadian family uses 350 litres of water a day. In Africa, the average is 20 litres and in Europe, 165 litres.

- A dairy cow needs to drink about four litres of water a day to produce one litre of milk.

- A tomato is about 95 per cent water.

- About 9,400 litres of water are used to make four car tires.

- About 1.4 billion litres of water are needed to produce a day's supply of the world's newsprint.

Sources: International Development Initiative of McGill University, Canada; Saint Paul Water Utility, Minnesota, USA

Lack of access to safe water and basic sanitation, by region, 1990-1996 (percent)

Region	People without access to safe water	People without access to basic sanitation
Arab States	21	30
Sub-Saharan Africa	48	55
South-East Asia and the Pacific	35	45
Latin America and the Caribbean	23	29
East Asia	32	73
East Asia (excluding China)	13	—
South Asia	18	64
Developing countries	29	58
Least developed countries	43	64

Source: *Human Development Report 1998*, New York, UNDP

Periods of complete renewal of the earth's water resources

Kinds of water	Period of renewal
Biological water	several hours
Atmospheric water	8 days
Water in river channels	16 days
Soil moisture	1 year
Water in swamps	5 years
Water storages in lakes	17 years
Groundwater	1 400 years
Mountain glaciers	1 600 years
World ocean	2 500 years
Polar ice floes	9 700 years

Source: *World Water Balance and Water Resources of the Earth*, Gidrometeoizdat, Leningrad, 1974 (in Russian)

A thirsty planet

We now have less than half the amount of water available per capita than we did 50 years ago. In 1950, world reserves, (after accounting for agricultural, industrial and domestic uses) amounted to 16.8 thousand cubic metres per person. Today, global reserves have dropped to 7.3 thousand cubic metres and are expected to fall to 4.8 thousand in just 25 years.

Scientists have developed many ways of measuring supplies and evaluating water scarcity. In the maps at right, "catastrophic" levels mean that reserves are unlikely to sustain a population in the event of a crisis like drought. Low supplies refer to levels which put in danger industrial development or ability to feed a population.

Just 50 years ago, not a country in the world faced catastrophic water supply levels. Today, about 35 per cent of the population lives under these conditions. By 2025, about two-thirds will have to cope with low if not catastrophic reserves. In contrast, "water rich" regions and countries—such as northern Europe, Canada, almost everywhere in South America, Central Africa, the Far East and Oceania—will continue to enjoy ample reserves.

The sharp declines reflect the soaring water demands of growing populations, agricultural needs and industrialization. In addition, nature has been far from even-handed. More than 40 per cent of the water in rivers, reservoirs and lakes is concentrated in just six countries: Brazil, Russia, Canada, the United States, China and India. Meanwhile just two per cent of river, reservoir and lake water is found in about 40 per cent of the world's land mass.

As a result, in 2025 Europe and the United States will have half the per capita reserves they did in 1950, while Asia and Latin America will have just a quarter of what they previously enjoyed. But the real drama is likely to hit Africa and the Middle East, where available supplies by 2025 may be only an eighth of what they were in 1950.

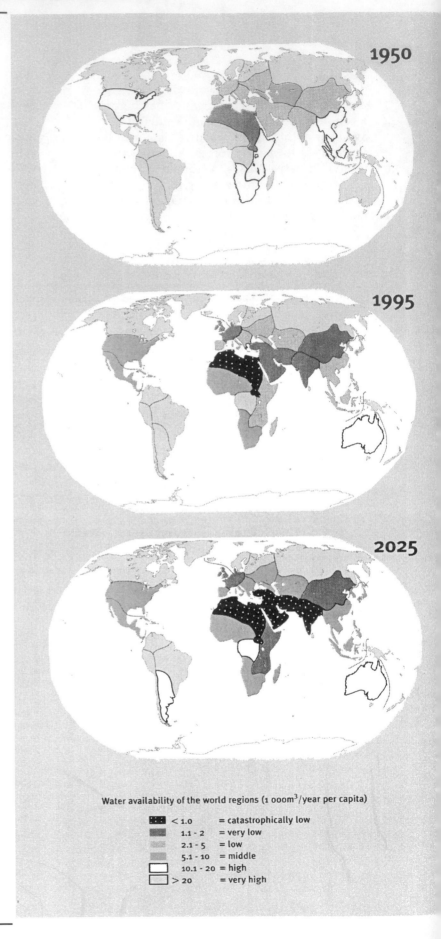

Water availability of the world regions (1 000m³/year per capita)

< 1.0	= catastrophically low
1.1 - 2	= very low
2.1 - 5	= low
5.1 - 10	= middle
10.1 - 20	= high
> 20	= very high

The Changing Geography of U.S. Hispanics, 1850–1990

by Terrence Haverluk

DEPARTMENT OF ECONOMICS AND GEOGRAPHY, UNITED STATES AIR FORCE ACADEMY, USAFA, COLORADO

In 1930, the majority of Hispanics were of Mexican descent and lived in the five Southwestern states of Arizona, California, Colorado, New Mexico, and Texas. After World War II, the Latino migrant stream began to diversify and include large numbers of Caribbeans, and Central and South Americans who generally settled in the Eastern states and California. Hispanics of non-Mexican origin now account for 36 percent of the U.S. Hispanic population. Mexican immigration has continued, and large numbers of Mexican Americans live in regions far from the border states. The U.S. Hispanic population has increased from approximately one million in 1930, to approximately 32 million in 1997. County maps chronicle the changing distribution and numbers of Hispanics from 1850 to 1990. Key words: Hispanic, Latino, Hispano, Tejano, Spanish-language radio stations.

Knowing where Hispanics settle and why is a necessary first step toward understanding a number of contemporary social issues such as illegal immigration, bilingual education, English-only laws, multiculturalism, and assimilation. The U.S. Hispanic population has increased from four million in 1960, to 32 million in 1997, and the United States is now the fifth largest Spanish speaking country in the world after Mexico, Spain, Argentina, and Colombia (U.S. Department of Commerce 1993). This so-called "browning of America," is a popular topic among educators and academics, but little longitudinal mapping exists that chronicles the changing geography of Hispanics.

Definition of Terms and Methods

The United States is the only country in the world with a Hispanic minority. The term Hispanic comes from the Latin word for Iberia, *Hispania*. Widespread use of the term began in the late 1970s when the U.S. Census Bureau adopted it to describe all persons in the United States who are descendant from Spain or from a Spanish-speaking country of the New World (Garcia 1996, 197). Hispanic is based more on history and geography than ethnicity because Hispanics may be of any race— African, Asian, European, or Native American—as long as they can trace their ancestry to Spain or one of Spain's colonies. The most numerous Hispanics are Mexicans, followed by Puerto Ricans, and Cubans. Hispanics not part of these groups are labeled "Other" Hispanics by the Census Bureau. "Other" Hispanics include Dominicans, Spaniards, Central and South Americans, and native Hispanics such as Tejanos, Hispanos, and Chicanos. Recently, the term Hispanic has been losing ground to Latino. *Latino* originated from within the social group it

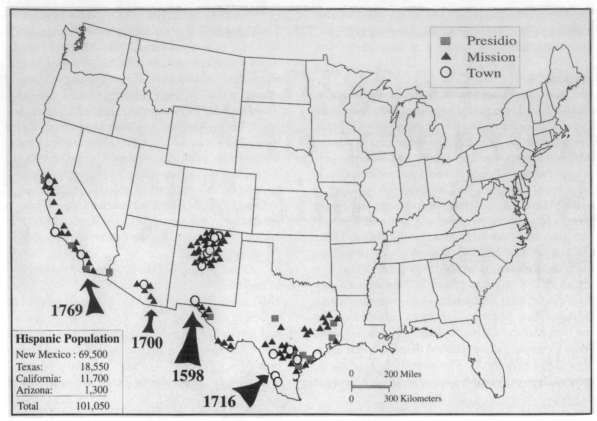

Figure 1. Hispanic population, 1850. *Source:* Martinez (1975), Carlson (1966), Simons and Hoyt (1992), Bannon (1979), and Bufkin (1975).

describes and is considered a more appropriate term. I use Latino and Hispanic interchangeably. The term Anglo-American, or simply Anglo, is commonly used to describe non-Hispanic whites and is used in this article.

This article presents a series of maps showing the distribution of U.S. Hispanics for 1850, 1900, 1930, 1960, and 1990. Accompanying the maps is an extensive historical geography that helps explain current and past Hispanic distributions. The 5 percent level is the lowest breakpoint used in the maps because it is an important political and social threshold. The 1975 amendment to the *Voting Rights Act* states that jurisdictions must provide bilingual ballots and bilingual election materials when 5 percent of its voting age population belongs to a single-language minority (Kusnet 1992, 15).

Hispanics in 1850

The first Europeans to settle permanently in what is now the United States were from Spain, not England. The Upper Rio Grande Valley, currently part of New Mexico, was settled by Juan de Oñate in 1598. In 1607, Santa Fe was founded, and in 1610 it became the capital of New Mexico, making it the oldest U.S. state capital (Carlson 1990).

Oñate's expedition was one of four broad *entradas* (entries) into the present boundaries of the United States. Figure 1 presents the dates, population, and settlement geography of the four entradas. The entradas were part of a Spanish strategy to provide a buffer from Russian and French advances and to Hispanicize Native Americans. The entradas were based on three components: 1) missions to christianize Native Americans, 2) towns for commerce and administration, and 3) *presidios* (forts) for protection.

The second entrada began a century after the Oñate expedition. Father Eusebio Kino's missionary work advanced into Pimería Alta, currently southern Arizona, in 1700. Father Kino helped establish the mission at San Xavier del Bac near present-day Tucson. The third entrada began in response to French settlements in Louisiana. Fearful of French advances, Spain established a presidio at Nacogdoches, in east Texas in 1716, followed by the presidio in San Antonio in 1718. The fourth and final entrada began in response to Russian settlements in the Pacific Northwest. Led by Father Junipero Serra, the Spanish established a string of missions along the California coast, beginning in San Diego, California, in 1769 (Nostrand 1970).

These four settlement areas constituted Spain's northern frontier; close to the hostile Apache and Comanche Indians, but far away from the civilization of Mexico City and the wealth of the silver mines on the Central Mexican Plateau. Because of this isolation, recruiting settlers was difficult. Although there were a few aristocratic Spaniards and Jesuits leading the entradas, the majority of the settlers were poor mestizos (Spanish and Indian), mulatos (Spanish and black), and coyotes (mestizo and Indian) (De la Teja 1995, 24).

Mexico gained its independence from Spain in 1821, a year that also marked a change in frontier strategy that included the secularization of the missions (Bolton 1921, 275). Lack of support for missionary activity reduced the population and importance of the missions, but more liberal Mexican trade policies opened up the Santa Fe Trail, leading to increased commercial activity between Mexico and the United States. Contact with an expansionist United States eventually led to one of the least remembered but most important wars in American history—the Mexican-American War.

Mexico's northern frontier stood in the way of American Manifest Destiny and the desire to control the continent from Atlantic to Pacific. In 1846, the U.S. Senate approved a declaration of war with Mexico and 17 months later General Winfield Scott "rode triumphantly through the [Mexico] City Square amid the deafening cheers of what was left of his army" (Eisenhower 1989, 342). Having won the battle for Mexico City, the victorious Americans began dictating terms. Mexico offered to

cede California and Texas, but not New Mexico. The Americans rejected this proposal and under the terms of the Treaty of Guadalupe Hidalgo in 1848 forced Mexico to cede all lands north of the Rio Grande and north of a horizontal line drawn one marine league south of the port of San Diego extending to the Rio Grande at El Paso. Later, in 1853, the Gadsden Purchase finalized the U.S.-Mexico boundary in southern Arizona.

The invasion, annexation, and purchase of the northern half of Mexico left a legacy of Spanish speakers, Spanish architecture, Mexican food, and Roman law throughout the Southwest. More importantly, United States expansionism created a minority population with strong historic, geographic, familial, and cultural links to Mexico that are evident today.

U.S. Hispanics, 1900

After the United States appropriated the northern half of Mexico in 1848, Anglos migrated westward, especially to the gold fields of California and to the farmland of east Texas. Figure 2 reveals that by 1900 California and east Texas had been completely overwhelmed by Anglos.

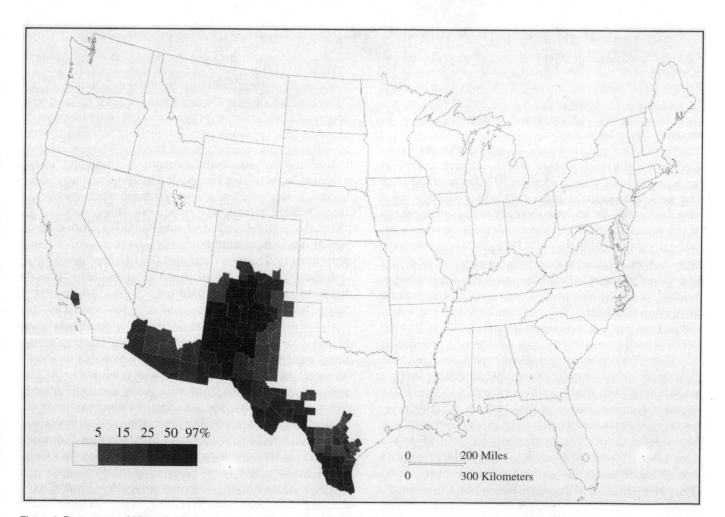

Figure 2. Percentage of Hispanics by county, 1900. *Source:* Nostrand (1980), De Leon and Jordan (1982), Hornbeck (1983), and Camarillo (1979).

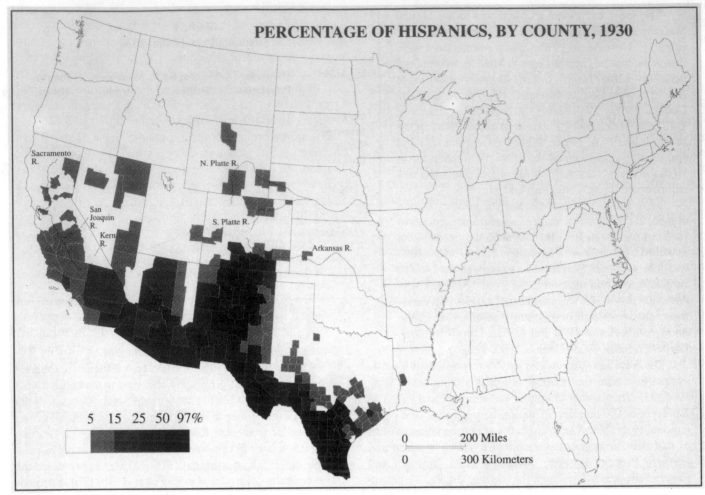

PERCENTAGE OF HISPANICS, BY COUNTY, 1930

Figure 3. Percentage of Hispanics by county, 1930. *Source:* U.S. Bureau of the Census, 1930; Haverluk 1993.

In California, only Santa Barbara County remained at least 5 percent Hispanic. In southern Arizona, Hispanics had become the minority in every county. Mexicans were recruited to work the copper mines throughout southern Arizona, and as a result, several Arizona counties had sizable Hispanic proportions in 1900.

In contrast to California, the percentage of Hispanics in New Mexico and Colorado expanded during this period. The establishment of U.S. military power in New Mexico controlled Ute, Apache, and Comanche raids, thereby facilitating the expansion of New Mexican Hispanics, called Hispanos, throughout New Mexico and into Colorado. Hispanos moved into southern Colorado and established the state's first town, San Luis, in 1851 (Figure 2). The year 1900 marked the demographic and geographic apogee of what Richard Nostrand (1980) and Alvar Carlson (1990) call the Hispano Homeland. The Hispano Homeland is a distinctive cultural area, the core of which is in north-central New Mexico around Taos and Santa Fe. Some unique cultural attributes in the Hispano Homeland include archaic Spanish words, long lots along irrigation ditches, and Penitente *moradas* (meeting houses).

Arreola (1993) identified a second Hispanic homeland in South Texas. After the Anglo take-over, many Texas Hispanics, called Tejanos, fled to Mexico or to the relative safety of South Texas below the Nueces River. The 1900 core of the Tejano Homeland was along the United States side of the Rio Grande in the cluster of 13 counties greater than 50 percent Tejano. Like the Hispano Homeland, the Tejano Homeland has distinctive characteristics, including a high percentage of central town plazas, unique festival celebrations, and a distinctive music.

Although the Hispano and Tejano homelands were able to maintain their demographic and cultural dominance after 1850, contact with Anglo-Americans altered the economy and their role in society. With the Anglos came American capital and a religious determination to develop the land. Anglos immediately began to integrate the northern Mexican frontier into the expanding industrial capitalism of the American West, first in California and Texas, and later in Arizona, New Mexico, and Colorado. Development meant clearing land of mesquite and cactus, building dams, digging irrigation canals, constructing railroads, and expanding vegetable and cotton production. The primary source of labor to accomplish these tasks consisted of Mexicans and the recently conquered Mexican-Americans. One Anglo rancher of South Texas put it this way:

> ...if it were not for those Meskins [sic], this place wouldn't be on the map. It is very true about the Anglo know-how, but without those Meskin [sic] hands no one could have built up the prosperity we have in this part of the nation (Spillman 1979, 22).

Anglo immigration, the number of farms, and tillable acres in the West increased rapidly after 1900 (Worster 1985). The pre-1850 linkages between the northern Mexican frontier (now the U.S. Southwest) and the Mexican interior continued during the American period.

U.S. Hispanics, 1930

The historic linkages between Mexico and the United States were an essential component to American economic expansion as people, money, and ideas flowed virtually unimpeded between the two countries. From 1900 to 1925, approximately 700,000 Mexicans migrated to the United States. Since there was no border patrol, crossing was safe and easy (Daniels 1990, 326). Figure 3 shows the expanded geography of Mexican migration. Between 1900 and 1930, Mexicans migrated to several newly developed irrigated valleys of the West: the sugar beet regions of the North and South Platte Rivers in Wyoming, Nebraska and Colorado; the fruit and vegetable regions of the Arkansas River in Colorado and Kansas; and the cotton, fruit, and vegetable areas of the Sacramento, San Joaquin and Kern Rivers in Central California.

The establishment of Hispanic communities in counties that did not traditionally have Hispanic populations created new migrant streams that made it easier for subsequent generations of Mexican migrants to relocate—it also made it easier for subsequent generations of Anglo farmers and industrialists to hire Mexicans and Mexican-Americans.

Several thousand Mexicans were also recruited to work the factories and the fields of the Midwest. By 1930, more than 30,000 Mexicans were working in the factories in the Chicago-Gary area, but unlike the West, where Mexicans were the primary source of labor, in the Midwest they were only one of several sources, and as a result, no Midwestern county had Hispanic populations of at least 5 percent in 1930 (Taylor 1932).

The mass movement of Mexicans to the United States spurred the Census Bureau to create a new racial category, *Mexican*, in 1930. Census enumerators were instructed that "all persons born in Mexico, who are not definitely white, Negro, Indian, Chinese, or Japanese should be returned as Mexican" (U.S. Bureau of the Census 1930, 27). The 1930 census enumerated 1,422,533 Mexicans, 90 percent of whom lived in the West, primarily in Texas. Table 1 provides a regional and ethnic break-

Table 1

Hispanic Population, 1930

Region**	Hispanic Population	Percent by Region	Mexican* Origin	Puerto Rican
West	1,478,535	92	1,477,273	1,262
Midwest	59,227	4	59,227	NA
South	6,908	1	6,908	NA
East	58,882	4	7,370	51,512
Total	1,603,522	100	1,550,778	52,774

NA = Not Available
* Includes Hispanos
** West: WA, OR, CA, ID, NV, MT, WY, CO, NM, UT, ND, SD, NE, KS, OK, TX. Midwest: MN, WI, MI, OH, IN, IL, IA, MO. South: AK, FL, LA, MS, AL, GA, SC, NC, TN, KY, VA, WV. East: MD, DE, DC, PA, NJ, NY, CT, RI, MA, NH, VT, ME.

Source: U.S. Bureau of the Census, 1930.

down of the Hispanic population in 1930 and reveals that the U.S. Hispanic population was still overwhelmingly Mexican and Western.

After 1900, Mexicans became the primary source of labor in the West as a result of several amendments to U.S. immigration law. In California, Chinese were excluded by the *Chinese Exclusion Act of 1882*, which was followed by a Gentlemen's Agreement in 1907 to curb Japanese immigration to the United States. These Asian exclusion acts were the first in a series of amendments that restricted immigration, culminating in the *Immigration Act of 1924* and the creation of the U.S. Border Patrol in 1925 (Daniels 1990, 328).

The *National Origins Act of 1924* established a quota system based on the national origin of immigrants (Yang 1995, 23). The number of immigrants admitted was based on 2 percent of the number of foreign-born persons of a given nationality in 1890. The year 1890 was chosen because prior to that date immigrants were primarily from western and northern Europe. Between 1890 and 1920, southern and eastern Europeans dominated the migrant stream. By basing admission on 2 percent of the 1890 population, the amendment effectively restricted southern and eastern European immigration, as well as most Asian immigration—but it also created labor shortages (Daniels 1990, 283).

The state department worked in cooperation with industry and agriculture to keep immigration from Mexico open and, by the 1920s, Mexicans became the most important source of immigrant labor in the West. Mexican labor, it was argued, would be easier to send home during recessions, but Mexico was close enough and labor plentiful enough that labor streams could be re-established when necessary (Daniels 1990, 309). The idea that

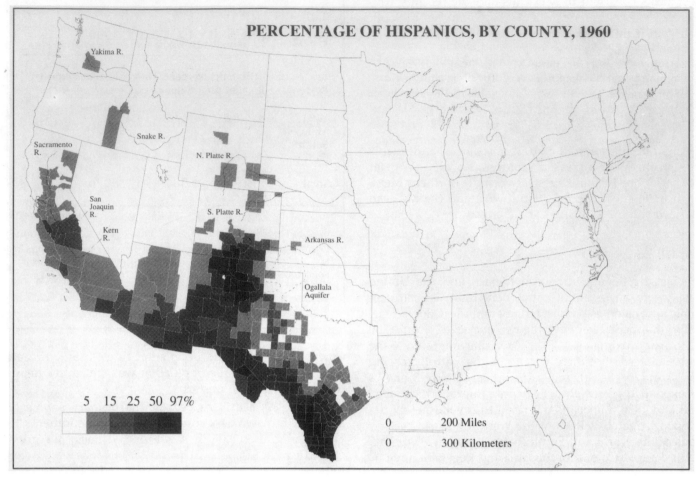

Figure 4. Percentage of Hispanics by county, 1960. *Source:* U.S. Bureau of the Census, 1960.

Mexican labor could be forcibly returned to Mexico during recessions is what I call the *repatriation strategy.*

The Great Depression led to the implementation of the repatriation strategy and thousands of Mexicans were forcibly returned to Mexico. The number of Mexicans in the United States declined throughout the 1930s and early 1940s, only to rebound again during and after World War II.

U.S. Hispanics, 1960

The end of the Depression and the onset of World War II led to labor shortages in agriculture and industry. Mexican workers were again seen as a partial solution to U.S. labor shortages. In 1942, the United States and Mexico established a guest worker system called the Bracero Program. The Bracero Program (from *brazo,* arm) guaranteed transportation, food, housing, and a minimum wage for braceros.

Figure 4 shows that the Bracero Program expanded Mexican settlement geography, this time to the cotton fields of the Texas south plains; the hops and orchard farms of the Yakima River Valley in Washington; and the vegetable growing areas of the Snake River Valley in Idaho. The Bracero Program was in effect between 1942

and 1964. During its peak in the mid-1950s, an average of 400,000 Mexicans entered the United States each year (Grebler et al. 1970, 176). Braceros were expensive, however. Many farmers bypassed the Bracero Program with its high cost and cumbersome bureaucracy and hired un-

Table 2

Hispanic Population, 1960

Region	Hispanic Population	Percent by Region	Mexican* Origin	Puerto Rican
West	3,561,668	78	3,519,318	42,350
Midwest	199,184	4	133,012	66,172
South	47,313	1	15,457	31,856
East	754,684	17	22,219	732,465
Total	4,562,849	100	3,690,006	872,843

* Includes Hispanos

Source: U.S. Bureau of the Census, 1960.

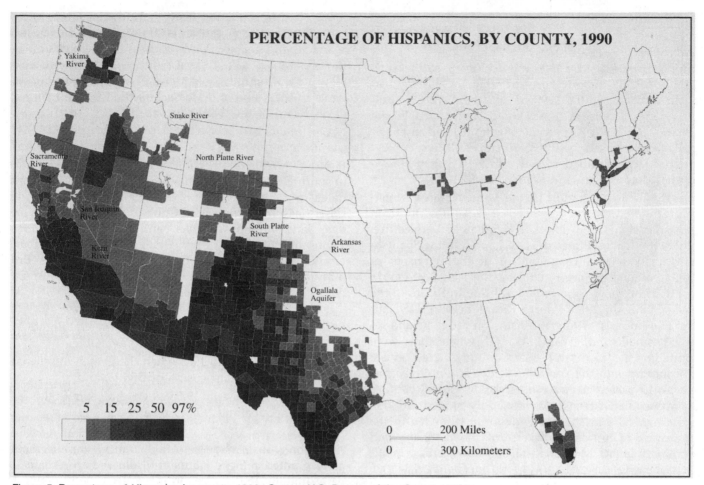

Figure 5. Percentage of Hispanics by county, 1990. *Source:* U.S. Bureau of the Census, 1993.

documented (illegal) Mexicans directly. Undocumented Mexicans were less expensive and easier to control than Braceros, and in 1953, over 800,000 undocumented Mexicans were apprehended along the border (Grebler et al. 1970, 68).

Figure 4 also reveals that the Hispano Homeland in New Mexico and Colorado contracted after 1900, while the Tejano Homeland was able to maintain its dominance. California's Hispanic proportions also increased, especially in the central valley around Fresno.

The period of high immigration in the 1950s was interrupted by a recession after the end of the Korean War, which again led to forcible repatriation. In 1954, the United States established a paramilitary organization called Operation Wetback, whose goal was to find and return undocumented workers. Operation Wetback was run by a retired Army general who successfully repatriated over 3.8 million undocumented Mexicans (many of whom were sent back more than once). Success at stemming the flow of undocumented Mexicans was only temporary because the western United States economy still relied on Mexicans in several economic sectors, especially agriculture. The end of the recession led to the termination of Operation Wetback, and in 1955, undocu-

mented Mexican migration returned to pre-Operation Wetback levels (Grebler 1970, 176).

The Bracero Program was terminated in 1964 because of widespread abuse on both sides of the border, but immigration continued—albeit illegally. American farmers found it cheaper and easier to hire undocumented aliens. American businesses who hired undocumented Mexicans were breaking no law and after the termination of the Bracero Program in 1964, use of Mexican labor continued to be widespread.

Table 2 reveals that the Mexican population doubled from 1930 to 1960, but that the Puerto Rican population increased 16 times. Widespread Puerto Rican immigration to the United States began in the 1950s with the introduction of inexpensive flights from Puerto Rico. The overwhelming number of Puerto Ricans settled in New York City resulting in the increasing importance of Latinos in the East. In 1930, 92 percent of Hispanics lived in the West, in 1960 only 78 percent lived there. Changing settlement destinations among Puerto Ricans, a revolution in Cuba, and changes in U.S. immigration laws initiated new Latino migrant streams, continuing the trend of Latino diversification seen in Table 2.

U.S. Hispanics 1990

Puerto Ricans

Figure 5 reveals that by 1990 Puerto Ricans had become a highly visible minority group along the eastern seaboard. According to the 1990 census, there were 2.6 million Puerto Ricans in the United States, (triple the number from 1960) 70 percent of whom lived in megalopolis. The majority of Puerto Ricans still live in New York, but large numbers also live in New Jersey, Massachusetts, Connecticut, and Pennsylvania.

Puerto Rican migration to the United States is relatively easy because they are U.S. citizens. Puerto Rican citizenship stems from another U.S. colonial adventure. In 1898, the United States defeated Spain in the Spanish-American War and took control of the remnants of Spain's colonial empire, including the island of Puerto Rico. In 1910, there were only 1,513 Puerto Ricans in the United States. In 1917, Puerto Ricans became U.S. citizens. The number of Puerto Ricans on the mainland remained small until World War II when they were recruited by the War Man Power Commission to alleviate labor shortages in the East and Midwest. Puerto Ricans were given preference over other labor sources because they were U.S. citizens. At first, only skilled workers were admitted, but the need for more labor led to the recruitment of railroad workers for the Baltimore and Ohio line in 1944 (Moldanado-Denis 1976, 111).

After World War II, spurred by the expanding U.S. economy and the availability of inexpensive commercial flights to the mainland, Puerto Rican immigration increased markedly. During the 1960s, Puerto Ricans migrating to the United States overwhelmingly chose New York City, but agricultural workers were recruited to several eastern and midwestern states, especially Ohio and Pennsylvania.

In Megalopolis, the term Hispanic is associated with Puerto Ricans or "Other" Hispanics. Puerto Rican food, music, and immigration history are distinct from Mexican. Puerto Rican music is Bomba and Plena, not Mariachi and Tejano. Puerto Rican folk architectural influences can be seen on *casitas* (little houses) in many of New York's community gardens. The majority of my students in Maryland and Massachusetts, for example, had never heard of the word Chicano.

Cubans

Figure 5 reveals another important cluster of Hispanics in south Florida. This population is primarily Cuban and dates from the Revolution of 1959. In that year, Fidel Castro assumed control of Cuba and expropriated private land holdings, banks, and industrial concerns. Thousands of mostly upper- and middle-class Cubans who opposed the regime fled to the United States. Prohias and Casals (1973, 12) identified three stages of Cuban migration to the United States:

1. between 1959 and 1962 when commercial flights between Cuba and the U.S. were available,
2. between 1962 and 1965 when the Cuban missile crisis led to the suspension of flights, and
3. between 1965 and 1973 when daily air flights between Cuba and Miami resumed.

All Cubans who made it to this country were immediately granted refugee status and allowed to remain legally. Initially, the U.S. government attempted to relocate Cubans to other parts of the United States to lessen the impact of Cubans on south Florida, but most relocated Cubans eventually moved back to south Florida.

Two more stages must now be added to the list:

4. between 1973 and 1994, when, for the most part, Cubans used boats to sail the 90 miles between Cuba and south Florida, and

---

Table 3

Hispanic Population, 1990

Region	Hispanic Population	Percent by Region	Mexican Origin	Puerto Rican	Cuban	"Other" Hispanics*
West	14,325,314	66	12,186,656	220,777	108,114	1,809,772
Midwest	1,524,611	7	1,018,797	251,735	32,561	221,518
South	2,116,532	10	346,672	321,626	709,360	738,874
East	3,739,781	17	188,706	1,826,466	196,179	1,528,430
Total	21,706,243	100	13,740,831	2,620,604	1,046,214	4,298,594

* Includes Hispanos, Dominicans, Central Americans, South Americans, and Hispanics who identified themselves as Spanish, Chicano, Tejano, Californio, and so on.

Source: U.S. Bureau of the Census, 1990.

Table 4

"Other" Hispanic Populations, 1990

Region	"Other" Hispanics*	Percent by Region	Central Americans	South Americans	Dominicans
West	1,039,198	36	771,649	256,317	11,232
Midwest	104,646	4	45,444	54,330	4,873
South	493,846	17	229,350	224,650	39,846
East	1,227,775	43	276,095	496,560	455,120
Total	2,865,465	100	1,322,538	1,031,857	511,071

* Includes Dominicans, Central Americans, and South Americans.

Source: U.S. Bureau of the Census, 1990.

5. after August 1994, when President Bill Clinton revoked automatic refugee status for Cubans, thereby pulling up the 36-year old Cuban welcome mat in Florida.

Between 1959 and 1994, approximately 715,000 Cubans successfully relocated to the United States (*Washington Post* 1994), most to Florida. In Florida, the word Hispanic is associated primarily with Cubans, whose language and traditions are distinct from Puerto Ricans and Mexicans.

Table 3 presents the 1990 population of the most numerous Hispanic groups—Mexicans, Puerto Ricans, and Cubans, as well as "Other" Hispanics. The percentage by region reveals a continuation of the decreasing demographic dominance of the West—from 92 percent in 1930, to 66 percent in 1990, and the corresponding increase of the south and east. Until 1990, the South had fewer Hispanics than the Midwest. Table 3 also reveals the importance of the "Other" Hispanic population, which in 1990 was larger than the Puerto Rican and Cuban populations combined.

Other Hispanics

Between 1924 and 1965, most Latin Americans were restricted from legally immigrating to the United States. As already mentioned, the exceptions were Puerto Ricans, who are U.S. citizens, Cuban refugees, and Mexican braceros. Until 1965, U.S. Hispanics were overwhelmingly from Mexico and the Caribbean. After 1965, the Latin American migrant stream expanded to include Central and South Americans.

The social revolution in the 1960s and world-wide condemnation of the restrictive *National Origins Act of 1924* led to the emendation of U.S. immigration law. The *Immigration and Nationality Act of 1965* phased out the quota system and its restrictions and placed in its stead a hemispheric allocation system that admitted 170,000 persons from the eastern hemisphere and 120,000 from the western hemisphere for an annual ceiling of 290,000.

Available slots were meted-out based on family reunification, which favored Mexico with its large U.S. population. In 1976, the system was amended to allow only 20,000 from one country, which expanded the Latin migrant stream to Central and South America.

In 1980, Congress established the *Refugee Act*, which allowed 50,000 refugees to enter annually. The *Refugee Act* facilitated the migration of thousands of Central Americans caught in U.S.-sponsored wars against communism, especially in Nicaragua, El Salvador, and Guatemala. In 1980, thousands of anticommunist Nicaraguans, like Cubans 20 years earlier, fled their country and relocated to the United States, primarily to Florida (79,056). Salvadorans (338,769) and Guatemalans (159,177) preferred California, but several thousand also migrated to New York and south Florida (Daniels 1990, 341). Even with these changes, however, Mexicans are still the largest legal immigrant group not only among Latin Americans, but among all legal immigrant groups (Yang, 1995, 24).

Central Americans, South Americans, and Dominicans are lumped together by the Census Bureau under the title "Other" Hispanics. About one-half of all Central Americans live in California (637,656), especially Los Angeles County (453,048). About three-fourths of all Dominicans live in New York City (357,868). South Americans also prefer New York (279,101), but also live in California (182,384), Florida (170,531), and New Jersey (126,286). "Other" Hispanics are the fastest growing segment of the Latino population in the East and the South. "Other" Hispanic are the majority Latino population in Rhode Island (Guatemalans), Maryland (Salvadorans), and the District of Columbia (Salvadorans and Dominicans). In New York and New Jersey "Other" Hispanics are now almost as numerous as Puerto Ricans. Unlike the West and Midwest, where Mexicans are the overwhelming majority, the eastern and southern Hispanic populations are more heterogeneous. West of the Mississippi, only San Francisco, Los Angeles, and Houston have substantial "Other" Hispanic populations. Table 4

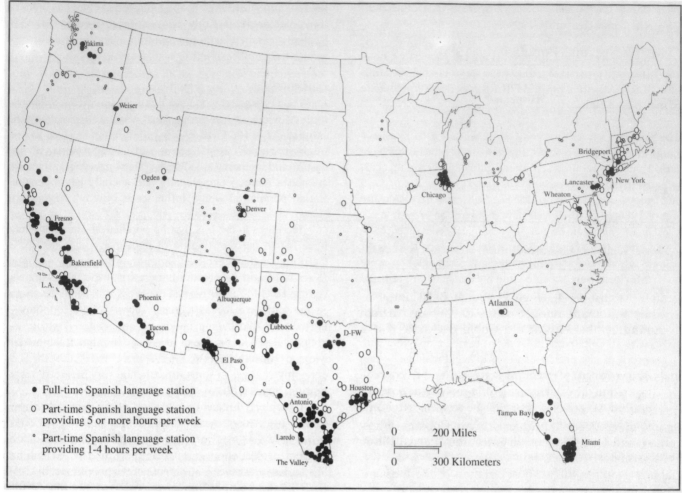

Figure 6. Spanish Language Radio Stations, 1990. Full-time formats in the West include Country and Western/Hispanic, bilingual Tejano, Hispanic/Anglo/Indian, Spanish music, Spanish hit radio, bilingual Hispanic, and Spanish. Full-time formats outside the West include Spanish news, Spanish urban Contemporary, Spanish, Spanish music, Spanish hit radio, Spanish/Caribbean, Spanish/Portuguese, Spanish talk, Hispanic, and Latin.
Source: *Broadcasting and Cable Yearbook, 1990.*

shows the distribution of the three principal "Other" Hispanic groups. The "Other" Hispanic population is essentially bicoastal, with about 90 percent of the population living along the northeastern seaboard, Florida, or California.

The Puerto Rican, Cuban, and "Other" Hispanic population is primarily eastern. In 1990, the East had 25 counties greater than 5 percent Hispanic, compared to zero in 1960. The same is true of Florida, which in 1960 was overwhelmingly Anglo, but in 1990 had 15 counties greater than 5 percent Hispanic, including Dade County, the only county east of the Mississippi greater than 50 percent Hispanic.

The Midwestern cluster of counties centered around Chicago is primarily Mexican (612,442 or about 70 percent of the Illinois Hispanic population). The large number of Mexicans in Chicago is a continuation of Mexican migration to Chicagoland factories that began during the 1920s. The Michigan Hispanic population is also primarily Mexican. Michigan was the second most important sugar beet producing state after Colorado in the 1920s, and thousands of Mexicans migrated to Michigan to

work beets (Taylor 1932). Michigan and other parts of the Midwest still rely on Mexican agricultural labor.

Even more dramatic than Hispanic increases in the East and South is the continued growth and geographic expansion of the western Hispanic population shown in Figure 5. The Hispanic population west of the Mississippi is still overwhelmingly of Mexican descent. In 1960, the Northwest had one county, Yakima, greater than 5 percent Hispanic. In 1990, there were 34 counties with a Hispanic population greater than 5 percent. The migrant linkages established between Mexico and the Northwest during the Bracero Program still exist, and some Yakima Valley communities, such as Toppenish and Sunnyside, are now majority Hispanic. The Snake River Valley in Idaho has almost as many counties greater than 5 percent Hispanic (14) as the East (25). The number of Hispanics in the Northwest is not as large as in Megalopolis, but their proportions and their influence on the landscape and the culture are perhaps greater.

In Texas, the Tejano Homeland has expanded northward and westward—in 1960 Texas had 17 counties greater than 50 percent Hispanic, in 1990 there were 34 counties greater than 50 percent Hispanic. There is now a continuous cluster of Hispanic majority counties 700 miles long from El Paso in extreme West Texas to Corpus Christi along the Gulf of Mexico. In 1990, these counties had a population of 2.2 million and is one reason that Texas overtook New York to become the second largest state in 1994. Based on current Hispanic growth rates, several more counties in south Texas and on the South Plains will be majority Hispanic by the year 2000.

In New Mexico and Colorado the Hispano homeland has lost some of its demographic dominance while the southern part of New Mexico has become increasingly Mexican. In California and Nevada, almost every county has seen sharp rises in the percentage of Hispanics since 1960, yet only one county, Imperial, along the Mexican border is greater than 50 percent Hispanic. Even though the western Hispanic population is not as dominant as it once was, 66 percent of all Hispanics live in the West and it is America's most ethnically diverse region.

Hispanic Influence

Hispanics, especially Mexicans, have historically had a large impact on the cultural and economic geography of the United States. The American cowboy and the ranching (*rancho*) industry owe much of their existence to the Mexican *vaquero* (buckaroo). Many of the tools and vocabulary of the cowboy originated in Mexico: lariat (*la reata*), chaps (*chaparejos*), mustangs (*mesteños*) and rodeo (*rodear*, to round up), to name a few. Mexicans have also influenced music and contributed much of the "Western" to Country & Western music. In south Florida, the "Miami sound" is influenced by Caribbean rhythms and the Spanish language (Roberts 1979). Mexican food, which has always been popular out West, is now common throughout the United States and salsa now outsells ketchup (Minneapolis Star and Tribune, 13 November 1992). The increasing number and wider distribution of Hispanics in the United States means that the Spanish language is more widespread now than it has been historically, and many phrases such as nada, hasta la vista, mañana, mano a mano, and no más have entered the popular vernacular.

Hispanics, like the U.S. Asian population, are a diverse group from many different countries. Unlike the Asian population, however, Hispanics share a common language. Spanish was the first European language in North America, and the use of Spanish has increased with the Hispanic population. Spanish is reinforced through Spanish language media, including magazines, newspapers, TV, and radio. Hispanics, however, seem to prefer radio to print and TV (Greenberg 1983). Full- and part-time Spanish-language radio stations (SLRS) were once primarily a border phenomenon but are now common throughout the United States, especially in the West. SLRS can be an important diagnostic feature of the relative importance of the Spanish language in a place (Figure 6).

Texas has the most SLRS in the country—twice as many as California, which has double the Hispanic population of Texas. The dominance of Texas in SLRS suggests that Spanish may be more important there than in California, and in fact, Texas Hispanics have higher Spanish language usage rates than California Hispanics (U.S. Bureau of the Census 1993). One-half of Texas' full-time SLRS are in the Tejano Homeland, also known as "The Valley," where Tejano music originated.

Tejano music is rooted in traditional, accordion-based norteño music of Mexico but also incorporates synthesizers, salsa rhythms, and even hip-hop. Tejano music began in south Texas and has spread throughout Texas and the West. Many western radio stations shown in Figure 6 play predominately Tejano music. In San Antonio and Los Angeles, the highest rated radio stations in 1993 were Tejano stations (Dallas Morning News, 1993). When the well-known Tejano singer Selena was gunned down by a disgruntled fan in 1995, the news spread rapidly through the network of radio stations, and for days afterward there were Selena tributes as far north as the Yakima Valley. The popularity of Tejano may partially explain why there are more full-time SLRS in Lubbock, Texas, with 51,000 Hispanics than in New York City with over one million Hispanics. The state of Washington, with 206,018 Hispanics, has almost as many full-time SLRS as the East with 3 million Hispanics. The reason for the greater concentration of SLRS in the West is the result of several factors:

1. In the Texas and New Mexico homelands, Hispanics have always been the dominant population, thereby legitimizing Hispanic culture from a very early date;

2. Proximity to the border reinforces the use of Spanish among Hispanics and even many non-Hispanics;

3. Most western Hispanics are of Mexican descent and share more similar tastes and values than the more heterogeneous eastern Hispanic population;

4. The ability to procure airway space is easier in Lubbock, Texas, than in New York City, because the airways are more crowded in New York City; and

5. The West is where most U.S Hispanics live. My own research has shown that non-Hispanics living in areas with large numbers of Spanish speakers are significantly more likely to speak Spanish than other non-Hispanics (Haverluk 1993).

South Florida and Chicago, which are primarily Cuban and Mexican respectively, also have large numbers of full-time SLRS. Atlanta's full-time SLRS is perhaps emblematic of the "New South."

The Future

Large increases in non-European immigrants since 1965 have fueled anti-immigration sentiment and latent racism that culminated in the *Immigration Reform and Control Act of 1986* and Proposition 187 (Prop. 187) in California, in 1994. The Act of 1986 established four new immigration provisions:

1. Amnesty for illegal aliens in the U.S. since 1982. Over 3 million illegal immigrants, 70 percent of whom were Mexican, were accepted into the amnesty program and are now in various stages of the legalization process.
2. Requirements that employers verify the eligibility of all newly hired employees. This provision is the reason why all employees must now provide employers with proof of citizenship or proof of legal residence upon hiring. Instead of inhibiting immigration, this provision has led to the establishment of a sophisticated underground network of fraudulent document providers. Employers only have to ask for documentation, they do not have to verify its authenticity. Many western farmers still rely on illegal aliens and comply with the letter, but not the spirit, of the law.
3. Sanctions against employers that knowingly hire illegals. This provision was designed to mete out tough fines to persons or businesses that knowingly hire illegals. Many western farmers and businesses have relied on illegal Mexican labor for decades and are unwilling to blow the whistle on their neighbor for hiring illegals. Furthermore, in order to get the necessary votes to pass the bill, western growers insisted that the Immigration and Naturalization Service (INS) could not conduct raids during harvest, effectively taking the teeth out of the amendment.
4. Provisions to allow agricultural workers to be recruited during times of labor shortages. Clearly the most ironic, even hypocritical, component of a bill designed to reduce immigration is the Replenishment of Agricultural Workers (RAW) component of the Act. RAW established a mechanism to authorize an additional 250,000 agricultural workers each year who could eventually be returned to Mexico—again the repatriation strategy. Ten years after its implementation, the Act of 1986 has not been very effective at stemming Latino immigration (Daniels 1990, 342).

The Act's ineffectiveness led to Proposition 187 in California, which proposes that all state services be denied to illegal immigrants and their children, even if they were born in the United States. This latter component of the bill is probably unconstitutional, and a California court has blocked its implementation. At the national level, there is debate in Congress to amend current immigration and amnesty laws to reduce the number of immigrants from all areas.

Conclusion

Until the 1960s, the U.S. Latino population was overwhelmingly Mexican and western, and so was the Hispanic cultural imprint. After 1965, changes in immigration laws led to new migrant streams from Central and South America to the eastern and southern United States. These new migrant streams created a more geographically and socially heterogeneous Latino population whose influence is now felt beyond the West.

Increases in the U.S. Latino population since the 1960s have led to the reinstitution of the repatriation strategy in California and a general anti-immigrant sentiment. Attempts to restrict Latin American immigration, especially Mexican immigration to California and the West, will be unsuccessful in the long-term because many sectors of its economy—agriculture, landscaping, child care, and janitorial services—are predicated on the use of low-wage immigrant labor from Mexico (Bustamante et al. 1992; Jones 1995). Unless the U.S. economy changes in some fundamental way Mexican immigration, legal or illegal, will continue.

Continued Latino immigration combined with higher Latino birth rates means that the Hispanic population is growing seven times faster than the non-Hispanic population. Current projections suggest that there will be 81 million Hispanics, constituting 30 percent of the population, by 2050 (U.S. Department of Commerce 1993). In cities such as San Antonio, Miami, and El Paso, Hispanics are already the majority; in Los Angeles and New York they are approaching majority status. Culturally, these communities are quite different from other U.S. cities: the nightly news is given in Spanish and English; newspapers are bilingual; Latino holidays are celebrated as often as traditional "American" holidays; specialty crops are planted and/or imported to provide traditional Latino foods; and Latino history, sometimes taught in Spanish, is presented along with Anglo American history.

Increasing numbers of Latinos in the United States have led to new attempts to restrict immigration and immigrant services. English Only and/or Official English laws have been passed by many states and are being discussed at the national level. Bilingual education is being challenged, while at the same time many colleges and universities are requiring the study of foreign languages which, increasingly, means Spanish. In many respects, the entire socio-political framework of the United States is affected, especially in the shaded areas of Figure 5. The reason for these social debates is that unlike many previous immigrant groups, Latinos are more likely to maintain their language and culture even while they learn English and absorb American culture (Matovina 1995). Unlike other immigrant groups that have maintained their language and culture, such as the Greeks and the Chinese, Latinos are much more numerous and their source areas are closer to the United States.

The long-term maintenance of Mexican culture in the West, for example, has allowed many non-Hispanics to embrace aspects of Latino culture such as Santa Fe style clothing, Southwestern architecture, and New Mexican cuisine. Whereas some Anglos feel threatened by the existence of a large non-Anglo cultural group in society and wish to foster the Americanization of Hispanics, other Anglos feel enriched by Latino contributions to society. How the United States deals with these tensions will be a continuing theme in American politics and society well into the future. Let us hope we develop along the Swiss, rather than the Balkan, model.

References

Arreola, D. D. 1993. The Texas-Mexican homeland. *Journal of Cultural Geography* 13(2)6–74.

Bannon, C. 1979. *The Spanish borderlands frontier, 1513–1821.* Albuquerque, NM: University of New Mexico Press.

Bolton, H. E. 1921. The Spanish borderlands. *The Chronicle of America Series,* No. 23. New Haven, CT: Yale University Press.

——. 1964. *Bolton and the Spanish borderlands.* Norman, OK: University of Oklahoma Press.

Broadcasting and Cable Yearbook. 1990. *Broadcasting yearbook.* Washington, DC: Broadcast Publishing.

Bustamante, J. A., C. W. Reynolds, and R. A. Hinojosa. 1992. *US-Mexican relations: Labor market interdependence.* Stanford, CA: Stanford University Press.

Camarillo, A. 1979. *Chicanos in a changing society: From Mexican pueblo to American barrio in Santa Barbara and Southern California, 1848–1930.* Cambridge, MA: Harvard University Press.

Carlson, A. W. 1966. *The historical geography of the Spanish settlement in the middle Rio Grande Valley, 1598–1821.* Unpublished research paper. Minneapolis, MN: University of Minnesota.

——. 1990. *The Spanish-American homeland: Four centuries in New Mexico's Rio Arriba.* Baltimore, MD: Johns Hopkins University Press.

Dallas Morning News. 1993. *Tejano Attracts the big labels.* 10 October, 9C.

Daniels, R. 1990. *Coming to America.* New York: Harper Collins.

De Leon, A., and Jordan. 1982. *The Tejano community, 1836–1900.* Albuquerque, NM: University of New Mexico Press.

Eisenhower, J. S. D. 1989. *So far from God: The U.S. war with Mexico, 1846–1848.* New York: Doubleday.

García, I. M. 1996. Backwards from Aztlán: Politics in the age of Hispanics. In *Chicanos and Chicanas in contemporary society,* ed. Roberto M. de Anda, Needham Heights, NJ: Allyn and Bacon.

Grebler, L. et al. 1970. *The Mexican-American people: The nation's second largest minority.* New York: Free Press.

Greenberg, B. 1983. *Mexican Americans and the mass media.* Norwood, NJ.

Haverluck, T. W. 1993. *The Regional Hispanization of the United States.* Unpublished doctoral dissertation. University of Minnesota.

Hornbeck, D. 1983. *California patterns: A geographical and historical atlas.* Mountain View, CA: Mayfield Publishing.

Jones, R. C. 1995. *Ambivalent journey, U.S. migration and economic mobility in north central Mexico.* Tucson, Az: University of Arizona Press.

Kusnet, D. 1992. *Voting rights in America: Continuing the quest for full participation.* Edited by Karen McGil Arrington and William L. Taylor. Washington, DC: Leadership Conference Education Fund.

Maldonado-Denis, M. 1976. *En las entranas: Un analasis sociohistorico de la emigracion Puertorriquena.* Habana, Cuba: Casa de las Americas.

Matovina, T. M. 1995. *Tejano religion and ethnicity: San Antonio 1821–1860.* Austin, TX: University of Texas Press.

Minneapolis Star and Tribune. 1992. *Salsa savvy: Market is hot for spicy Mexican sauce.* 13 November 1992, E1.

Nostrand, R. L. 1970. The Hispanic-American horderland: Delimitation of an American cultural region. *Annals of the Association of American Geographers* 60(4):638–661.

—— 1980. The Hispano homeland in 1900. *Annals of the Association of American Geographers* 70(3):382–396.

Roberts, J. S. 1979. *The Latin tinge: The impact of Latin American music on the United States.* Oxford University Press.

Simons, H., and C. Hoyt. 1992. *Hispanic Texas: A historical guide.* Austin, TX: University of Texas Press.

Spillman, R. C. 1979. *Hispanic population patterns in southern Texas, 1850–1970.* Research paper number 57. Eugene, OR: University of Oregon.

Taylor, P. S. 1932. *Mexican labor in the United States.* Berkeley, CA: University of California Press.

United States Bureau of the Census. 1990. *1990 social and economic characteristics.* Washington, DC: U.S. Government Printing Office.

——. 1960. *Persons of Spanish surname 1963.* Subject Reports PC(2)-1D. Washington, DC: U.S. Government Printing Office.

——. 1930. *Race and foreign origin 1933.* Washington, DC: U.S. Government Printing Office.

United States Department of Commerce. 1993. *We the Americans . . . Hispanics.* Washington, DC: Bureau of the Census.

The Washington Post. 1994. *Recurring waves of Cuban refugees.* 20 August.

Worster D. 1985 *Rivers of empire: Water, aridity and the growth of the American West.* New York: Pantheon.

Yang P. Q. 1995. *Post-1965 immigration to the United States: Structural determinants.* Westport: Praeger.

RUSSIA'S POPULATION SINK

In the former heart of the Soviet empire, deaths are far outpacing births.

Toni Nelson

Toni Nelson is a staff researcher at the Worldwatch Institute

In Nadvoitsy, a small Russian town near the Finnish border, an estimated 4,000 children have been poisoned by fluoride, which replaces calcium in the body, leaving its victims with blackened, rotting teeth and weakened bones. Although the town's aluminum plant no longer dumps fluoride into unlined landfills, the contamination persists because neither the authorities nor the company can afford a full-fledged clean-up. Today 5 to 10 percent of the town's kindergartners continue to exhibit signs of fluorosis.

Nadvoitsy's experience provides a glimpse into the myriad problems facing the countries of the Former Soviet Union (FSU). Years of environmental contamination have combined with economic instability to push the region into a public health crisis, and several FSU countries are now experiencing the most dramatic peacetime population decline in modern history. In Russia, which has more than half the FSU's population, the situation may be at its worst. As the country's birth rate falls and its death rate climbs, the population is expected to shrink by some 9 million between 1992 and 2005. More important, perhaps, is the rising incidence of birth defects and other health problems whose effects may linger for generations.

Russia's demographic decline began in the mid-1980s, well before the collapse of the Soviet Union in 1991 (see graph). Total live births in Russia dropped from a peak of 2.5 million in 1987 to 1.4 million in 1994, while total deaths climbed from 1.5 million to 2.3 million over the same period. The year 1994 brought the most precipitous decline on record, with deaths exceeding births by more than 880,000 and the population falling by 0.6 percent (excluding immigration, which compensated for two-thirds of the decline). Life expectancy, which provides the best general measure of a country's health conditions, also dropped sharply between 1987 and 1994, from 65 to 57

years for men, and from 75 to 71 years for women. This decline has no precedent in industrialized societies; Russian male life expectancy is now the lowest of all developed countries.

Russia's deteriorating social and ecological conditions have had serious consequences for the country's children as well. Infant mortality has climbed to at least 20 deaths per 1,000 live births, although some experts suggest the figure could be as high as 30 per 1,000—more than three times the U.S. rate and double that of Costa Rica, one of the most advanced developing countries. Birth defects occur in 11 percent of newborns, and 60 percent exhibit symptoms of allergies or the deficiency disease known as rickets, caused by a lack of vitamin D. Children's health tends to decline throughout childhood; scarcely one-fifth of Russia's children can be considered healthy by the end of their school years.

Maternal health, and the health of women of reproductive age in general, is also declining—a trend that will almost certainly intensify problems with infant health. Gynecological pathologies have been found in 40 to 60 percent of women in their child-bearing years, and even girls in their early teens are showing signs of reproductive abnormalities. Fully 75 percent of Russian women experience complications during pregnancy, and the death rate during childbirth is 50 per 1,000 births—more than six times the U.S. rate. Only 45 percent of Russian births qualify as normal by Western medical standards.

The factors underlying these trends are complex and numerous, but most can be traced to some combination of environmental contamination and economic instability. In part, the fertility decline is a matter of simple demographics: the number of marriages has decreased, and there are fewer women of childbearing age in the population due to a brief decline in births after World War II.

From *World Watch* magazine, January/February 1996, pp. 22–23. © 1996 by the Worldwatch Institute, Washington, DC. Reprinted by permission.

But life in the FSU is still haunted by the abrupt transition from communism, which provided work and housing for nearly everyone, to a capitalist system driven by competition and characterized by insecurity. The uncertain economic situation has prompted many couples to forgo childbearing, according to Carl Haub, Director of Information and Education at the Population Reference Bureau. In a survey of 3,000 Russian women in 1992, 75 percent cited insufficient income as a factor discouraging childbearing. As Haub observed in a 1994 report, "The recent birth dearth, not surprisingly, is a direct result of the collapse of the economy and a general lack of confidence in the future."

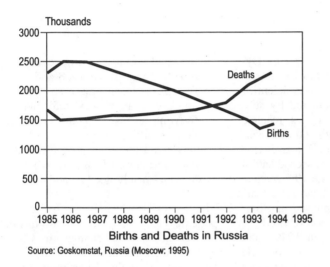

Births and Deaths in Russia

Source: Goskomstat, Russia (Moscow: 1995)

The rise in Russia's death rate illustrates even more dramatically the extent to which the country's social fabric is fraying. The incidence of stress-related conditions such as heart attacks and alcoholism has risen sharply. When the death rate jumped from 12.1 per 1,000 persons to 14.5 between 1992 and 1993, for example, three-quarters of the increase was due to cardiovascular disease, accidents, murder, suicide, and alcohol poisoning. According to Gennadi I. Gerasimov, a former spokesman for Mikhail Gorbachev, 100,000 Russians have killed themselves in the past two years—a suicide rate more than three times

that of the United States. A sharp rise in alcoholism is causing increases in both alcohol poisoning and fatal accidents at work. And lack of adequate health care has exacerbated the problem—treatable conditions are increasingly fatal, and people often die after a relatively mild heart attack.

The death rate also has a clear environmental connection. According to Murray Feshbach, a professor of demography at Georgetown University in Washington, D.C., and an expert on Russia's environment, pollution plays a role in 20 to 30 percent of the country's deaths. In the most affected areas, that percentage is much higher. Feshbach cites the example of Nikel, a town of 21,000 on the Norweigan border in Russia's extreme north. For employees at the local smelter, life expectancy is just 34 years; for the town in general, it is 44 years.

Contamination, in one form or another, is undermining public health throughout the FSU. Water in as many as 75 percent of the region's rivers and lakes is unfit for drinking, according to Alexey Yablokov, chairman of a Russian Federation commission on the environment and a former environmental advisor to both Gorbachev and Boris Yeltsin. And at least 50 percent of tap water fails to meet sanitary norms for microbial, chemical, and other forms of pollution. Waterborne infectious diseases are epidemic (see "Environmental Intelligence," November/December 1995), and children are often the principal victims. Yablokov has classified about 15 percent of Russia's land area as ecological disaster zones.

Presumably, economic improvements will eventually stabilize Russia's demographic picture. But economic recovery alone will not erase the effects of decades of gross environmental abuse. Russia's public health crisis is a lesson in the human costs of unbridled development—a lesson that may have particular relevance for the most rapidly developing countries. In China, for example, where the economy has grown by 57 percent in the past four years, coal accounts for more than three-quarters of the commercial energy supply. Yet, nearly 1 million people a year may already be dying from lung diseases related to air pollution—and China plans to double its coal use over the next two decades.

China's changing land

Population, food demand and land use in China

Gerhard K. Heilig

China's land-use changes directly affect the country's capacity to generate sufficient food supplies. Losses of arable land due to natural disasters, agricultural restructuring, and infrastructure expansion might lead to food production deficits in the future.

China's food prospects, however, are also of geostrategic and geopolitical relevance to the West. Food deficits in China might destabilize the country and jeopardize the process of economic and political reform. Despite China's strong commitment to self-sufficiency, the country may become a major importer of (feed) grain. This is of great economic interest to large grain exporters such as the United States, France, or Australia.

There is also the danger that climate change might affect China's vegetation cover—especially the large grasslands, which are important for the country's livestock production. An increased frequency of catastrophic weather events could trigger massive floods or extended droughts that would affect major agricultural areas in various parts of China. These trends pose serious risks for the country's future food security.

What are the major trends in China's land-use changes?

There are few places in the world where people have changed the land so intensively, and over such a long period, as in China. The Loess Plateau of Northern China, for instance, was completely deforested in pre-industrial times. The Chinese started systematic land reclamation and irrigation schemes, converting large areas of natural land into rice pad-

dies, as far back as the early Han Dynasty, in the fourth and third centuries BC. This process, which was scientifically planned and coordinated by subsequent dynastic bureaucracies, reached its first climax in the 11th and 12th centuries. Another period of massive land modification followed in the second half of the 18th and first half of the 19th century.

The LUC project, however, deals with a more recent phase—approximately since the foundation of the People's Republic of China in 1949—and in particular the period since economic reforms began in 1978. Although there has been widespread speculation in the scientific literature about losses of cultivated land in China due to pollution and urban expansion, very little hard data was available when the LUC project started. Through its collaboration with Chinese partners and other sources, LUC has received highly detailed land-use and land-cover data for China. This information includes statistical data, mapped information and remote sensing data. Table 1 presents some results from LUC's analyses of the most recent trends, based on new surveys from the Chinese State Land Administration:

- Conversion of cropland into horticulture was the most important factor in land-use change in China in recent years. Between 1988 and 1995 farmers converted some 1.2 million hectares (ha) of cropland into horticulture. This conversion is a positive trend, indicating a growing market orientation of Chinese agriculture.
- Manifold construction activities (roads, settlements, industry and mining) di-

minished China's cultivated land by 980,000 ha between 1988 and 1995.
- The third most important type of land-use change was reforestation, diverting almost 970,000 ha of previously cultivated land.
- Between 1988 and 1995, China also lost some 850,000 ha due to natural disasters—mainly flooding and droughts.

These data clearly indicate that anthropogenic factors are mainly responsible for recent land-use changes in China. The growing demand for meat, fish, fruit, and vegetables drives much of the agricultural restructuring, such as conversion to horticultural land and fishponds. Environmental programs are promoting reforestation, and the expansion of infrastructure is a function of rapid economic development and urbanization.

What are important drivers of land-use change in China?

Proximate determinants, such as those discussed above, are just the last step in a chain of causation that triggers land-use change in China. There are, however, three fundamental factors behind these trends:

- Population growth. Most recent projections from the United Nations Population Division (see Figure 1) assume that China's population will increase to some 1.49 billion people by 2025 and then slowly decline to 1.48 by 2050.

From *Options*, Summer 1999, pp. 18-20. © 1999 by the International Institute of Applied Systems Analysis. Reprinted by permission.

Table 1: Increase, decrease and net-change of cultivated land in China by region, 1988 – 1995 (in hectares).

| | Increase | | | | Decrease Conversion | | | | | | |
| | | | Re-use of abandoned land | Conv. from agricultural land | Construct. (1) | to horticulture | to forest land | to grassland | to fishponds | Disasters | Net-Change |
	Reclamation	Drainage									
North	289.733	18.233	106,028	41,080	–229.825	–298.595	–111.895	–41.255	–7.979	–105,535	–340.020
Northeast	396.867	21.991	39.957	25.852	–109,662	–87.417	–156.501	–90.631	–9.426	–250.871	–219.842
East	43.250	21.147	34.284	44.628	–242.824	–178.677	–24.765	–1.416	–45.624	–29.328	–379.324
Central	79.943	7.665	13.159	26.529	–91.802	–101.858	–111.782	–2.689	–61.519	–41.456	–283.771
South	320.001	22.686	18.870	96.681	–117.023	–116.966	–81.465	–10.361	–87.096	–56.534	–11.206
Southwest	358.704	8.687	40.260	90.450	–92.583	–114.235	–161.770	–93.864	–7.902	–132.915	–105.169
Plateau	28.044	980	3.135	3.935	–7.927	–163	–363	–11.874	–1	–486	15.281
Northwest	681.902	19.798	93.428	84.270	–88.588	–325.316	–321.214	–296.605	–6.322	–239.252	–397.900
TOTAL	2.198.444	121.177	349.121	413.465	–980.235	–1.223.229	–969.756	–548.694	–225.868	–856.377	–1.721.951

Source: State Land Administration, Statistical Information on the Land of China in 1995. Beijing, 1996. And equivalent reports for 1988 to 1994.

Note: (1) "Construction" includes all kinds of infrastructure, industrial areas and residential areas. In the original data tables this category includes all construction by state-owned units (cities, towns, mining and factories, railways, highways, water reservoirs, public buildings) and constructions by rural communities (rural roads, township and village enterprises, rural water reservoirs, offices, education and sanitation, rural private resident housing).

- Income growth. By all measures, China had spectacular economic growth rates in recent years. The number of people in poverty declined by some 200 million.
- Urbanization. Although China still has a large agricultural population, most experts predict a rapid increase in the number and size of towns and cities in the future. This urbanization is associated with changes in labor-force participation and lifestyle.

Population growth increases the demand for food and thus leads to intensification of agriculture and expansion of cultivated land. However, as the LUC analysis shows, there is a high error range in projecting China's population for more than 20 to 30 years. Its huge initial size amplifies even very small changes in fertility and mortality. For instance, between estimates made in 1994 and 1998, the United Nations Population Division had to revise its population projection for China for the year 2050 by 128 million. This uncertainty is unavoidable, and must be taken into account in predicting food demand and other population-related changes.

There is also considerable uncertainty concerning China's future economic development. Current trends indicate that China will experience a rapid transition from an agricultural to an industrial and service society. This would lead to expansion of infrastructure and areas of settlement, encroaching on valuable cropland, especially around urban-industrial agglomerations in coastal provinces.

An increase of productivity in both industry and agriculture could also lead to massive unemployment—particularly among the agricultural population. China already has an excess agricultural labor force of at least 120 million, which will increase the potential for rural-urban migration. Policy measures could probably slow down urbanization to some extent, but most projections assume a massive

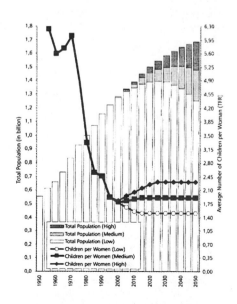

Figure 1: Population projection for China.

increase in urban population from the current 370 million to 800 million people. All these trends would—directly or indirectly—influence land use in China. With growing income and urban lifestyle, for instance, Chinese consumers will further change their diet (see Figure 2).

According to FAO estimates, there was already a massive increase in the domestic supply of meat from 6.6 to 47.9 million metric tons between the mid-1960s and the mid-1990s—with a significant effect on crop production patterns. The harvest of maize (the major feed crop) increased from 25 to 113 million tons between the mid-1960s and mid-1990s. Use of cereals for feeding animals grew from 12 million tons in 1964–66 to 107 million tons in 1994–96. Similar trends can be observed with other food commodities. Between the mid-1960s and mid-1990s, China's vegetable production increased from 37.9 to 182.7 million tons, and the production of fish increased from 3.3 to 24.3 million tons. These consumer-driven changes in agricultural production caused most of the land-use change in China during the last few decades.

Some general observations resulting from LUC's research

LUC's various models and analyses indicate that China is certainly facing a critical phase of its development during the next three decades. In that period, the demographic momentum will inevitably lead to further population growth, in the order of some 260 million people. Their demand for food, water, and shelter will greatly increase pressure on China's land and water resources. The country's urbanization and economic development will amplify this pressure. Without effective control of pollution and farmland loss, China faces serious problems of land degradation and productivity decline that might threaten the country's food security.

The LUC project has analyzed how global climate change might affect natural vegetation, agriculture, and water management systems in China. This is essential for a long-term water and land resource strategy. China will only be able to cope with its serious problems of flooding and drought if it takes into account the interactions between human and natural factors. For instance, flood damage has increased in recent years because farmers moved into areas that should have been reserved for flood

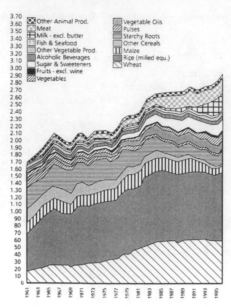

Figure 2: Per capita food consumption by commodity.

control. The serious water deficit in downstream areas of the Yellow River is amplified by a massive increase in (rather inefficient) agricultural water consumption upstream. As the LUC analysis on *Human Impact on Yellow River Water Management* (IIASA IR 98-016) has shown. China will need intense water management activities to secure a reliable water supply.

Despite recent progress, great data deficits and uncertainties still exist concerning both biophysical and socioeconomic conditions and trends. While LUC was able—with the help of Chinese colleagues—to improve, cross-check, and update several (georeferenced) data sets on China (climate, soils, water, cultivated land, grassland, natural vegetation, population, etc.), project staff still see the need for additional sources of information. In particular, remote sensing data could improve land-use analyses and planning for certain regions. Due to the rapid change in China, available statistical or mapped data are not always up to date. For instance, the conversion of arable land due to infrastructure expansion and urban sprawl in recent years could be analyzed more accurately with satellite images.

Helping the world's poorest

Jeffrey Sachs, a top academic economist, argues that rich countries must mobilise global science and technology to address the specific problems which help to keep poor countries poor

IN OUR Gilded Age, the poorest of the poor are nearly invisible. Seven hundred million people live in the 42 so-called Highly Indebted Poor Countries (HIPCS), where a combination of extreme poverty and financial insolvency marks them for a special kind of despair and economic isolation. They escape our notice almost entirely, unless war or an exotic disease breaks out, or yet another programme with the International Monetary Fund (IMF) is signed. The Cologne Summit of the G8 in June was a welcome exception to this neglect. The summiteers acknowledged the plight of these countries, offered further debt relief and stressed the need for a greater emphasis by the international community on social programmes to help alleviate human suffering.

The G8 proposals should be seen as a beginning: inadequate to the problem, but at least a good-faith prod to something more useful. We urgently need new creativity and a new partnership between rich and poor if these 700m people (projected to rise to 1.5 billion by 2030), as well as the extremely poor in other parts of the

Jeffrey Sachs is director of the Centre for International Development and professor of international trade at Harvard University. A prolific writer, he has also advised the governments of many developing and East European countries.

world (especially South Asia), are to enjoy a chance for human betterment. Even outright debt forgiveness, far beyond the G8's stingy offer, is only a step in the right direction. Even the call to the IMF and World Bank to be more sensitive to social conditions is merely an indicative nod.

A much more important challenge, as yet mainly unrecognised, is that of mobilising global science and technology to address the crises of public health, agricultural productivity, environmental degradation and demographic stress confronting these countries. In part this will require that the wealthy governments enable the grossly underfinanced and underempowered United Nations institutions to become vibrant and active partners of human development. The failure of the United States to pay its UN dues is surely the world's most significant default on international obligations, far more egregious than any defaults by impoverished HIPCS. The broader American neglect of the UN agencies that assist impoverished countries in public health, science, agriculture and the environment must surely rank as another amazingly misguided aspect of current American development policies.

The conditions in many HIPCS are worsening dramatically, even as global science and technology create new surges of wealth and well-being in the richer countries. The problem is that, for myriad reasons, the technological gains in wealthy countries do not readily diffuse to the poorest ones. Some

barriers are political and economic. New technologies will not take hold in poor societies if investors fear for their property rights, or even for their lives, in corrupt or conflict-ridden societies. *The Economist's* response to the Cologne Summit ("Helping the Third World", June 26th) is right to stress that aid without policy reform is easily wasted. But the barriers to development are often more subtle than the current emphasis on "good governance" in debtor countries suggests.

Research and development of new technologies are overwhelmingly directed at rich-country problems. To the extent that the poor face distinctive challenges, science and technology must be directed purposefully towards them. In today's global set-up, that rarely happens. Advances in science and technology not only lie at the core of long-term economic growth, but flourish on an intricate mix of social institutions—public and private, national and international.

Currently, the international system fails to meet the scientific and technological needs of the world's poorest. Even when the right institutions exist—say, the World Health Organisation to deal with pressing public health disasters facing the poorest countries—they are generally starved for funds, authority and even access to the key negotiations between poor-country governments and the Fund at which important development strategies get hammered out.

Reprinted with permission from *The Economist*, August 14, 1999, pp. 17-20. © 1999 by The Economist, Ltd. Distributed by The New York Times Special Features.

The ecology of underdevelopment

If it were true that the poor were just like the rich but with less money, the global situation would be vastly easier than it is. As it happens, the poor live in different ecological zones, face different health conditions and must overcome agronomic limitations that are very different from those of rich countries. Those differences, indeed, are often a fundamental cause of persisting poverty.

Let us compare the 30 highest-income countries in the world with the 42 HIPCS (see table below). The rich countries overwhelmingly lie in the world's temperate zones. Not every country in those bands is rich, but a good rule of thumb is that temperate-zone economies are either rich, formerly socialist (and hence currently poor), or geographically isolated (such as Afghanistan and Mongolia). Around 93% of the combined population of the 30 highest-income countries lives in temperate and snow zones. The HIPCS by contrast, include 39 tropical or desert societies. There are only three in a substantially temperate climate, and those three are landlocked and therefore geographically isolated (Laos, Malawi and Zambia).

Not only life but also death differs between temperate and tropical zones. Individuals in temperate zones almost everywhere enjoy a life expectancy of 70 years or more. In the tropics, however, life expectancy is generally much shorter. One big reason is that populations are burdened by diseases such as malaria, hookworm, sleeping sickness and schistosomiasis, whose transmission generally depends on a warm climate. (Winter may be the greatest public-health intervention in the world.) Life expectancy in the HIPCS averages just 51 years, reflecting the interacting effects of tropical disease and poverty. The economic evidence strongly suggests that short life expectancy is not just a result of poverty, but is also a powerful cause of impoverishment.

All the rich-country research on rich-country ailments, such as cardiovascular diseases and cancer, will not solve the problems of malaria. Nor will the biotechnology advances for temperate-zone crops easily transfer to the conditions of tropical agriculture. To address the special conditions of the HIPCS, we must first understand their unique problems, and then use our ingenuity and co-operative spirit to create new methods of overcoming them.

Modern society and prosperity rest on the foundation of modern science. Global capitalism is, of course, a set of social institutions—of property rights, legal and political systems, international agreements, transnational corporations, educational establishments, and public and private research institutions—but the prosperity that results from these institutions has its roots in the development and applications of new science-based technologies. In the past 50 years, these have included technologies built on solid-state physics, which gave rise to the information-technology revolution, and on genetics, which have fostered breakthroughs in health and agricultural productivity.

Science at the ecological divide

In this context, it is worth noting that the inequalities of income across the globe are actually exceeded by the inequalities of scientific output and technological innovation. The chart below shows the remarkable dominance of rich countries in scientific publications and, even more notably, in patents filed in Europe and the United States.

The role of the developing world in one sense is much greater than the chart indicates. Many of the scientific and technological breakthroughs are made by poor-country scientists working in rich-country laboratories. Indian and Chinese engineers account for a significant proportion of Silicon Valley's workforce, for example. The basic point, then, holds even more strongly: global science is directed by the rich countries and for the rich-country markets, even to the extent of mobilising much of the scientific potential of the poorer countries.

The imbalance of global science reflects several forces. First, of course, science follows the market. This is especially true in an age when technological leaps require expensive scientific equipment and well-provisioned research laboratories. Second, scientific advance tends to have in-

creasing returns to scale: adding more scientists to a community does not diminish individual marginal productivity but tends to increase it. Therein lies the origin of university science departments, regional agglomerations such as Silicon Valley and Route 128, and mega-laboratories at leading high-technology firms including Merck, Microsoft and Monsanto. And third, science requires a partnership between the public and private sectors. Free-market ideologues notwithstanding, there is scarcely one technology of significance that was not nurtured through public as well as private care.

If technologies easily crossed the ecological divide, the implications would be less dramatic than they are. Some technologies, certainly those involving the computer and other ways of managing information, do indeed cross over, and give great hopes of spurring technological capacity in the poorest countries. Others—especially in the life sciences but also in the use of energy, building techniques, new materials and the like—are prone to "ecological specificity". The result is a profound imbalance in the global production of knowledge: probably the most powerful engine of divergence in global well-being between the rich and the poor.

Consider malaria. The disease kills more than 1m people a year, and perhaps as many as 2.5m. The disease is so heavily concentrated in the poorest tropical countries, and overwhelmingly in sub-Saharan Africa, that nobody even bothers to keep an accurate count of clinical cases or deaths. Those who remember that richer places such as Spain, Italy, Greece and the southern United States once harboured the disease may be misled into thinking that the problem is one of social institutions to control its transmission. In fact, the sporadic transmission of malaria in the sub-tropical regions of the rich countries was vastly easier to control than is its chronic transmission in the heart of the tropics. Tropical countries are plagued by ecological conditions that produce hundreds of infective bites per year per person.

Different ecologies 1995	HIPCs* (42)	Rich countries (30)
GDP per person, PPP$†	1,187	18,818
Life expectancy at birth, years†	51.5	76.9
Population by ecozones, % in:		
tropical	55.6	0.7
dry	17.6	3.7
temperate and snow	12.5	92.6
highland	14.0	2.5

Source: J. Sachs *Highly indebted poor countries †Unweighted averages

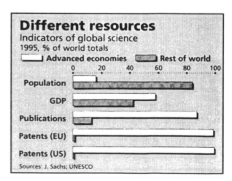

Different resources
Indicators of global science
1995, % of world totals
Advanced economies / Rest of world

Population
GDP
Publications
Patents (EU)
Patents (US)

Sources: J. Sachs; UNESCO

Mosquito control does not work well, if at all, in such circumstances. It is in any event expensive.

Recent advances in biotechnology, including mapping the genome of the malaria parasite, point to a possible malaria vaccine. One would think that this would be high on the agendas of both the international community and private pharmaceutical firms. It is not. A Wellcome Trust study a few years ago found that only around $80m a year was spent on malaria research, and only a small fraction of that on vaccines.

The big vaccine producers, such as Merck, Rhône-Poulenc's Pasteur-Mérieux-Connaught and SmithKline Beecham, have much of the in-house science but not the bottom-line motivation. They strongly believe that there is no market in malaria. Even if they spend the hundreds of millions, or perhaps billions, of dollars to do the R&D and come up with an effective vaccine, they believe, with reason, that their product would just be grabbed by international agencies or private-sector copycats. The hijackers will argue, plausibly, that the poor deserve to have the vaccine at low prices—enough to cover production costs but not the preceding R&D expenditures.

The malaria problem reflects, in microcosm, a vast range of problems facing the HIPCS in health, agriculture and environmental management. They are profound, accessible to science and utterly neglected. A hundred IMF missions or World Bank health-sector loans cannot produce a malaria vaccine. No individual country borrowing from the Fund or the World Bank will ever have the means or incentive to produce the global public good of a malaria vaccine. The root of the problem is a much more complex market failure: private investors and scientists doubt that malaria research will be rewarded financially. Creativity is needed to bridge the huge gulfs between human needs, scientific effort and market returns.

Promise a market

The following approach might work. Rich countries would make a firm pledge to purchase an effective malaria vaccine for Africa's 25m newborn children each year if such a vaccine is developed. They would even state, based on appropriate and clear scientific standards, that they would guarantee a minimum purchase price—say, $10 per dose—for a vaccine that meets minimum conditions of efficacy, and

perhaps raise the price for a better one. The recipient countries might also be asked to pledge a part of the cost, depending on their incomes. But nothing need be spent by any government until the vaccine actually exists.

Even without a vast public-sector effort, such a pledge could galvanise the world of private-sector pharmaceutical and biotechnology firms. Malaria vaccine research would suddenly become hot. Within a few years, a breakthrough of profound benefit to the poorest countries would be likely. The costs in foreign aid would be small: a few hundred million dollars a year to tame a killer of millions of children. Such a vaccine would rank among the most effective public-health interventions conceivable. And, if science did not deliver, rich countries would end up paying nothing at all.

Malaria imposes a fearsome burden on poor countries, the AIDS epidemic an even weightier load. Two-thirds of the world's 33m individuals infected with the HIV virus are sub-Saharan Africans, according to a UN estimate in 1998, and the figure is rising. About 95% of worldwide HIV cases are in the developing world. Once again, science is stopping at the ecological divide.

Rich countries are controlling the epidemic through novel drug treatments that are too expensive, by orders of magnitude, for the poorest countries. Vaccine research, which could provide a cost-effective method of prevention, is dramatically underfunded. The vaccine research that is being done focuses on the specific viral strains prevalent in the United States and Europe, not on those which bedevil Africa and Asia. As in the case of malaria, the potential developers of vaccines consider the poor-country market to be no market at all. The same, one should note, is true for a third worldwide killer. Tuberculosis is still taking the lives of more than 2m poor people a year and, like malaria and AIDS, would probably be susceptible to a vaccine, if anyone cared to invest in the effort.

The poorer countries are not necessarily sitting still as their citizenry dies of AIDS. South Africa is on the verge of authorising the manufacture of AIDS medicines by South African pharmaceutical companies, despite patents held by American and European firms. The South African government says that, if rich-country firms will not supply the drugs to the South African market at affordable prices (ones that are high enough to meet marginal production costs but do not include the patent-generated monop-

oly profits that the drug companies claim as their return for R&D), then it will simply allow its own firms to manufacture the drugs, patent or no. In a world in which science is a rich-country prerogative while the poor continue to die, the niceties of intellectual property rights are likely to prove less compelling than social realities.

There is no shortage of complexities ahead. The world needs to reconsider the question of property rights before patent rights allow rich-country multinationals in effect to own the genetic codes of the very foodstuffs on which the world depends, and even the human genome itself. The world also needs to reconsider the role of institutions such as the World Health Organisation and the Food and Agriculture Organisation. These UN bodies should play a vital role in identifying global priorities in health and agriculture, and also in mobilising private-sector R&D towards globally desired goals. There is no escape from such public-private collaboration. It is notable, for example, that Monsanto, a life-sciences multinational based in St Louis, Missouri, has a research and development budget that is more than twice the R&D budget of the entire worldwide network of public-sector tropical research institutes. Monsanto's research, of course, is overwhelmingly directed towards temperate-zone agriculture.

People, food and the environment

Public health is one of the two distinctive crises of the tropics. The other is the production of food. Poor tropical countries are already incapable of securing an adequate level of nutrition, or paying for necessary food imports out of their own export earnings. The HIPC population is expected to more than double by 2030. Around one-third of all children under the age of five in these countries are malnourished and physically stunted, with profound consequences throughout their lives.

As with malaria, poor food productivity in the tropics is not merely a problem of poor social organisation (for example, exploiting farmers through controls on food prices). Using current technologies and seed types, the tropics are inherently less productive in annual food crops such as wheat (essentially a temperate-zone crop), rice and maize. Most agriculture in the equatorial tropics is of very low productivity, reflecting the fragility of most tropical soils at high temperatures combined with heavy rainfall.

High productivity in the rainforest eco-zone is possible only in small parts of the tropics, generally on volcanic soils (on the island of Java, in Indonesia, for example). In the wet-dry tropics, such as the vast savannahs of Africa, agriculture is hindered by the terrible burdens of unpredictable and highly variable water supplies. Drought and resulting famine have killed millions of peasant families in the past generation alone.

Scientific advances again offer great hope. Biotechnology could mobilise genetic engineering to breed hardier plants that are more resistant to drought and less sensitive to pests. Such genetic engineering is stymied at every point, however. It is met with doubts in the rich countries (where people do not have to worry about their next meal); it requires a new scientific and policy framework in the poor countries; and it must somehow generate market incentives for the big life-sciences firms to turn their research towards tropical foodstuffs, in co-operation with tropical research centres. Calestous Juma, one of the world's authorities on biotechnology in Africa, stresses that there are dozens, or perhaps hundreds, of underused foodstuffs that are well adapted to the tropics and could be improved through directed biotechnology research. Such R&D is now all but lacking in the poorest countries.

The situation of much of the tropical world is, in fact, deteriorating, not only because of increased population but also because of long-term trends in climate. As the rich countries fill the atmosphere with increasing concentrations of carbon, it looks ever more likely that the poor tropical countries will bear much of the resulting burden.

Anthropogenic global warming, caused by the growth in atmospheric carbon, may actually benefit agriculture in high-latitude zones, such as Canada, Russia and the northern United States, by extending the growing season and improving photosynthesis through a process known as carbon fertilisation. It is likely to lower tropical food productivity, however, both because of increased heat stress on plants and because the carbon fertilisation effect appears to be smaller in tropical ecozones. Global warming is also contributing to the increased severity of tropical climatic disturbances, such as the "one-in-a-century" El Niño that hit the tropical world in 1997–98, and the "one-in-a-century" Hurricane Mitch that devastated Honduras and Nicaragua a year ago. Once-in-a-century weather events seem to be arriving with disturbing frequency.

The United States feels aggrieved that poor countries are not signing the convention on climatic change. The truth is that these poor tropical countries should be calling for outright compensation from America and other rich countries for the climatic damages that are being imposed on them. The global climate-change debate will be stalled until it is acknowledged in the United States and Europe that the temperate-zone economies are likely to impose heavy burdens on the already impoverished tropics.

New hope in a new millennium

The situation of the HIPCS has become intolerable, especially at a time when the rich countries are bursting with new wealth and scientific prowess. The time has arrived for a fundamental re-thinking of the strategy for co-operation between rich and poor, with the avowed aim of helping the poorest of the poor back on to their own feet to join the race for human betterment. Four steps could change the shape of our global community.

First, rich and poor need to learn to talk together. As a start, the world's democracies, rich and poor, should join in a quest for common action. Once again the rich G8 met in 1999 without the presence of the developing world. This rich-country summit should be the last of its kind. A G16 for the new millennium should include old and new democracies such as Brazil, India, South Korea, Nigeria, Poland and South Africa.

Second, rich and poor countries should direct their urgent attention to the mobilisation of science and technology for poor-country problems. The rich countries should understand that the IMF and World Bank are by themselves not equipped for that challenge. The specialised UN agencies have a great role to play, especially if they also act as a bridge between the activities of advanced-country and developing-country scientific centres. They will be able to play that role, however, only after the United States pays its debts to the UN and ends its unthinking hostility to the UN system.

We will also need new and creative institutional alliances. A Millennium Vaccine Fund, which guaranteed future markets for malaria, tuberculosis and AIDS vaccines, would be the right place to start. The vaccine-fund approach is administratively straightforward, desperately needed and within our technological reach. Similar efforts to merge public and private science activities will be needed in agricultural biotechnology.

Third, just as knowledge is becoming the undisputed centrepiece of global prosperity (and lack of it, the core of human impoverishment), the global regime on intellectual property rights requires a new look. The United States prevailed upon the world to toughen patent codes and cut down on intellectual piracy. But now transnational corporations and rich-country institutions are patenting everything from the human genome to rainforest biodiversity. The poor will be ripped off unless some sense and equity are introduced into this runaway process.

Moreover, the system of intellectual property rights must balance the need to provide incentives for innovation against the need of poor countries to get the results of innovation. The current struggle over AIDS medicines in South Africa is but an early warning shot in a much larger struggle over access to the fruits of human knowledge. The issue of setting global rules for the uses and development of new technologies—especially the controversial biotechnologies—will again require global co-operation, not the strong-arming of the few rich countries.

Fourth, and perhaps toughest of all, we need a serious discussion about long-term finance for the international public goods necessary for HIPC countries to break through to prosperity. The rich countries are willing to talk about every aspect except money: money to develop new malaria, tuberculosis and AIDS vaccines; money to spur biotechnology research in food-scarce regions; money to help tropical countries adjust to climate changes imposed on them by the richer countries. The World Bank makes mostly loans, and loans to individual countries at that. It does not finance global public goods. America has systematically squeezed the budgets of UN agencies, including such vital ones as the World Health Organisation.

We will need, in the end, to put real resources in support of our hopes. A global tax on carbon-emitting fossil fuels might be the way to begin. Even a very small tax, less than that which is needed to correct humanity's climate-deforming overuse of fossil fuels, would finance a greatly enhanced supply of global public goods. No better time to start than as the new millennium begins.

AE Article Review Form

We encourage you to photocopy and use this page as a tool to assess how the articles in **Annual Editions** expand on the information in your textbook. By reflecting on the articles you will gain enhanced text information. You can also access this useful form on a product's book support Web site at **http://www.dushkin.com/ online/.**

NAME: DATE:

TITLE AND NUMBER OF ARTICLE:

BRIEFLY STATE THE MAIN IDEA OF THIS ARTICLE:

LIST THREE IMPORTANT FACTS THAT THE AUTHOR USES TO SUPPORT THE MAIN IDEA:

WHAT INFORMATION OR IDEAS DISCUSSED IN THIS ARTICLE ARE ALSO DISCUSSED IN YOUR TEXTBOOK OR OTHER READINGS THAT YOU HAVE DONE? LIST THE TEXTBOOK CHAPTERS AND PAGE NUMBERS:

LIST ANY EXAMPLES OF BIAS OR FAULTY REASONING THAT YOU FOUND IN THE ARTICLE:

LIST ANY NEW TERMS/CONCEPTS THAT WERE DISCUSSED IN THE ARTICLE, AND WRITE A SHORT DEFINITION:

ANNUAL EDITIONS revisions depend on two major opinion sources: one is our Advisory Board, listed in the front of this volume, which works with us in scanning the thousands of articles published in the public press each year; the other is you—the person actually using the book. Please help us and the users of the next edition by completing the prepaid article rating form on this page and returning it to us. Thank you for your help!

ANNUAL EDITIONS: Geography 00/01

ARTICLE RATING FORM

Here is an opportunity for you to have direct input into the next revision of this volume. We would like you to rate each of the 43 articles listed below, using the following scale:

1. Excellent: should definitely be retained
2. Above average: should probably be retained
3. Below average: should probably be deleted
4. Poor: should definitely be deleted

Your ratings will play a vital part in the next revision.
So please mail this prepaid form to us just as soon as you complete it.
Thanks for your help!

RATING

ARTICLE

1. Rediscovering the Importance of Geography
2. The Four Traditions of Geography
3. The American Geographies
4. World Prisms: The Future of Sovereign States and International Order
5. Human Domination of Earth's Ecosystems
6. The Role of Science in Policy: The Climate Change Debate in the United States
7. California Fumes Over Oil: Offshore-Drilling Decision Gives New Life to an Old Battle
8. Lead in the Inner Cities
9. The Season of El Niño
10. The Great Climate Flip-Flop
11. Human Imprint on Climate Change Grows Clearer
12. Beyond the Valley of the Dammed
13. A River Runs through It
14. Sprawling, Sprawling
15. Past and Present Land Use and Land Cover in the USA
16. Operation Desert Sprawl
17. Can This Swamp Be Saved? Bold Everglades-Protection Strategies May Revive the River of Grass
18. A Greener, or Browner, Mexico?
19. The Importance of Places, or, a Sense of Where You Are
20. The Rise of the Region State
21. Europe at Century's End: The Challenge Ahead
22. Russia's Fragile Union

RATING

ARTICLE

23. Kosovo: America's Balkan Problem
24. Sea Change in the Arctic
25. Greenville: From Back Country to Forefront
26. Flood Hazards and Planning in the Arid West
27. Does It Matter Where You Are?
28. Transportation and Urban Growth: The Shaping of the American Metropolis
29. Amtrak Plans High-Speed Trains as Plane Alternative
30. GIS Technology Reigns Supreme in Ellis Island Case
31. Teaching about Karst Using U.S. Geological Survey Resources
32. County Buying Power, 1987–97
33. Do We Still Need Skyscrapers?
34. Indian Gaming in the U.S.: Distribution, Significance and Trends
35. Before the Next Doubling
36. The Population Surprise
37. The End of Cheap Oil
38. Gray Dawn: The Global Aging Crisis
39. A Rare and Precious Resource
40. The Changing Geography of U.S. Hispanics, 1850–1990
41. Russia's Population Sink
42. China's Changing Land: Population, Food Demand and Land Use in China
43. Helping the World's Poorest

(Continued on next page)

NO POSTAGE
NECESSARY
IF MAILED
IN THE
UNITED STATES

BUSINESS REPLY MAIL
FIRST-CLASS MAIL PERMIT NO. 84 GUILFORD CT

POSTAGE WILL BE PAID BY ADDRESSEE

Dushkin/McGraw-Hill
Sluice Dock
Guilford, CT 06437-9989

ABOUT YOU

Name _____   Date _____

Are you a teacher? ☐ A student? ☐
Your school's name _____

Department _____

Address _____   City _____   State _____   Zip _____

School telephone # _____

YOUR COMMENTS ARE IMPORTANT TO US !

Please fill in the following information:
For which course did you use this book?
_____

Did you use a text with this *ANNUAL EDITION*? ☐ yes ☐ no
What was the title of the text?
_____

What are your general reactions to the *Annual Editions* concept?
_____

Have you read any particular articles recently that you think should be included in the next edition?
_____

Are there any articles you feel should be replaced in the next edition? Why?
_____

Are there any World Wide Web sites you feel should be included in the next edition? Please annotate.
_____

May we contact you for editorial input? ☐ yes ☐ no
May we quote your comments? ☐ yes ☐ no
_____